LONDON MATHEMATICAL SOCIETY LECTURE NOTE SERIES

Managing Editor: Professor M. Reid, Mathematics Institute,
University of Warwick, Coventry CV4 7AL, United Kingdom

The titles below are available from booksellers, or from Cambridge University Press at
http://www.cambridge.org/mathematics

London Mathematical Society Lecture Note Series: 425

Geometry, Topology, and Dynamics in Negative Curvature

Edited by

C. S. ARAVINDA
TIFR Centre for Applicable Mathematics, Bangalore

F. T. FARRELL
Tsinghua University, Beijing

J.-F. LAFONT
Ohio State University, Columbus

CAMBRIDGE
UNIVERSITY PRESS

CAMBRIDGE
UNIVERSITY PRESS

University Printing House, Cambridge CB2 8BS, United Kingdom

One Liberty Plaza, 20th Floor, New York, NY 10006, USA

477 Williamstown Road, Port Melbourne, VIC 3207, Australia

314-321, 3rd Floor, Plot 3, Splendor Forum, Jasola District Centre, New Delhi - 110025, India

79 Anson Road, #06-04/06, Singapore 079906

Cambridge University Press is part of the University of Cambridge.

It furthers the University's mission by disseminating knowledge in the pursuit of education, learning and research at the highest international levels of excellence.

www.cambridge.org
Information on this title: www.cambridge.org/9781107529007

First published 2016

A catalogue record for this publication is available from the British Library

ISBN 978-1-107-52900-7 Paperback

Contents

Contributors

Jayadev S. Athreya
Department of Mathematics, University of Washington,
Padelford Hall, Seattle, WA 98915 U.S.A.
E-mail: jathreya@uw.edu

Igor Belegradek
School of Mathematics, Georgia Institute of Technology,
Atlanta, GA 30332-0160 U.S.A.
E-mail: ib@math.gatech.edu

Marc Bourdon
Laboratoire Painlevé,
UMR CNRS 8524, Université de Lille 1,
59655 Villeneuve d'Ascq, FRANCE
E-mail: bourdon@math.univ-lille1.fr

David Constantine
Department of Mathematics and Computer Science,
Wesleyan University, Middletown, CT 06459 U.S.A.
E-mail: dconstantine@wesleyan.edu

Gerhard Knieper
Faculty of Mathematics, Ruhr University Bochum,
44780 Bochum, GERMANY
Email: gerhard.knieper@rub.de

Peter A. Linnell
Department of Mathematics, Virginia Tech,
Blacksburg, VA 24061-0123 U.S.A.
E-mail: plinnell@math.vt.edu

Mahan Mj
RKM Vivekananda University, Belur Math, WB-711 202, INDIA
Email: mahan.mj@gmail.com; mahan@rkmvu.ac.in

Hee Oh
Hee Oh
Mathematics Department, 10 Hillhouse Ave, P.O. Box 208283
Yale University, New Haven, CT 06520 U.S.A., and
Korea Institute for Advanced Study, Seoul, KOREA
E-mail: hee.oh@yale.edu

Nimish Shah
Department of Mathematics, The Ohio State University,
Columbus, OH 43210 U.S.A.
E-mail: shah@math.osu.edu

Jouni Parkkonen
Department of Mathematics and Statistics, P.O. Box 35
40014 University of Jyväskylä, FINLAND.
E-mail: jouni.t.parkkonen@jyu.fi

Frédéric Paulin
Département de mathématique, UMR 8628 CNRS, Bât. 425
Université Paris-Sud, 91405 ORSAY Cedex, FRANCE
E-mail: frederic.paulin@math.u-psud.fr

Anne Thomas
School of Mathematics and Statistics F07
University of Sydney NSW 2006, AUSTRALIA
E-mail: anne.thomas@sydney.edu.au

Preface

The geodesic flow on the unit tangent bundle of a closed surface of constant negative curvature is one of the earliest examples of an ergodic dynamical system. This was first proven by G. A. Hedlund in 1936. Soon afterwards, it was reproved by E. Hopf, who also generalized to the case of closed surfaces of variable negative curvature. Hopf's proof already indicated the relevance of negative curvature to the ergodicity of the geodesic flow. About 20 years later, Hopf's theorem was generalized by Anosov to geodesic flows on unit tangent bundles of higher dimensional closed negatively curved manifolds.

Around the same time, combining certain basic results in geometry and topology, it was observed that the fundamental group of a closed manifold M of negative curvature determines M up to homotopy equivalence. Mostow's celebrated strong rigidity theorem (1968) showed that, within the class of closed locally symmetric spaces M of non-compact type of dimension ≥ 3, the fundamental group $\pi_1(M)$ determines M up to isometry (possibly after scaling the metric by a positive constant). Further study of this rigidity phenomenon led to two important generalizations. On the one hand, Margulis established his super-rigidity theorem in higher rank. Using the harmonic map techniques of Eells-Sampson, versions of the super-rigidity theorem were established for certain rank 1 cases by Corlette, Jost-Yau and Mok-Siu-Yeung. On the other hand, it led Ballmann-Brin-Eberlein to introduce, in the mid 1980's, the notion of geometric rank. Generalizing a notion that existed for locally symmetric spaces, this culminated in the rank rigidity theorem due, independently, to Ballmann and Burns-Spazier. Around this time, Gromov realized that many results of this nature could be formulated and proved, in a synthetic way, in the more general setting of metric spaces of non-positive curvature. The study of this broader class of spaces resulted in certain new rigidity phenomena (such as quasi-isometric rigidity). It also allowed these techniques to be applied to the study of certain infinite groups. Research in this direction has since exploded and created the whole new field of geometric group theory.

Concurrently, investigations continued on how the fundamental group $\pi_1(M)$ of a closed negatively curved manifold M constrained the topology and geometry of M. Farrell-Jones proved, in the late 1980s, the remarkable topological rigidity theorem, which showed that in dimensions $n \neq 3, 4$, the fundamental group $\pi_1(M)$ determines M up to homeomorphism. They also constructed examples of closed negatively curved manifolds which are homeomorphic but not diffeomorphic, thereby showing that smooth rigidity, as conjectured by Lawson-Yau, fails in general.

Besson-Courtois-Gallot reintroduced dynamical ideas to study, on a given closed locally symmetric manifold M, the special role played by the locally symmetric metric within the broader class of negatively curved metrics. They defined the notion of a natural map, and in the early 1990s used the natural map to show that, among all negatively curved metrics, the locally symmetric ones minimize the topological entropy.

Farrell-Jones-Ontaneda-Raghunathan constructed exotic PL-structures on certain closed real hyperbolic manifolds. These constructions have exposed certain limitations to the otherwise successful analytic methods (such as the harmonic map, the natural map and the Ricci flow) in taking the study of closed negatively curved manifolds further.

With many important questions still waiting to be resolved, one of the aims of the GTDNC conference was to bring together mathematicians working in different aspects of negative curvature, and to discuss some of the recent highlights of the field.

India has had its own share of participation in these exciting developments. Some of the leading players visited India (especially TIFR Mumbai) over the years, inspiring significant research activity on some of these topics, and forging great collaborations. A testimony to the fact that research in these areas flourished in India is that some of these mathematicians made multiple visits to India. Several of these visitors made extended visits and gave lecture courses at TIFR. Notes from some of these courses were written up by young students at TIFR, and appeared in the TIFR Lecture notes series. These visits also inspired articles and books that appeared elsewhere too, notable among these is the foundational article of Gromov that heralded the beginnings of geometric group theory. Atle Selberg made the first of his three visits to India in 1956, Armand Borel was a frequent visitor, Marcel Berger visited a couple of times, Dan Mostow visited in 1968, Gromov visited in 1984-85, Margulis visited in

1989, Tom Farrell first visited in 1993 and returned to India several times after that. In recent times Besson has been a regular visitor. Hamilton, Yau and Gabai have also delivered special lecture series.

The ICM 2010 provided a great occasion to get together again in India and discuss various facets of the fascinating area of negative curvature. The ICM 2010 satellite conference *Geometry, Topology and Dynamics in Negative Curvature* was held in the beautiful ambiance of the Raman Research Institute in Bangalore during 2-7 August, 2010. The conference was a big event, with 35 one hour lectures, and two open problem sessions. We had 20 participants from abroad, 23 participants from Bangalore, and 24 participants from other cities throughout India. A large number of photos and videos of talks from the conference are posted on the conference webpage:

<*http://www.icts.res.in/archive/program/details/91/*>

We hope that the articles in these proceedings provide some snippets of the conference.

C.S. Aravinda
TIFR Centre for Applicable Mathematics
P O Box 6503, Sharadanagara
G K V K Post, Bangalore - 560 065, INDIA.
e-mail: aravinda@math.tifrbng.res.in

F.T. Farrell
Department of Mathematical Sciences, Binghamton University
Binghamton, New York 13902-6000
e-mail: farrell@math.binghamton.edu

J.-F. Lafont
Department of Mathematics, The Ohio State University
231 West 18th Avenue, Columbus, OH 43210-1174, U.S.A.
e-mail: jlafont@math.ohio-state.edu

Acknowledgments

We, as a part of the organizing committee, thoroughly enjoyed working on this event, and would like to take this opportunity to thank all the people who helped make the conference, and these proceedings, possible. Specifically, we would like to acknowledge the help of

- our distinguished Scientific Committee, consisting of M. Davis, P. B. Eberlein, R. S. Kulkarni, P. Ontaneda, G. Prasad, M. S. Raghunathan, and R. J. Spatzier, for their assistance in selecting a truly stellar list of speakers.
- S. K. Roushon and J. Samuel, for helping with the organization of the conference.
- our distinguished speakers for their interesting talks.
- the contributors to this volume, for taking the time to prepare these articles.
- our anonymous referees, for their time and effort.

Particular thanks are due to the staff of the Raman Research Institute, for their wonderful hospitality in hosting the conference. The conference would not have been possible without the generous funding received from the International Centre for Theoretical Sciences, Bangalore (who recognized our conference as an official ICTS event), and from the National Science Foundation, USA (under grant DMS-1016098).

1

Gap distributions and homogeneous dynamics

JAYADEV S. ATHREYA[1]

Abstract

We survey the use of dynamics of $SL(2, \mathbb{R})$-actions to understand gap distributions for various sequences of subsets of $[0, 1)$, particularly those arising from special trajectories of various two-dimensional dynamical systems. We state and prove an abstract theorem that gives a unified explanation for some of the examples we present.

1 Introduction

The study of the distribution of gaps in sequences is a subject that arises in many different contexts and has connections with many different areas of mathematics, including number theory, probability theory, and spectral analysis. In this paper, we study gap distributions from the perspective of dynamics and geometry, exploring examples connected with the dynamics of $SL(2, \mathbb{R})$-actions on moduli spaces of geometric objects, in particular the space of lattices and the space of translation surfaces.

The inspiration for this article is the a quote from the beautiful paper of Elkies-McMullen [8], referring to their explicit computation of the gap distribution of the sequence of fractional parts of $\sqrt{n}$, using the dynamics of the $SL(2, \mathbb{R})$-action on the space of *affine unimodular lattices* in $\mathbb{R}^2$.

"... *the uniform distribution of lattices explains the exotic distribution of gaps.*"

Indeed, the main results of our paper, Theorem 7, Theorem 10, and Theorem 11, give unified explanations of several examples of 'exotic' gap

[1] J.S.A. partially supported by NSF CAREER grant DMS 1351853, NSF grant DMS 1069153, and grants DMS 1107452, 1107263, 1107367 "RNMS: GEometric structures And Representation varieties" (the GEAR Network).

distributions via uniform distribution on various moduli spaces of geometric objects.

1.1 Equidistribution, randomness, and gap distributions

Suppose that for each positive integer k, we are given a finite list of points $F(k) \subset [0, 1)$, where by a *list*, we mean a finite non-decreasing sequence of real numbers where N_k denotes the number of terms in the kth sequence $F(k)$. We write

$$F(k) = \left\{ F_k^{(0)} \leq F_k^{(1)} \leq \ldots F_k^{(N_k)} \right\},$$

and we assume $N_k \to \infty$ as $k \to \infty$. In many situations, we are interested in the 'randomness' of the sequence of lists $\{F(k)\}_{k=1}^{\infty}$. A first test of 'randomness' is whether the lists $F(k)$ *uniformly distribute* in $[0, 1)$, that is the measures $\Delta_k = \frac{1}{N_k} \sum_{j=0}^{N_k} \delta_{F_k^{(j)}}$ converge weak-* to Lebesgue measure, i.e., for any $0 \leq a \leq b \leq 1$,

$$\lim_{k \to \infty} \Delta_k(a, b) = b - a. \tag{1}$$

A more refined question (not necessarily dependent on (1)) is to examine the distribution of *gaps* for the sequences $F(k)$. That is, form the associated *normalized gap sets*

$$G(k) := \left\{ N_k \left(F_k^{(i+1)} - F_k^{(i)} \right) : 0 \leq i < N_k \right\}, \tag{2}$$

and given $0 \leq a < b \leq \infty$, what is the behavior of

$$\lim_{k \to \infty} \frac{|G(k) \cap (a, b)|}{N_k}? \tag{3}$$

If the sequence $F(k)$ is 'truly random', that is, given by

$$F(k) = \{X_{(0)} \leq X_{(1)} \leq \ldots \leq X_{(k)}\},$$

where the $\{X_{(i)}\}$ are the order statistics generated by independent, identically distributed (i.i.d.) uniform $[0, 1)$ random variables $\{X_n\}_{n=0}^{\infty}$, it is an exercise in probability theory to show that the gap distribution converges to a *Poisson process* of intensity 1. Precisely, for any $t > 0$,

$$\lim_{k \to \infty} \frac{|G(k) \cap (t, \infty)|}{N_k} = e^{-t} \tag{4}$$

However, many sequences that arise 'in nature' satisfy an equidistribution property but do not have Poissonian gaps. Following [8], we call such gap distributions *exotic*. In this paper, we discuss in detail some examples of exotic gap distributions, which, moreover, can be calculated (or at least shown to exist) using methods arising from homogeneous dynamics, in particular dynamics of $SL(2, \mathbb{R})$ actions on appropriate moduli spaces. In particular, the results we discuss share a similar philosophy; the sets $F(k)$ are associated to sets of angles or slopes of a discrete set of vectors in $\mathbb{R}^2$, and the gap distribution is studied by appropriate linear renormalizations, which can be viewed as part of an $SL(2, \mathbb{R})$ action on an appropriate moduli space of geometric objects. The main novelty of this paper is the statement of three meta-theorems (Theorem 7, Theorem 10, and Theorem 11), which give unified explanations of some of these examples by linking them to uniform distribution on various moduli spaces and which we expect can be used for future applications.

1.2 Organization of the paper

This paper is organized as follows: in the remainder of this introduction we state results about our main (previously studied) examples: the Farey sequences $\mathcal{F}(Q)$ (Section 1.4); slopes for lattice vectors (Section 1.4); and saddle connection directions for translation surfaces (Section 1.5). We also briefly discuss the space of affine lattices and $\{\{\sqrt{n}\}\}_{n \geq 1}$ in Section 1.6. In Section 2, we state the main results Theorem 7, Theorem 10, and Theorem 11. We describe how to use these results to explain our examples in Section 3-Section 4, and prove the theorems in Section 5. Finally, in Section 6, we pose some natural questions suggested by our approach.

1.3 Acknowledgements

Parts of this work are based on joint work with J. Chaika, Y. Cheung, and S. Lelievre. I thank them for their stimulating intellectual partnership. Discussions with F. Boca, M. Boshernitzan, A. Eskin, J. Marklof, and A. Zaharescu helped the author to see the common structural features in this circle of ideas, and without them this paper would not have been possible. In particular, the stunning paper [14] of J. Marklof and A. Strombergsson has inspired not only this paper but the papers [2, 3, 4]. I would also like to thank the Polish Academy of Sciences for their hospitality, and in particular Piotr Przytycki for arranging a visit to Warsaw where many of the ideas in this paper were clarified. Finally, thanks are due to the organizers of the ICM Satellite Conference on Geometry, Topology, and Dynamics in Negative Curvature for the opportunity to speak at the meeting and also to contribute to this volume. We would also like to thank the anonymous referee for their careful reading and many remarks which clarified the exposition of this paper.

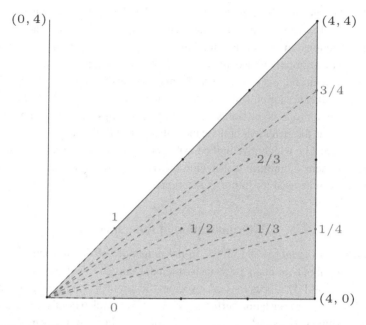

Figure 1 The triangle T_4. Primitive integer vectors are given by dashed lines, and are labeled by their slopes.

1.4 Farey sequences

Consider the integer lattice $\mathbb{Z}^2$. If we imagine an observer sitting at the origin 0, the 'visible' points in $\mathbb{Z}^2$ correspond to the set of *primitive* vectors, that is, integer vectors which are not integer multiples of other integer vectors. If we consider slopes of vectors (as opposed to angles), it is natural to consider the set of vectors with slopes in $[0, 1]$. The set of slopes of (primitive) integer vectors with horizontal component at most Q intersected with the interval $[0, 1]$ gives the *Farey sequence* of level Q. More simply, $\mathcal{F}(Q)$ consists of the set of fractions in between 0 and 1 with denominator at most Q. We write

$$\mathcal{F}(Q) := \left\{ \gamma_0 = \frac{0}{1} < \gamma_1 = \frac{1}{Q} < \gamma_2 \ldots < \gamma_i = \frac{p_i}{q_i} < \ldots \gamma_N = \frac{1}{1} \right\}$$

Here, $N = N(Q) = \sum_{i=1}^{Q} \varphi(i)$ is the cardinality of $\mathcal{F}(Q)$. By the above discussion, these correspond to the slopes of primitive integer vectors $\binom{q_i}{p_i}$ in the (closed) triangle T_Q with vertices at $(0,0)$, $(Q, 0)$, and (Q, Q). That is, it is bounded above by the line $\{y = x\}$, below by the x-axis, and on the right by the line $\{x = Q\}$. The triangle T_4 is shown in Figure 1.

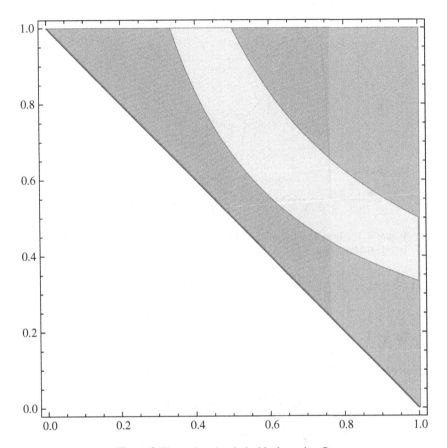

Figure 2 The region $A_{a,b}$ in inside the region Ω.

The sequences $\mathcal{F}(Q)$ equidistribute in $[0, 1]$ (by, for example, Weyl's criterion [22]). We denote by $\mathcal{G}(Q)$ the set of normalized gaps between Farey fractions, that is,

$$\mathcal{G}(Q) = \left\{ N(Q)(\gamma_{i+1} - \gamma_i) = \frac{N(Q)}{q_i q_{i+1}} : 0 \le i < N(Q) \right\}.$$

The limiting distribution for $\mathcal{G}(Q)$ is given by the following beautiful theorem of R. R. Hall, and illustrated in Figures 2 and 3 [2]. Let

$$\Omega := \{(u, v) \in [0, 1]^2 : u + v > 1\},$$

---

[2] Color versions of the Figures in this paper are available on the author's webpage.

Jayadev S. Athreya

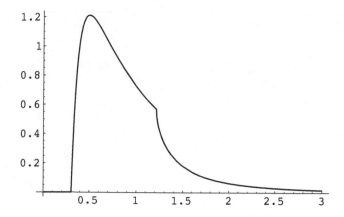

Figure 3 The limiting distribution of gaps for Farey fractions and, (appropriately rescaled) lattice slopes.

and for $0 \le a < b \le \infty$, let

$$A_{a,b} = \left\{ (u, v) \in \Omega : b^{-1} < uv < a^{-1} \right\}, \qquad (5)$$

and set $\tilde{A}_{a,b} := A_{\frac{\pi^2}{3}a, \frac{\pi^2}{3}b}$.

Theorem 1 *[11, R.R.Hall] Fix $0 \le a < b < \infty$. Then*

$$\lim_{Q \to \infty} \frac{|\mathcal{G}(Q) \cap (a, b)|}{N(Q)} = 2|\tilde{A}_{a,b}|.$$

Differentiating the cumulative distribution function

$$F_{Hall}(t) := |\tilde{A}_{0,t}|,$$

one can compute the probability distribution function $P_{Hall}(t)$ so that

$$\int_a^b P_{Hall}(t)dt := 2|\tilde{A}_{a,b}|.$$

We call this distribution (and any scalings) *Hall's distribution*. The graph of $P_{Hall}(t)$ is given in Figure 3, which is drawn from [6]. The points of non-differentiability $\frac{3}{\pi^2}$ and $\frac{12}{\pi^2}$ correspond to the transitions when the hyperbola $\left\{ xy = \frac{3}{\pi^2}t^{-1} \right\}$ enters the region Ω $\left(t = \frac{3}{\pi^2} \right)$ and when it hits the line $x + y = 1$ $\left(t = \frac{12}{\pi^2} \right)$. In Section 3, we will, following [4], give a proof of Hall's theorem inspired by the work of F. Boca, C. Cobeli, and A. Zaharescu [5]. They created a map $T : \Omega \to \Omega$, now known as the *BCZ map*, and used equidistribution properties of periodic orbits of this map to obtain many statistical results on $\mathcal{F}(Q)$ and $\mathcal{G}(Q)$. In [4], the author and Y. Cheung showed that these results could be obtained by studying the horocycle flow on the space $X_2 = SL(2, \mathbb{R})/SL(2, \mathbb{Z})$ of unimodular lattices in $\mathbb{R}^2$.

Geometry of Numbers

One can also study the behavior of an arbitrary unimodular lattice Λ. Let $\Lambda \subset \mathbb{R}^2$ be a unimodular lattice, and suppose Λ does not have vertical vectors. Let $\{s_1 < s_2 < \ldots < s_n < \ldots\}$ denote the slopes of the vectors (written in increasing order) in the vertical strip $V_1 = \{(u, v)^T : u \in (0, 1], v > 0\}$. Here, and below, we use $(u, v)^T$ to denote the *column vector* $\begin{pmatrix} u \\ v \end{pmatrix}$, as our matrices act on the left. Let

$$G_N(\Lambda) = \{s_{n+1} - s_n : 0 \leq n \leq N\}$$

denote the set of gaps in this sequence. Note that in this setting, we do not need to normalize, as our sequence is not contained in $[0, 1)$. Then we have that the limiting distribution of G_N is also given by Hall's distribution. That is:

Theorem 2 *[4] Let $0 \leq a < b \leq \infty$. Then*

$$\lim_{N \to \infty} \frac{1}{N} |G_N(\Lambda) \cap (a, b)| = 2|A_{a,b}|.$$

1.5 Saddle Connections

We saw above that the Farey sequence could be interpreted geometrically as slopes of primitive integer vectors in $\mathbb{R}^2$. Primitive integer vectors also correspond to (parallel families) of closed geodesics on the torus $\mathbb{R}^2/\mathbb{Z}^2$, which can also be interpreted as closed billiard trajectories in the square $[0, 1/2]^2$. A natural generalization would be to try and understand similar families of trajectories for higher-genus surfaces, and/or for billiards in more complex polygons More precisely, let P be a Euclidean polygon with angles in $\pi \mathbb{Q}$. The billiard dynamical system on P is given the (frictionless) motion of a point mass at unit speed with elastic collisions with the sides, satisfying the law of geometric optics: *angle of incidence = angle of reflection*. A *generalized diagonal* for the polygon P is a trajectory for the billiard flow that starts at one vertex of P and ends at another vertex. Since the group Δ_P generated by reflections in the sides of P is finite, the *angle* of a trajectory is well defined in $S^1 \cong S^1/\Delta_P$. The natural gap distribution question that arises in this context is:

Question 3 What is the limiting distribution of the gaps between angles of generalized diagonals (normalized in terms of the length)?

More generally, one can ask about the limiting distribution for gaps for *saddle connections*) in the more general setting of translation surfaces. A translation surface is a pair (M, ω), where M is a Riemann surface and ω a holomorphic 1-form.

A saddle connection is a geodesic γ in the flat metric induced by ω, connecting two zeros of ω. To each saddle connection γ associate the holonomy vector $\mathbf{v}_\gamma = \int_\gamma \omega \in \mathbb{C}$. The set of holonomy vectors $\Lambda_{sc}(\omega)$ is a discrete subset of $\mathbb{C} \cong \mathbb{R}^2$, and varies equivariantly under the natural $SL(2, \mathbb{R})$ action on the set of translation surfaces. Motivated by such concerns, and inspired by the work of Marklof-Strombergsson [14], the author and J. Chaika [2] studied the gap distribution for *saddle connection directions*. The relationship between flat surfaces and billiards in polygons is given by a natural *unfolding* procedure, which associates to each (rational) polygon P a translation surface (X_P, ω_P). The main result of [2] used the dynamics of the $SL(2, \mathbb{R})$ action on the moduli space Ω_g of genus g translation surfaces to show that generically, a limiting distribution exists.

More precisely, given $R > 0$, let

$$F_R^\omega := \{\arg(\mathbf{v}) : \mathbf{v} \in \Lambda_\omega \cap B(0, R)\} \tag{6}$$

denote the set of directions of saddle connections of length at most R. Masur [16] showed that the counting function $N(\omega, R) := |F^\omega(R)|$ grows quadratically in R for any ω. Denote the associated normalized gap set by $G^\omega(R)$.

Theorem 4 *([2, Theorem 1.1]) For almost every (with respect to Lebesgue measure on Ω_g) translation surface ω, there is a limiting distribution for the gap set $G^\omega(R)$. Moreover, this distribution has support at 0, that is, for almost every $\omega \in \Omega_g$, and for any $\epsilon > 0$,*

$$\lim_{R \to \infty} \frac{|G^\omega(R) \cap (0, \epsilon)|}{N(\omega, R)} > 0. \tag{7}$$

Lattice Surfaces

The support at 0 in Theorem 4 is in contrast to the setting of the torus, where, as seen in Figure 3, there a gap between 0 and $3/\pi^2$. This gap at 0 is, in some sense, due to the symmetry of the torus- if we think of the $SL(2, \mathbb{R})$ action on the moduli space X_2 of flat tori, the stabilizer of any point is (conjugate to) $SL(2, \mathbb{Z})$. More generally, It was shown in [2] that if ω is a *lattice surface* (i.e., the stabilizer of the flat surface ω under the $SL(2, \mathbb{R})$ action is a lattice) that the limiting distribution for gaps has no support at 0.

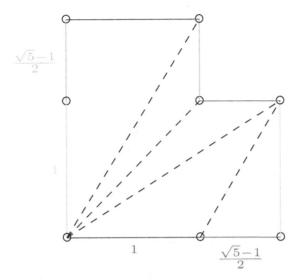

Figure 4 The Golden L. The long sides of the L each have length $\frac{1+\sqrt5}{2}$.

While it was in principle possible to compute the limiting distribution using the techniques in [2], the more geometric nature of the techniques in [4] and the use of horocycle flows on moduli spaces can be generalized to the setting of lattice surfaces to give a roadmap for explicitly calculating the limiting distribution of gaps. In joint work [3] with J. Chaika and S. Lelievre, we proved Theorem 5 on the gap distribution for the golden L, which is a surface of genus 2 with one double zero, displayed in Figure 4.

Theorem 5 *[3] There is an explicit limiting gap distribution for the set of slopes (equivalently, angles) for saddle connections on the golden L. The probability distribution function is differentiable except at a set of eight points.*

Remark: The limiting and empirical distributions are shown in Figure 5, drawn from [3]. We refer the reader to [3] for the precise formulas for the limiting distribution.

1.6 Visible affine lattice points

Another natural generalization of the Farey sequence is to consider *affine lattices*, that is, translates of lattices by some fixed vector. We write

$$\Lambda = M\mathbb{Z}^2 + \mathbf{v},$$

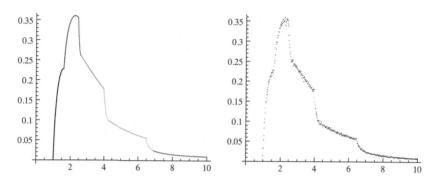

Figure 5 The limiting and empirical distributions for gaps of saddle connection slopes on the golden L.

where $M \in SL(2, \mathbb{R})$ and $\mathbf{v} \in \mathbb{R}^2$ (really $\mathbf{v}$ is well-defined up to the lattice $M\mathbb{Z}^2$, so we think of it as an element of the torus $\mathbb{R}^2/M\mathbb{Z}^2$). Marklof-Strombergsson [14] used dynamics on the space of affine lattices $\tilde{X}_2 = SL(2, \mathbb{R}) \ltimes \mathbb{R}^2/SL(2, \mathbb{Z}) \ltimes \mathbb{Z}^2$ to study the gap distribution for the angles of visible affine lattice points. They in fact considered much more general problems, studying the distribution of visible affine lattice points in higher dimensions, but for the purposes of this paper, we focus on their two-dimensional results.

Consider the set of angles of lattice points of length at most R, that is,

$$F_\Lambda(R) := \{\arg \mathbf{w} : \mathbf{w} \in \Lambda \cap B(0, R)\}.$$

To calculate the associated gap distribution P_Λ, the key is to estimate the probability of finding multiple lattice points in 'thinning' wedges. Given $\sigma > 0, \theta \in [0, 2\pi)$ and $R > 0$ consider the wedge

$$A_R^\theta(\sigma) := \{\mathbf{w} \in \mathbb{R}^2 : \mathbf{w} \in B(0, R), \arg(\mathbf{w}) \in (\theta - \sigma R^{-2}, \theta + \sigma R^{-2})\},$$

shown in Figure 6. Here, the factor of R^{-2} corresponds to the normalizing factor $\frac{1}{N}$ above, since the cardinality of $F_\Lambda(R)$ is on the order of R^2. The gap distribution will be given by (the second derivative) of the limiting probability

$$p_{\Lambda,0}(\sigma) = \lim_{R \to \infty} \lambda(\theta : A_R^\theta(\sigma) \cap \Lambda = \emptyset)$$

that this wedge does not affine lattice points. This follows from the fact that if we let $P_\Lambda(t)$ denote the probability distribution function of the limiting gap

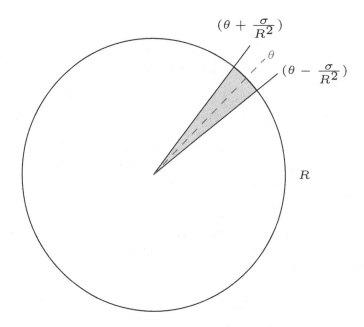

Figure 6 The wedge $A_R^\theta(\sigma)$.

distribution, we have

$$p_{\Lambda,0}(\sigma) = \sigma - \int_0^\sigma \int_0^t P_\Lambda(s)\,ds\,dt \qquad (8)$$

To compute $p_{\Lambda,0}(\sigma)$, note that rotating the region $A_R^\theta(\sigma)$ by the element

$$r_{-\theta} = \begin{pmatrix} \cos\theta & \sin\theta \\ -\sin\theta & \cos\theta \end{pmatrix}, \qquad (9)$$

and scaling by

$$g_t = \begin{pmatrix} e^{-t/2} & 0 \\ 0 & e^{t/2} \end{pmatrix}, \qquad (10)$$

(with $t = 2\log R$) we obtain (approximately) the triangle $T(\sigma)$ with vertices at $(0, 0)$ and $(1, \pm\sigma)$. Thus, the probability that (for a randomly chosen θ) the region A_R^θ does not contain points from Λ corresponds to the integral

$$\int_0^{2\pi} h_\sigma(g_t r_\theta \Lambda)\,d\theta,$$

where $h_\sigma : \tilde{X}_2 \to \{0, 1\}$ is given by

$$h_\sigma(\Lambda) = \begin{cases} 1 & \Lambda \cap T(\sigma) \neq \emptyset \\ 0 & \text{otherwise} \end{cases} \tag{11}$$

To understand the limits (as R, or equivalently $t \to \infty$) of the integrals $\int_0^{2\pi} h_\sigma(g_t r_\theta \Lambda) d\theta$, we need to apply tools from Ratner's theory of unipotent flows. It can be shown that these integrals converge to $\int_{\tilde{X}_2} h_\sigma d\mu$, where the limit measure μ can be shown to be invariant under a unipotent subgroup $N \subset SL(2, \mathbb{R}) \ltimes \mathbb{R}^2$, following work of Shah [20]. Using Ratner's Theorem [18], the possibilities for the measure μ are either:

- The Haar probability measure $\tilde{\mu}_2$
- A probability measure supported on the set of torsion points $\tilde{X}_2[n]$, that is, the support of the measure is restricted to affine lattices $M\mathbb{Z}^2 + \mathbf{v}$ whose translation vector $\mathbf{v}$ satisfies $n\mathbf{v} \in M\mathbb{Z}^2$.

Depending on the properties of the initial affine lattice Λ, torsion-supported measures can (and do) occur. As a consequence, the limiting gap distributions (which can all be explicitly computed) differ depending on the initial lattice, and in particular, whether the initial translation vector $\mathbf{v}$ is a torsion point of the torus $\mathbb{R}^2/M\mathbb{Z}^2$ or not, and at what level [14, Theorem 1.3].

$\sqrt{n}$ mod 1

Extraordinarily, the gap distribution for generic affine lattices Λ (that is, those for which the vector $\mathbf{v}$ is not torsion, and so for which the limiting measure is Haar) coincides with the gap distribution for the sequence of fractional parts of $\{\sqrt{n}\}$. The reason they coincide is that both are coming from equidistribution of certain homogeneous trajectories with respect to the Haar measure on the space $\tilde{X}_2$. We strongly urge the reader to look at the beautiful paper [8] for further details on the connection of $\sqrt{n}$ mod 1 to homogeneous dynamics.

2 Meta-Theorems on Gap Distributions

As discussed in the introduction, the common thread that runs through these results is the creation of associated discrete sets of $\mathbb{R}^2$ and appropriate dynamical systems, which turns gap distribution questions into questions of equidistribution of orbits on certain moduli spaces.

2.1 Setup

Our setup is very similar to that of [1]. Let X be a locally compact metric space with a continuous $SL(2, \mathbb{R})$-action, and a $SL(2, \mathbb{R})$-equivariant assignment

$$x \longmapsto \Lambda_x$$

that associates to each point $x \in X$ a countable, discrete subset $\Lambda_x \subset \mathbb{R}^2 \backslash \{0\}$. Equivariance, in this context, means that

$$\Lambda_{gx} = g\Lambda_x$$

for all $g \in SL(2, \mathbb{R})$, $x \in X$. Our theorems will connect the gap distributions for slopes and angles in the set Λ_x to the orbit of the point x under various subgroups of the $SL(2, \mathbb{R})$ action.

2.2 Slope gaps

The key idea of our meta-theorems comes from a simple observation on the behavior of slopes under (vertical) shears (a corresponding discussion can be made with inverse slopes and horizontal shears). Let $\mathbf{u} = (u_1, u_2)^T$, $\mathbf{v} = (v_1, v_2)^T \in \mathbb{R}^2$. Let $s_{\mathbf{u}} = \frac{u_2}{u_1}$ and $s_{\mathbf{v}} = \frac{v_2}{v_1}$ denote their slopes. Let

$$h_s = \begin{pmatrix} 1 & 0 \\ -s & 1 \end{pmatrix}. \tag{12}$$

Observation For any $s \in \mathbb{R}$, he difference in slopes between $h_s\mathbf{u}$ and $h_s\mathbf{v}$ is the same as the difference in slopes between $\mathbf{u}$ and $\mathbf{v}$, that is,

$$s_{h_s\mathbf{u}} - s_{h_s\mathbf{v}} = s_{\mathbf{u}} - s_{\mathbf{v}}.$$

This follows from the (even simpler) observation that for any $\mathbf{u} \in \mathbb{R}^2$,

$$s_{h_s\mathbf{u}} = s_{\mathbf{u}} - s.$$

Thus, if we have a set of slopes $\mathbb{S}(\Lambda) = \{s_1 < s_2 < \dots s_n < \dots\}$ of a family of vectors $\Lambda \subset \mathbb{R}^2$, the behavior of the associated gap set $\Gamma(\Lambda) = \{s_{i+1} - s_i : 1 \le i \le \infty\}$ is *invariant* under the action of h_s, that is,

$$\Gamma(h_s\Lambda) = \Gamma(\Lambda).$$

In particular, by considering the flow h_s at the times $s_1, s_2, \dots$, we can recover the gap $s_{i+1} - s_i$ by sampling the smallest positive element of $\mathbb{S}(h_{s_i}\Lambda)$.

Minkowski properties, short vectors, and vertical trips

To state our theorem on slopes, we require a few definitions.

Symmetry. Λ_x is centrally symmetric, $-\Lambda_x = \Lambda_x$.

Minkowski Property. For all $x \in X$, there is a $c = c(x)$ such that for any convex, centrally symmetric set $K \subset \mathbb{R}^2$ of volume at least c,

$$K \cap \Lambda_x \neq \emptyset.$$

The minimal such constant $c(x)$ is *invariant* under the action of $SL(2, \mathbb{R})$.

Vertically and horizontally short points. $x \in X$ is *vertical* (respectively *horizontal*) if Λ_x contains a vertical (resp. horizontal) vector. Given a constant η, x is *η-vertically short* (resp. *η-horizontally short*) if Λ_x contains a vertical (resp. horizontal) vector of length at most η.

Vertical Strip Property. For $\eta > 0$, et $V_\eta := \{(u, v)^T : 0 < u \leq \eta, v > 0\}$ denote the vertical strip of width η in $\mathbb{R}^2$. Suppose x is horizontal. Then $\Lambda_x \cap V_\eta$ is non-empty for any $\eta > 0$.

Exceptional points. x is *η-exceptional* if it is $\frac{\eta}{4c(x)}$-vertically short, that is, if it contains a vertical vector of length at most $\frac{\eta}{4c(x)}$. We denote by X_η the set of non-η-exceptional points.

Horocycles and transversals

Since h_s fixes vertical vectors, X_η is an h_s-invariant set for any $\eta > 0$.

Lemma 6 *Let $\eta > 0$, and let $\Omega_\eta \subset X$ denote the set of η-horizontally short points. Then Ω_η is a transversal to the h_s action on the set X_η. That is, for every $x \in X_\eta$, the h_s-orbit of x intersects Ω_η in a non-empty, discrete set of times.*

We prove this lemma in Section 5. This lemma yields the existence of the induced return map $T_\eta : \Omega_\eta \to \Omega_\eta$ and return time function $R_\eta : \Omega_\eta \to \mathbb{R}^+$, given by

$$R_\eta(x) = \min\{s > 0 : h_s x \in \Omega_\eta\}$$

and $T^t(x) = h_{R(x)}x$. In particular, any finite h_s-invariant measure ν on X_η can be decomposed as $d\nu = d\tilde\nu ds$, where $\tilde\nu$ is a T_η-invariant probability measure on Ω_η and we identify X_η with the suspension space

$$\{(x, s) : x \in \Omega_\eta, 0 \leq s \leq R_\eta(x)\}/ \sim,$$

where $(x, R_\eta(x)) \sim (T_\eta(x), 0)$.

Long horocycles

Define

$$\mathbb{S}_\eta(x) := \{0 \le s_1^\eta(x) < s_2^\eta(x) < \ldots < s_n^\eta(x) < \ldots\} \tag{13}$$

Let

$$\Gamma_N^\eta(x) := \{1 \le i \le N : s_{i+1}^\eta(x) - s_i^\eta(x)\} \tag{14}$$

denote the set of the first N slope gaps. Here, we do not normalize, since the slopes s_i^t are going to ∞ in all of our applications. Given $x \in X$, and $S > 0$, let $\sigma_{x,S}$ denote the Lebesgue probability measure $d\sigma_{x,S} = \frac{1}{S}ds$ on the segment $\{h_s x : 0 \le s \le S\}$.

Theorem 7 *Let $t > 0$. Suppose $\sigma_{x,S} \to \nu$ as $S \to \infty$ (here, and below, convergence of measures is in the weak-$\star$ topology), where ν is a finite measure supported on X_η, and $d\nu = d\tilde{\nu}ds$ as above. Then for any $0 \le a \le b \le \infty$, we have*

$$\lim_{N \to \infty} \frac{|\Gamma_N^\eta(x) \cap [a, b]|}{N} = \tilde{\nu}\left(R_\eta^{-1}([a, b])\right). \tag{15}$$

That is, the slope gap distribution is given by the distribution of the return time function R_η.

This theorem has the following immediate corollary:

Corollary 8 *Let ν be a ergodic, h_s-invariant probability measure supported on X_η. Then for ν-almost every x,*

$$\lim_{N \to \infty} \frac{|\Gamma_N^\eta(x) \cap [a, b]|}{N} = \tilde{\nu}\left(R_\eta^{-1}([a, b])\right).$$

Long closed horocycles

We can also state a theorem for x with periodic h_s-orbits. $x \in X$ is h_s-*periodic* with period $s_0 > 0$ if $h_{s_0}x = x$, and $h_s x \ne x$ for all $0 < s < s_0$.

Lemma 9 *Let x be h_s-periodic. Then for every $\eta > 0$ there is an $N_0 = N_0(\eta)$ such that for any $N \ge N_0$,*

$$\Gamma_N^\eta(x) = \Gamma_{N_0}^\eta(x).$$

That is, the set of gaps repeats.

We prove the lemma in Section 5. Let $g_t = \mathrm{diag}(e^{t/2}, e^{-t/2})$ be as in (10). Then we have the conjugation relation

$$g_t h_s g_{-t} = h_{se^{-t}}. \tag{16}$$

Thus, if x is h_s-periodic with period s_0, $g_{-t}x$ is $h_{s_0 e^t}$-periodic, since

$$h_{s_0 e^t} g_{-t} x = g_{-t} h_{s_0} x = g_{-t} x.$$

If $x \in X_\eta$, then $g_{-t}x \in X_{e^{t/2}\eta}$ for any $t > 0$, since g_{-t} expands by $e^{t/2}$ in the vertical direction. Let $\rho_{x,T}$ denote the Lebesgue probability measure $d\rho_{x,T} = \frac{ds}{e^T s_0}$ supported on the periodic orbit

$$\{h_s g_{-T} x : 0 \le s \le e^T s_0\}.$$

Theorem 10 *Suppose $\rho_{x,T} \to \nu$ as $T \to \infty$, with ν a finite measure supported on X_{η_0} for some $\eta_0 > 0$ (in fact, it will be supported on any X_η), and $d\nu = d\tilde\nu ds$, with $\tilde\nu$ the T_{η_0}-invariant probability measure on X_{η_0}. Then*

$$\lim_{\eta \to \infty} \frac{\left| \left(\frac{\eta}{\eta_0}\right)^2 \Gamma^\eta_{N_0(\eta)}(x) \cap [a,b] \right|}{N_0(\eta)} = \tilde\nu\left(R_{\eta_0}^{-1}([a,b])\right). \tag{17}$$

The right hand choice of this limit does not depend on the choice of η_0, which can be chosen to be any positive number.

2.3 Angle gaps and thinning wedges

In this section, we work with angles, as opposed to slopes. Let $x \in X$, and let

$$\Theta_x(R) := \{0 < \theta_1 < \theta_2 < \ldots < \theta_N < 2\pi\}$$

denote the set of angles of vectors in the set $\Lambda_x \cap B(0,R)$, where $B(0,R) = \{\mathbf{v} \in \mathbb{R}^2 : \|\mathbf{v}\|_2 \le R\}$ is the ball of radius R in $\mathbb{R}^2$, and $N = N(R)$ is the cardinality of the set of distinct angles. Arguing as in Section 1.6, the gap distribution for $\Theta_x(R)$ can be reduced to studying the limiting probability

$$p_0(x,\sigma) = \lim_{R \to \infty} \lambda(\theta : A_\theta^\sigma(R) \cap \Lambda_x = \emptyset),$$

where λ denotes the Lebesgue probability measure on S^1, $\sigma > 0$ is a fixed parameter, and

$$A_R^\theta(\sigma) := \{\mathbf{v} \in \mathbb{R}^2 : \mathbf{v} \in B(0,R), \arg(\mathbf{v}) \in (\theta - \sigma R^{-2}, \theta + \sigma R^2)\}, \tag{18}$$

as shown in Figure 6. Let $\lambda_{x,R}$ denote the Lebesgue probability measure $d\lambda_{x,R} = \frac{d\theta}{2\pi}$ supported on the orbit $\{g_t r_\theta x\}_{0 \leq \theta < 2\pi}$, with $t = 2 \ln R$.

Theorem 11 *Suppose $\lambda_{x,R} \to v$ as $R \to \infty$, where v is a probability measure on X. Then*

$$p_0(x, \sigma) = v(x \in X : \Lambda_x \cap T(\sigma) = \emptyset), \qquad (19)$$

where $T(\sigma)$ is the triangle with vertices at $(0,0)$ and $(1, \pm\sigma)$. Moreover, if we define

$$p_i(x, \sigma) := \lim_{R \to \infty} \lambda(\theta : |A_\theta^\sigma(R) \cap \Lambda_x| = i),$$

we have

$$p_i(x, \sigma) = v(x \in X : |\Lambda_x \cap T(\sigma)| = i). \qquad (20)$$

3 Farey sequences

We show how to use Theorem 10 to prove Hall's Theorem 1, and how to use Theorem 7 to prove the geometry of numbers result Theorem 2. We follow the exposition in [4], where it was shown how these theorems can be obtained from results on equidistribution of long orbits of the group $\{h_s\}$ on the space of unimodular lattices $X_2 = G/\Gamma$, where $G = SL(2, \mathbb{R})$ and $\Gamma = SL(2, \mathbb{Z})$. Following the notation in Section 2, our space X is given by the space X_2, the assignment of a discrete set is given by

$$g SL(2, \mathbb{Z}) \longmapsto g\mathbb{Z}_{prim}^2,$$

and we see immediately that the Minkowski and symmetry conditions of Section 2.2 are satisfied. It is also possible to verify the vertical strip condition explicitly (see [4]). We note that $\mathbb{Z}^2$ is h_s-periodic with period 1, since $h_1\mathbb{Z}^2 = \mathbb{Z}^2$. Given $\eta > 0$, the set of slopes in V_η is the set of non-negative rational numbers with denominator at most $Q = \lfloor \eta \rfloor$. That is, we can write

$$\mathbb{S}_N^\eta(\mathbb{Z}^2) := \{n + \mathcal{F}(Q) : n \geq 0\},$$

where n ranges over the non-negative integers. Thus, the gap set $\Gamma_N^\eta(\mathbb{Z}^2)$ corresponds to the set of gaps in the Farey sequence $\mathcal{F}(Q)$, and has cardinality $N(Q)$. We recall that Sarnak [19] proved that the measures $\rho_{\mathbb{Z}^2, T}$ equidistribute with respect to the Haar measure probability μ_2. That is,

$$\rho_{\mathbb{Z}^2, T} \to \mu_2.$$

Thus, to complete the proof of Theorem 1 (assuming Theorem 10), we need to describe the distribution of the return time function (with respect to the disintegrated measure) on the transversal

$$\Omega_\eta := \{\Lambda \in X_2 : \Lambda \text{ is } \eta - \text{horizontally short}\},$$

for some choice of $\eta > 0$.

3.1 Description of transversal and return map

The main result of [4] is to explicitly describe the transversal Ω_1 (more generally, Ω_η) and the associated return map and return time function. It was shown that Ω_η is in bijective correspondence with the set

$$\{(a, b) \in \mathbb{R}^2 : a, b \in (0, \eta], a + b > \eta\},$$

via the map

$$(a, b) \mapsto \Lambda_{a,b} = \begin{pmatrix} a & b \\ 0 & a^{-1} \end{pmatrix} \mathbb{Z}^2$$

and that the roof function is given by

$$R_\eta(a, b) = \frac{1}{ab}.$$

In particular, it does not depend on η. If we set $\eta = 1$, the return map is the *BCZ map* (named for its creators, Boca-Cobeli-Zaharescu [5]), given by

$$T_1(a, b) = \left(b, -a + \left\lfloor \frac{1 + a}{b} \right\rfloor \right).$$

More generally,

$$T_\eta(a, b) = \left(b, -a + \left\lfloor \frac{\eta + a}{b} \right\rfloor \right).$$

T_η is piecewise linear, and all the linear maps have determinant 1, and so the Lebesgue probability measure $\frac{2}{\eta^2} da\,db$ on Ω_b is T_η-invariant, and that $da\,db\,ds$ is, up to scaling, Haar measure on X_2 (which gives full measure to the set $X_{2,\eta}$ for any $\eta > 0$). The distribution of the roof function R can be seen to be independent of η (see Section 5.1 for an explanation of a general self-similarity phenomenon), so we can choose $\eta = 1$, obtaining the result of Theorem 1 up to the natural normalization factor $\frac{3}{\pi^2}$, which occurs since in the statement of Theorem 1 we normalize the gaps with a factor of $N(Q) \sim \frac{3}{\pi^2} Q^2$, as opposed to the factor Q^2 in the statement of Theorem 10. □

3.2 Proof of Theorem 2

The above computations are also crucial in the proof of Theorem 2 assuming Theorem 7. The other crucial step in the proof is Dani's measure classification [7] result for h_s-invariant measures on X_2, which states that any ergodic h_s-invariant probability measure on X_2 is either supported on a periodic orbit or Haar measure. A lattice is h_s-periodic if and only if it has a vertical vector. To show that the existence of a vertical vector implies periodicity, note that such lattices can be written as

$$\Lambda = \begin{pmatrix} 0 & -a^{-1} \\ a & b \end{pmatrix} \mathbb{Z}^2,$$

and it is a direct verification to show

$$h_{a^2}\Lambda = \begin{pmatrix} 0 & -a^{-1} \\ a & b-a \end{pmatrix} \mathbb{Z}^2 = \begin{pmatrix} 0 & -a^{-1} \\ a & b-a \end{pmatrix} \begin{pmatrix} 1 & 1 \\ 0 & 1 \end{pmatrix} \mathbb{Z}^2 = \Lambda.$$

On the other hand, if Λ is h_s-periodic, it must be *divergent* under g_t, which is equivalent to the existence of a vertical vector. This measure classification and characterization, combined with the Birkhoff ergodic theorem, yields:

Lemma 12 *Let $\Lambda \in X_2$ be a lattice without vertical vectors. Then the measures $\sigma_{\Lambda,S}$ converge to the Haar probability measure μ_2.*

Thus, applying Theorem 7, we see that for any $\eta > 0$,

$$\lim_{N\to\infty} \frac{1}{N}|\Gamma_N^\eta(\Lambda) \cap (a,b)| = \frac{2}{\eta^2}|\{(u,v) \in \Omega_\eta : R_\eta(u,v) \in (a,b)\}| = 2|A_{a,b}|$$

where $A_{a,b}$ is as in (5), and $|\cdot|$ is Lebesgue measure $dxdy$. The statement of Theorem 2 is the above statement for $\eta = 1$. □

4 Saddle connections

In this section, we sketch a proof of Theorem 4, following a similar strategy to the outline in Section 1.6, and attempting to indicate how it can be seen as a consequence of Theorem 11. We will particularly focus on the conclusion that for generic tranlsation surfaces, the gap distribution has support at 0. Recall notation: Ω_g is the moduli space of holomorphic 1-forms on compact genus g Riemann surfaces, and a saddle connection γ on a surface $(M,\omega) \in \Omega_g$ is a geodesic (with respect to the flat metric given by the one-form) connecting two zeros of ω with no zeros in its interior. The holonomy vector $\mathbf{v}_\gamma \in \mathbb{C} \equiv \mathbb{R}^2$

associated to γ is given by

$$\mathbf{v}_\gamma = \int_\gamma \omega.$$

The set of holonomy vectors Λ_ω is a discrete subset of $\mathbb{R}^2$, and we are interested in the set of *small gaps* between directions of vectors. Precisely, letting $F_R^\omega :=$ $\{\arg(\mathbf{v}) : \mathbf{v} \in \Lambda_\omega \cap B(0, R)\}$ denote the set of directions, we are interested in the normalized gap set $G^\omega(R)$, and in particular the size of the set $G^\omega(R) \cap (0, \epsilon)$ for $\epsilon > 0$. Note that having a small gap is equivalent to having two directions in a 'thin wedge'.

4.1 $SL(2, \mathbb{R})$-action

A point $(M, \omega) \in \Omega_g$ determines (and is determined by) an atlas of charts from the surface punctured at the zeros of ω to the plane $\mathbb{C}$, given by integration of ω. The transition maps for these charts are translations, and ω is given by dz in these coordinates. This also gives a natural $SL(2, \mathbb{R})$ action on Ω_g via linear postcomposition with charts. There is ([15], [21]) a natural $SL(2, \mathbb{R})$-invariant probability measure on Ω_g, known as the Masur-Veech measure, which we denote μ_g.

Equivariance

The assigment $(M, \omega) \mapsto \Lambda_\omega$ is an $SL(2, \mathbb{R})$-equivariant assignment: for any $h \in SL(2, \mathbb{R})$,

$$\Lambda_{h\omega} = h\Lambda_\omega.$$

4.2 Renormalization

As above, we consider the wedges $A_R^\theta(\sigma)$. To understand the support of the distribution at 0, we consider probability that for a randomly chosen θ there are at least 2 points in the thinning wedge $A_R^\theta(\sigma)$. This will give us the probability that our gap set has gaps of size at most σ. As above, renormalizing by g_t and r_θ, we can write this probability as

$$\int_0^{2\pi} h_{2,\sigma}(g_t r_\theta \Lambda) d\theta,$$

where $h_{2,\sigma} : \Omega_g \to \{0, 1\}$ is given by

$$h_{2,\sigma}(\omega) = \begin{cases} 1 & |\Lambda_\omega \cap T(\sigma)| \geq 2 \\ 0 & \text{otherwise} \end{cases} \tag{21}$$

4.3 Equidistribution

Since for $g \geq 1$ the space Ω_g is not a homogeneous space, one cannot use Ratner-type technology to classify the possible limit measures for the measures $\lambda_{\omega, R}$ supported on $\{g_t r_\theta \omega\}_{0 \leq \theta < 2\pi}$. However, Eskin-Masur [9, Theorem 1.5] used a general ergodic result of Nevo to show that for μ_g-almost every starting point ω, the measures converge (in an appropriately smoothed sense) to μ_g, that is, the main assumption of Theorem 11 is satisfied, with $\nu = \mu_g$. It is worth noting that *there is no known example of a point $\omega \in \Omega_g$ for which there does not exist a measure ν satisfying the hypothesis of Theorem 11.* Using Nevo's theorem, the author and J. Chaika [2] showed:

$$\lim_{t \to \infty} \int_0^{2\pi} h_{2,\sigma}(g_t r_\theta \Lambda) d\theta = \int_{\Omega_g} h_{2,\sigma} d\mu_g,$$

Finally, a measure estimate on the set of surfaces with simultaneous short saddle connections due to [17] (see also [10, Lemma 7.1]) shows that the integral on the right hand side is positive for any $\sigma > 0$.

4.4 Slopes

For certain measures, Theorem 7 and Corollary 8 be applied in this setting, as the Minkowski condition is satisfied by [13, Theorem 1]. However, to give gap distributions for slopes of surfaces in the support of h_s-invariant measures on Ω_g, one still has to verify the vertical strip condition for the support of these measures, and then, identify the transversal, the roof function, and the associated measure $\tilde{\nu}$ on the transversal in order to get explicit formulas. If we have a *closed* $SL(2, \mathbb{R})$ orbit, then the points in that orbit are lattice surfaces, where it is possible to carry out this program.

4.5 Lattice Surfaces

If (M, ω) is a lattice surface (for example the golden L), with stabilizer $\Gamma \subset SL(2, \mathbb{R})$, we restrict our parameter space X to from Ω_g to the $SL(2, \mathbb{R})$-orbit of (M, ω), which can be identified with $SL(2, \mathbb{R})/\Gamma$ via

$$g(M, \omega) \longmapsto g\Gamma.$$

It is possible (see [3]) to verify the vertical strip condition in this setting. Since the stabilizer Γ must be non-uniform, we can assume that (M, ω) is h_s-periodic, that is, that Γ contains an element h_{s_0} for some $s_0 > 0$. Applying Theorem 10, we obtain that the limiting distribution of the slope gaps for vectors in Λ_ω (in

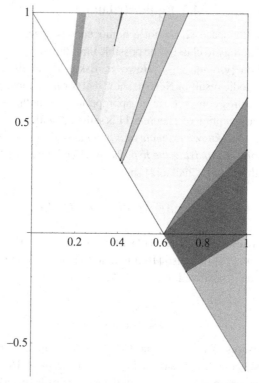

Figure 7 The transversal X_1 to the horocycle flow h_s for the $SL(2, \mathbb{R})$-orbit of the golden L. The colors indicate different behaviors of the return map.

the first quadrant) is given by the distribution of the return time function R for h_s to the transversal $X_1 := \{g\Gamma \in X : \Lambda_{g\omega} \text{ is } 1\text{-horizontally short}\}$. In [3], the transversal X_1, the return time function, and the return map were explicitly computed for the translation surface (M, ω) associated to the golden L. The transversal, in this case, can be identified with the triangle with vertices at $(0, 1)$, $(1, 1)$, and $\left(1, \frac{1-\sqrt{5}}{2}\right)$, shown in Figure 7. In principal, with some effort, this program can be carried out for any lattice surface.

5 Abstract Homogeneous Dynamics

In this section, we prove our main results Theorem 7, Theorem 10, and Theorem 11. We prove Theorem 7 and Theorem 10 in Section 5.1 and Theorem 11 in Section 5.2.

5.1 Horocycle Flows

We recall notation: X is our parameter space, equipped with an $SL(2, \mathbb{R})$-action, and given $x \in X$ we have the associated discrete set $\Lambda_x \subset \mathbb{R}^2$, satisfying:

Equivariance. For any $g \in SL(2, \mathbb{R})$, $\Lambda_{gx} = g\Lambda_x$.

Symmetry. Λ_x is centrally symmetric, that is $-\Lambda_x = \Lambda_x$

Minkowski Property. For all $x \in X$, there is a $c = c(x)$ such that for any convex, centrally symmetric set $K \subset \mathbb{R}^2$ of volume at least c,

$$K \cap \Lambda_x \neq \emptyset.$$

Vertical Strip Property. Fix $\eta > 0$. Let $V_\eta := \{(u, v)^T : 0 < u \leq \eta, v > 0\}$ denote the vertical strip in $\mathbb{R}^2$. Suppose x is horizontal. Then $\Lambda_x \cap V_\eta$ is non-empty.

Recall that we say that $x \in X$ is *vertical* (respectively *horizontal*) if Λ_x contains a vertical (resp. horizontal) vector, and that given a constant η, we say that x is *η-vertically short* (resp. *η-horizontally short*) if Λ_x contains a vertical (resp. horizontal) vector of length at most η. We say that x is *η-exceptional* if it is $\frac{\eta}{4c(x)}$-vertically short, that is, if it contains a vertical vector of length at most $\frac{\eta}{4c(x)}$. We denote by X_η the set of non-η-exceptional points. Note that X_η and $X \backslash X_\eta$ are both h_s-invariant sets.

Existence of return map

In this subsection, we prove Lemma 6, which gives the existence of a well-defined transversal Ω_η and return map R_η for the flow h_s on the invariant set X_η, for any fixed $\eta > 0$. Recall that

$$\Omega_\eta = \{x \in X : x \text{ is } \eta - \text{horizontally short}\},$$

and

$$V_\eta = \{(u, v)^T \in \mathbb{R}^2 : u \in (0, \eta), v > 0\}.$$

Then for any $x \in X_\eta$, the set $\mathbb{S}_\eta(x)$ of slopes of vectors of Λ_x in V_η defined in (13) corresponds precisely to the set of times

$$\{s > 0 : h_s x \in \Omega_\eta\},$$

and this set is non-empty. This follows from the observation that if $(u, v)^T \in V_\eta \cap \Lambda_x$, then, setting $s = \frac{v}{u}$ to be the slope of the vector, we have

$$h_s(u, v)^T = (u, 0)^T,$$

with $0 < u \le \eta$, that is, $h_s x \in \Omega_\eta$. The set of slopes $\mathbb{S}_\eta(x)$ is *contained* in the set of return times $\{s > 0 : h_s x \in \Omega_\eta\}$. For the reverse containment, suppose $h_s x \in \Omega_\eta$. Then there is a vector $(u, 0)^T \in \Lambda_{h_s x}$ with $0 < u \le \eta$, so $h_{-s}(u, 0)^T = (u, su) \in \Lambda_x \cap V_\eta$, so $s \in \mathbb{S}_\eta(x)$ as desired. To show that this set is discrete, note that if there was an accumulation point, say s_0, there would be a sequence of distinct vectors in $\Lambda_x \cap V_\eta$ accumulating along the line of slope s_0, so by passing to a subsequence, we would obtain a sequence of distinct vectors in $\Lambda_x \cap V_\eta$ converging to a point along the line segment $\{(u, s_0 u) : 0 < u \le \eta\}$, a contradiction to the discreteness of Λ_x.

Finally, to show that this set is non-empty, we need to use the Minkowski and vertical strip properties. We first show that for any $x \in X_\eta$, there is an $s \in \mathbb{R}$ so that $h_s x \in \Omega_\eta$. To show this, let

$$A = [-\eta, \eta] \times \left[-\frac{c}{4\eta}, \frac{c}{4\eta} \right],$$

where $c = c(x)$ is as in the Minkowski property. Then A is convex, centrally symmetric, and has volume c. Since we assume $x \in X_\eta$, Λ_x does not contain a vertical vector of length at most $\frac{c}{4\eta}$, so we have a vector $(v_1, v_2)^T \in A \cap \Lambda_x$, with $v_1 \neq 0$. Since Λ_x is centrally symmetric, we can assume $v_1 > 0$, and setting $s = \frac{v_2}{v_1}$, we have $(v_1, 0) \in h_s x$, so $h_s x \in \Omega_\eta$ as desired.

Next, we will use the vertical strip property to show that for any $\omega \in \Omega_\eta$, there is an $s' > 0$ so that $h_{s'}\omega \in \Omega_\eta$. For $\omega \in \Omega_\eta$, we can use the vertical strip property to find $(u, v)^T \in \Lambda_\omega \cap V_\eta$, and setting $s' = \frac{v}{u}$, we have our result. This completes the proof of Lemma 6. $\qquad\square$

Self-Similarity

We let $R_\eta : \Omega_\eta \to \mathbb{R}^+$ be the return time function

$$R_\eta(\omega) = \min\{s > 0 : h_s \omega \in \Omega_\eta\},$$

and $T_\eta : \Omega_\eta \to \Omega_\eta$ be the return map

$$T_\eta(\omega) = h_{R_\eta(\omega)}\omega.$$

Given two positive parameters $\eta_1, \eta_2 > 0$, the maps T_{η_1} and T_{η_2} are related by the action of the diagonal subgroup g_t, via the conjugation relation $g_t h_s g_{-t} = h_{se^{-t}}$. Precisely, letting $T = 2\log\frac{\eta_1}{\eta_2}$, we have that $g_T \Omega_{\eta_2} = \Omega_{\eta_1}$, since the horizontal vectors of length (at most) η_2 get expanded by $e^{T/2} = \frac{\eta_1}{\eta_2}$ to vectors of length (at most) η_1, and the slopes of the vectors get scaled by $e^T = \left(\frac{\eta_1}{\eta_2}\right)^2$. This yields the self-similarity relations

$$T_{\eta_1}(g_T \omega) = g_T T_{\eta_2}(\omega),$$

and

$$R_{\eta_1}(g_T \omega) = \left(\frac{\eta_1}{\eta_2} \right)^2 R_{\eta_2}(\omega)$$

for any $\omega \in \Omega_{\eta_2}$.

Generic points

We now prove Theorem 7. Recall notation: $\sigma_{x,S}$ denotes the Lebesgue probability measure $\frac{ds}{S}$ on the orbit $\{h_s x : 0 \leq s \leq S\}$. Our assumption is that

$$\sigma_{x,S} \to \nu$$

as $S \to \infty$, where convergence is in the weak-$\star$ topology (in fact, all of our convergence of measures will be in this topology, so we do not mention it again), and that ν gives measure 0 to the set of vertical x. We want to show (15), that is, for any $0 \leq a \leq b \leq \infty$,

$$\lim_{N \to \infty} \frac{|\Gamma_N^\eta(x) \cap [a,b]|}{N} = \tilde{\nu} \left(R_\eta^{-1}([a,b]) \right),$$

where

$$\Gamma_N^\eta(x) = \{1 \leq i \leq N : s_{i+1}^\eta(x) - s_i^\eta(x)\}$$

is the set of the first N gaps of $\mathbb{S}_\eta(x)$, and $\tilde{\nu}$ is the measure on Ω_η such that $d\nu = d\tilde{\nu}\, ds$. Let $\omega_0 = h_{s_1^\eta(x)} \in \Omega_\eta$ be the first hitting point of Ω_η for the orbit $\{h_s x\}_{s>0}$. Then the assumption $\sigma_{x,S} \to \nu$ implies that the measures $\sigma_{\omega_0,N}$ given by

$$\sigma_{\omega_0,N} := \frac{1}{N} \sum_{i=0}^{N-1} \delta_{T_\eta^i(\omega_0)}$$

satisfy

$$\sigma_{\omega_0,N} \to \tilde{\nu}$$

as $N \to \infty$. Let $\chi_{a,b}$ denote the indicator function of the set $R_\eta^{-1}([a,b])$. Then we claim

$$\frac{|\Gamma_N^\eta(x) \cap [a,b]|}{N} = \sigma_{\omega_0,N}(\chi_{a,b}),$$

since

$$R_\eta(T^i(\omega)) = s_{i+1}^\eta(x) - s_i^\eta(x),$$

which follows from the proof of Lemma 6. That is,

Observation The gap distribution is a Birkhoff sum for the roof function R_η along the orbit of the map T_η

Then, applying the convergence $\sigma_{\omega_0,N} \to \tilde{\nu}$ to the bounded function $\chi_{a,b}$, we have our result. The above argument also shows how Corollary 8 follows from the Birkhoff ergodic theorem applied to the system $(\Omega_\eta, T_\eta, \tilde{\nu})$ and the function $\chi_{a,b}$. $\qquad\qquad\qquad\qquad\qquad\qquad\qquad\qquad\qquad\qquad\qquad\qquad\square$

Periodic points

To prove Theorem 10, we first need to prove Lemma 9, that is, we need to show if x is h_s-periodic, then the set of slope gaps eventually stabilizes. Let s_0 be the (minimal) period of x, so $h_{s_0}x = x$. Then $h_{s_0}\Lambda_x = \Lambda_x$. Let

$$M = \min\{n > 0 : s_n^\eta(x) \geq s_0\}.$$

Then

$$\mathbb{S}_\eta(h_{s_0}x) = \{s_M^\eta(x) - s_0 < s_{M+1}^\eta(x) - s_0 < s_{M+2}^\eta(x) - s_0$$
$$< \ldots < s_{M+i}^\eta(x) - s_0 < \ldots\},$$

but $h_{s_0}x = x$, so

$$s_{M+i}^\eta(x) - s_0 = s_{i+1}^\eta(h_{s_0}x) = s_{i+1}^\eta(x),$$

and thus

$$s_{i+1}^\eta(x) - s_i^\eta(x) = s_{M+i}^\eta(x) - s_{M+i-1}^\eta(x),$$

that is, the gaps have period $M - 1$. Setting $N_0 = M - 1$, we have, as desired, for any $N \geq N_0$,

$$\Gamma_N^\eta(x) = \Gamma_{N_0}^\eta(x).$$

This completes the proof of Lemma 9. To complete the proof of Theorem 10, we note that fixing an arbitrary $\eta_0 > 0$, periodic orbits for h_s naturally correspond to periodic orbits for the return map T_{η_0}. The above argument shows the periodic orbit of T_{η_0} associated to the orbit

$$\{h_s g_{-T}x : 0 \leq s \leq e^T s_0\}$$

has length $N_0(\eta)$, where $T = 2\log(\frac{1}{\eta_0})$. By moving x by h_s, and applying Lemma 6, we can assume $x \in \Omega_{\eta_0}$. We can then define the probability measure $\sigma_{x,\eta}$ supported on the periodic orbit for T_{η_0} on Ω_{η_0} induced by the probability measure $\rho_{x,T} = \frac{ds}{s_0 e^T}$ supported on the orbit $\{h_s g_{-T} x : 0 \le s \le e^T s_0\}$, via

$$\sigma_{x,\eta} := \frac{1}{N_0(\eta)} \sum_{i=0}^{N_0(\eta)-1} \delta_{T_{\eta_0}^i(x)}.$$

Recall from Section 5.1 that g_T scales V_η to V_{η_0} and multiplies slopes (and gaps of slopes) of vectors by the factor $\left(\frac{\eta}{\eta_0}\right)^2$. Thus we have

$$\frac{\left|\left(\frac{\eta}{\eta_0}\right)^2 \Gamma_{N_0(\eta)}^\eta(x) \cap [a,b]\right|}{N_0(\eta)} = \sigma_{x,\eta}(\chi_{a,b}),$$

and applying our convergence assumption, $\rho_{x,T} \to \nu$, which implies

$$\sigma_{x,\eta} \to \tilde{\nu},$$

we have our result. $\qquad\qquad\qquad\qquad\qquad\qquad\qquad\qquad\qquad\qquad\qquad$ □

5.2 Circle Limits

The proof of Theorem 11 is essentially already contained in [14, Section 9] and in [2, Section 4], which we discussed in Section 1.6 and Section 4. We sketch the argument once again: we define, for $x \in X, \sigma > 0$,

$$p_i(x,\sigma) := \lim_{R \to \infty} \lambda(\theta : |A_\theta^\sigma(R) \cap \Lambda_x| = i).$$

By the renormalization argument described in Section 1.6, we have, for $R >> 0$,

$$|\Lambda_x \cap A_\theta^\sigma(R)| \approx |g_t r_{-\theta} \Lambda_x \cap T(\sigma)|,$$

where $t = 2 \log R$. Thus,

$$p_i(x,\sigma) \approx \lambda_{x,R}(\{y \in X : |\Lambda_y \cap T(\sigma)| = i\}).$$

Applying our convergence assumption $\lambda_{x,R} \to \nu$, we have our result. $\qquad$ □

6 Further Questions

We collect some further questions and speculations. We discuss the space of translation surfaces in Section 6.1, the space of affine lattices in Section 6.2, and speculate wildly about other problems in Section 6.3.

6.1 Translation surfaces

It would be intriguing to push the machinery developed in this note further than the set of lattice surfaces. In particular, it would be interesting to check which $SL(2, \mathbb{R})$ (or h_s)-invariant measures on Ω_g are supported on the set of surfaces satisfying the vertical strip condition, and then to attempt to explicitly identify the transversal and the return time function (and the associated distribution). It would be particularly nice to do this for the Masur-Veech measure on Ω_g, and to understand if there were any 'continuity' properties of the gap distributions- namely, given a sequence of surfaces $\omega_n \to \omega$, do the gap distributions of saddle connection directions converge in any appropriate sense? Also, are there any surfaces $\omega \in \Omega_g$ which do have 'truly random' behavior, that is, an exponential distribution of gaps? We conjecture that this is not possible.

6.2 Affine Lattices

A nice test case, which in fact corresponds to an $SL(2, \mathbb{R})$-invariant subset of Ω_2 (see [2, Section 6]), is the collection of *affine unimodular lattices*, as discussed in Section 1.6. An explicit description of the (4-dimensional) transversal

$$\Omega_\eta \subset \tilde{X}_2 = \left(SL(2, \mathbb{R}) \ltimes \mathbb{R}^2\right) / \left(SL(2, \mathbb{Z}) \ltimes \mathbb{Z}^2\right),$$

and the return time function, would recover the (2-dimensional) results of [14] explicitly using unipotent flows.

6.3 Other gap distributions

As discussed in the introduction, although our paper was inspired by [8], our machinery does not seem to immediately give the gap distribution for the sequence $\{\{\sqrt{n}\}\}_{n \geq 0}$. It would be a nice application to formulate the results of loc. cit. in the language of our machine. A more ambitious, but probably very difficult project would be to try and understand the apparent exponential distribution of the gaps in $\{\{n^\alpha\}\}_{n \geq 0}$ for other $0 < \alpha < 1$. Another sequence of great interest in applications (see [12, Section 9]) would be the gaps in the sequence of squares of lengths of lattice vectors $\{|a + b\tau|^2 : a, b \in \mathbb{Z}\}$ for any fixed $\tau \in \mathbb{C}$, which are also conjectured to be exponential. In fact, we would be very excited to see any application of our technology which led to a non-exotic (i.e., exponential) distribution of gaps.

Jayadev S. Athreya
Department of Mathematics,
University of Washington
Padelford Hall, Seattle WA 98195, U.S.A.
E-mail: jathreya@uw.edu

References

[1] J. S. Athreya, *Cusp excursions on parameter spaces*, J. London Math. Soc. 87 (2013), no. 3, 741-765.

[2] J. S. Athreya and J. Chaika, *On the distribution of gaps for saddle connection directions*, Geom. Funct. Anal. 22 (2012), no. 6, 1491-1516.

[3] J. S. Athreya, J. Chaika, and S. Lelievre, *The distribution of gaps for saddle connections on the golden L*, Contemporary Mathematics, Vol. 631 (2015), 47-62.

[4] J. S. Athreya and Y. Cheung, *A Poincaré section for horocycle flow on the space of lattices*, Int. Math. Res. Notices (2014), no. 10, 2643-2690.

[5] F. Boca, C. Cobeli, and A. Zaharescu, *A conjecture of R. R. Hall on Farey points*. J. Reine Angew. Math. 535 (2001), 207 - 236.

[6] F. P. Boca and A. Zaharescu, *Farey fractions and two-dimensional tori*, in *Noncommutative Geometry and Number Theory* (C. Consani, M. Marcolli, eds.), Aspects of Mathematics E37, Vieweg Verlag, Wiesbaden, 2006, pp. 57-77.

[7] S. G. Dani, *On uniformly distributed orbits of certain horocycle flows*. Ergodic Theory Dynamical Systems 2 (1982), no. 2, 139 - 158 (1983).

[8] N.D. Elkies and C.T. McMullen, *Gaps in $\sqrt{n}$ mod 1 and ergodic theory*. Duke Math. J. 123 (2004), 95-139.

[9] A. Eskin and H. Masur, *Asymptotic Formulas on Flat Surfaces*, Ergodic Theory and Dynam. Systems, v.21, 443-478, 2001.

[10] A. Eskin, H. Masur, and A. Zorich, *Moduli spaces of abelian differentials: the principal boundary, counting problems, and the Siegel-Veech constants.* Publ. Math. Inst. Hautes Etudes Sci. No. 97 (2003), 61–179.

[11] R. R. Hall, *A note on Farey series.* J. London Math. Soc. (2) 2 1970 139 - 148.

[12] C. Hodgson and H. Masai, *On the number of hyperbolic 3-manifolds of a given volume*, Contemporary Mathematics, Vol. 597 (2013), 295-320.

[13] P. Hubert and T. Schmidt, *Diophantine approximation on Veech surfaces*, Bull. Soc. Math. France 140 (2012), no. 4, 551-568.

[14] J. Marklof and A. Strombergsson, *Distribution of free path lengths in the periodic Lorentz gas and related lattice point problems.* Ann. of Math. 172 (2010), 1949-2033.

[15] H. Masur, *Interval exchange transformations and measured foliations.* Ann. of Math. (2) 115 (1982), no. 1, 169–200.

[16] H. Masur, *The growth rate of trajectories of a quadratic differential*, Ergodic Theory Dynam. Systems 10 (1990), no. 1, 151-176.

[17] H. Masur and J. Smillie, *Hausdorff dimension of sets of nonergodic measured foliations*. Ann. of Math. (2) 134 (1991), no. 3, 455–543.

[18] M. Ratner, *Invariant measures and orbit closures for unipotent actions on homogeneous spaces*. Geom. Funct. Anal. 4 (1994), no. 2, 236-257.

[19] P. Sarnak, *Asymptotic behavior of periodic orbits of the horocycle flow and Eisenstein series*. Comm. Pure Appl. Math. 34 (1981), no. 6, 719 - 739.

[20] N. Shah, *Limit distributions of expanding translates of certain orbits on homogeneous spaces*. Proc. Indian Acad. Sci. (Math Sci) 106(2), (1996), pp. 105–125.

[21] W. Veech, *Gauss measures for transformations on the space of interval exchange maps*. Ann. of Math. (2) 115 (1982), no. 1, 201–242.

[22] H. Weyl, *Über die Gleichverteilung von Zahlen mod Eins*, Math.Ann.77(1916), 313-352. 103

2

Topology of open nonpositively curved manifolds

Igor Belegradek[1]

Abstract

This is a survey of topological properties of open, complete nonpositively curved manifolds which may have infinite volume. Topics include topology of ends, restrictions on the fundamental group, as well as a review of known examples.[2]

Among many monographs and surveys on aspects of nonpositive curvature, none deals with topology of open complete nonpositively curved manifolds, and this paper aims to fill the void. Most of the material discussed here is not widely-known. A number of questions is posed, ranging from naive to hopelessly difficult. Proofs are supplied when there is no explicit reference, and as always, the author is solely responsible for mistakes.

This survey has a narrow focus and does not discuss topological properties of

- hyperbolic 3-manifolds [161, 112, 51],
- open negatively pinched manifolds [19, 26],
- non-Riemannian nonpositively curved manifolds [61],
- compact nonpositively curved manifolds [78, 124],
- higher rank locally symmetric spaces [125, 73, 28, 87] and their compactifications [35],

which are covered in the above references. We choose to work in the Riemannian setting which leads to some simplifications, even though many results hold in a far greater generality, and the references usually point to the strongest

[1] I.B. partially supported by NSF grant DMS-1105045
[2] **Keywords:** nonpositive curvature, discrete group, open manifold, ends, negative curvature, finite volume, rank one.
AMS codes: Primary 53C20.

results available. Special attention is given to rank one manifolds, such as manifolds of negative curvature. Nonpositively curved manifolds are aspherical so we focus on groups of finite cohomological dimension (i.e. fundamental groups of aspherical manifolds), or better yet, groups of type F (i.e. the fundamental groups of compact aspherical manifolds with boundary).

Conventions: unless stated otherwise, manifolds are smooth, metrics are Riemannian, and sectional curvature is denoted by K.

Acknowledgments: The author is grateful for NSF support (DMS-1105045). Thanks are due to Jim Davis, Denis Osin, Yunhui Wu, and the referee for correcting misstatements in the earlier version.

1 Flavors of negative curvature

A *Hadamard manifold* is a connected simply-connected complete manifold of $K \leq 0$. By the Cartan-Hadamard theorem, any Hadamard manifold is diffeomorphic to a Euclidean space. Thus any complete manifold of $K \leq 0$ is the quotient of a Hadamard manifold by a discrete torsion-free isometry group (torsion-freeness can be seen geometrically: any finite isometry group of a Hadamard manifold fixes the circumcenter of its orbit, or topologically: a nontrivial finite group has infinite cohomological dimension so it cannot act freely on a contractible manifold).

A Hadamard manifold is a *visibility* manifold if any two points at infinity can be joined by a geodesic. A *flat* in X is a convex subset isometric to a Euclidean space. A *flat half plane* in X is a convex subset isometric to a Euclidean half plane. An *infinitesimal flat* of a geodesic is the space of parallel Jacobi fields along a geodesic. A geodesic is called *rank one* if its infinitesimal flat is one-dimensional. A complete manifold of $K \leq 0$ *has rank one* if it contain a complete (i.e. defined for all times) geodesic of rank one.

The following conditions on a Hadamard manifold X represent various manifestations of negative curvature:

(1) K is bounded above by a negative constant,
(2) X is Gromov hyperbolic,
(3) X is a visibility manifold,
(4) $K < 0$,
(5) X contains no flat half plane,
(6) X contains a complete geodesic that does not bound a flat half plane.

The implications $(1) \Rightarrow (4) \Rightarrow (5) \Rightarrow (6)$ and $(1) \Rightarrow (2) \Rightarrow (3) \Rightarrow (5) \Rightarrow$ (6) are immediate from definitions except for $(2) \Rightarrow (3)$ which can be found in [40, Lemma VIII.3.2].

Proposition 1 *Every other implication fails in dimension two.*

Remark There are of course analogs of (5) or (6) such as "no flat strip" or "no flat plane", see also various axioms on [72], but the above list is what comes up most often.

Remark Kleiner [118] proved that a Hadamard manifold has rank one if and only if it contains a complete geodesic that does not lie in a two-dimensional flat (see [11, Proposition IV.4.4] for the case when $\mathrm{Iso}(X)$ satisfies the duality condition, e.g. contains a lattice). Since (6) is intermediate between the two conditions, it is equivalent to them.

Proof of Proposition 1 That (2) $\not\Rightarrow$ (4) follows by doubling along the boundary any nonpositively curved compact surface of negative Euler characteristic whose metric is cylindrical near the boundary; the universal cover is hyperbolic because hyperbolicity is a quasiisometry invariant, but it contains a flat strip.

To show that (3) $\not\Rightarrow$ (2) recall a consequence of the Gauss-Bonnet theorem, mentioned on [13, page 57], that a 2-dimensional Hadamard manifold X is visibility if and only if for some $p \in X$ the total curvature of every sector bounded by two rays that start at p is infinite. Any smooth nonpositive function K on $[0, \infty)$ can be realized as the curvature of a rotationally symmetric metric $dr^2 + f(r)^2 d\theta$ on $\mathbb{R}^2$, namely, f is a unique solution of $f'' + Kf = 0$, $f(0) = 1$, $f'(0) = 1$. Note that $f(r) \geq r$ by Sturm comparison for ordinary differential equations, and the total curvature equals $2\pi \int_0^\infty Kf dr$. Let I_i be a sequence of disjoint compact subintervals of $(0, \infty)$ such that I_i has length i, and let K be a smooth negative function on $[0, \infty)$ that equals $-\frac{1}{i}$ on I_i. The associated rotationally symmetric metric has infinite total curvature on every sector from the origin, so it is visibility, and it contains arbitrary large regions of curvature $-\frac{1}{i}$, which contain triangles violating δ-hyperbolicity for any δ.

That (4) $\not\Rightarrow$ (3) is also shown in [72, Example 5.10]: modifying the argument of the previous paragraph yields a rotationally symmetric metric with $K < 0$ and finite total curvature.

To see that (6) $\not\Rightarrow$ (5) start from a finite volume complete hyperbolic metric on a punctured torus, and modify the metric near the end to a complete metric of $K \leq 0$ that is a cylinder outside a compact set. The cylinder lifts to a flat half plane so (5) fails, but any geodesic through a point of $K < 0$ satisfies (6).

The other non-implications are formal consequence of the above implications and non-implications. $\square$

Remark The non-implications in the above proof are justified by two dimensional examples, and it seems that similar examples in higher dimensions can be produced via iterated warping with $\cosh r$, i.e. replacing X with $\mathbb{R} \times_{\cosh r} X$, which contains X as a totally geodesic submanifold. For some non-implications we need to insert the warped product as a convex subset in a closed manifold of $K \leq 0$ following Ontaneda, see [140]. The warping clearly preserves conditions (1), (4), (5), (6).

Question 2 Does the warping with $\cosh r$ preserve (2) and (3)?

One wants to understand the relations among (1)–(6) when $\mathrm{Iso}(X)$ is large.

Example 3 If $\mathrm{Iso}(X)$ is cocompact, then clearly (1) $\Leftrightarrow$ (4), and also (2) $\Leftrightarrow$ (3) $\Leftrightarrow$ (5); in fact, Gromov hyperbolicity is equivalent to uniform visibility for proper CAT(0) spaces [40, Proposition III.3.1], and visibility is equivalent to the non-existence of a 2-flat for proper CAT(0) spaces with cocompact isometry groups [40, Theorem II.9.33]. For 2-dimensional Hadamard manifolds with cocompact isometry group (5) $\Leftrightarrow$ (6) because if X contains a flat half plane and has cocompact isometry group, then X is isometric to $\mathbb{R}^2$. On the other hand, (6) $\not\Rightarrow$ (5) in dimensions > 2 with examples given by the *Heintze manifolds*, obtained by chopping of cusps of a finite volume complete hyperbolic manifold, changing the metric near the cusp a metric of $K \leq 0$ with totally geodesic flat boundary, and doubling along the boundary; the boundary lifts to a flat of dimension > 1, while any geodesic through a point of $K < 0$ has rank one.

Example 4 If $\dim(X) = 2$ and $\mathrm{Iso}(X)$ contains a lattice, then either X is isometric to $\mathbb{R}^2$ or X is visibility [70, Proposition 2.5] so that (6) $\Leftrightarrow$ (3).

Example 5 If $\dim(X) > 2$, then (6) $\not\Rightarrow$ (3); indeed Nguyen Phan [139] and Wu [164] show that the universal covers of the finite volume manifolds of $K < 0$ constructed in [81] are not visibility, so (4) $\not\Rightarrow$ (3).

Example 6 Eberlein showed [71] that the ends of a finite volume complete manifold with bounded nonpositive curvature and visibility universal cover are π_1-injectively embedded. This is not the case for Buyalo's example [45] of a finite volume complete 4-manifold of $-1 < K < 0$; thus its universal cover is not visibility.

Remark Let us compare (3) and (4) in higher dimensions. Non-visibility of X can be checked by studying its two-dimensional totally geodesic submanifolds, cf. [139, 164], while proving visibility of X gets much harder, and indeed, it is a strong restriction on X with a variety of consequences for the geometry of horoballs, Tits boundary, and isometry groups. By contrast, the condition "$K < 0$" is easy to verify but has few implications. Either condition implies that every complete geodesic in X has rank one.

2 Manifolds of dimensions two and three

A classification of open connected 2-manifolds goes back to Kerékjártó [117]; see [150, 88, 157] for more recent accounts.

Richards [150, Theorem 3] proved that every open surface is obtained from S^2 by removing a closed totally disconnected subset, and then removing a finite or countable family of disjoint closed disks and attaching handles or Möbius bands along boundaries of the disks.

Remark Any closed totally disconnected subset $T \subset S^2$ can be moved by an ambient homeomorphism to a subset of the standard Cantor set. (Like any compact totally disconnected metric space, T is homeomorphic to a subset Q of the Cantor set in S^2. The homeomorphism type of a planar surface is determined by the homeomorphism type of its the space of ends [150, Theorem 1], which for $S^2 \setminus T$, $S^2 \setminus Q$ are identified with T, Q, respectively. Hence there is a homeomorphism $S^2 \setminus T \to S^2 \setminus Q$, which extends to a homeomorphism of the end compactifications $S^2 \to S^2$ mapping T to Q).

The uniformization theorem equips any connected surface with a constant curvature metric, which can be chosen hyperbolic for open surfaces:

Theorem 7 *Any open connected 2-manifold admits a complete metric of constant negative curvature.*

Proof Given an open manifold M, we equip it with a complete Riemannian metric, and pull the metric back to the universal cover $\widetilde{M}$, where $\pi_1(M)$ acts isometrically, and hence by preserving the conformal class of the metric. By the uniformization theorem $\widetilde{M}$ is conformal to the hyperbolic plane $\mathbf{H}^2$ or to $\mathbb{C}$. A self-diffeomorphism of $\mathbf{H}^2$ that preserve the conformal class of the hyperbolic metric is an isometry of $\mathbf{H}^2$. A self-diffeomorphism of $\mathbb{C}$ that preserves the conformal class of the Euclidean metric is of the form $z \to az + b$ or $z \to a\bar{z} + b$, and hence it either is an isometry (i.e. $a\bar{a} = 1$), or its square has a fixed point (in fact, $z \to az + b$ fixes $\frac{b}{1-a}$ and similarly for the square of $z \to a\bar{z} + b$). The deck-transformation $\pi_1(M)$-action is free, so it preserves the hyperbolic or the Euclidean metric. Thus M admits a complete metric of constant curvature -1 or 0. Finally, a flat open 2-manifold is $\mathbb{R}^2$, an annulus, or a Möbius band, so it also admits a complete hyperbolic metric. $\square$

Most of what is known on the geometrization of an open 3-manifold requires it to have finitely generated fundamental group, or better yet to be the interior of a compact manifold.

Theorem 8 *The interior of any compact aspherical 3-manifold with nonempty boundary admits a complete metric of* $-1 \leq K \leq 0$.

Proof We refer to [112] for the terminology used in this proof. Let N be a compact aspherical 3-manifold. Asphericity implies that N contains no essential 2-sphere or projective plane. There is a decomposition of N along incompressible 2-sided tori and Klein bottles into Seifert and atoroidal pieces, see [34], which extends previous proofs by Johannson and Jaco-Shalen to the non-orientable case. Each piece is π_1-injectively embedded and hence aspherical.

Atoroidal pieces contain no essential annuli or Möbius bands whose boundary circles lie on the components of ∂N of zero Euler characteristic (for the orientable case see e.g. [102, Lemma 1.16], and the non-orientable case follows by passing to the orientation cover for an essential annulus or Möbius band stays essential in the orientation cover, and a virtually Seifert piece is Seifert).

Define the pared manifold structure on every atoroidal piece N by letting its parabolic locus P be the union of boundary components of zero Euler characteristic. Thurston's hyperbolization theorem [112, Theorem 1.43] gives (N, P) a geometrically finite hyperbolic structure whose parabolic subgroups are of rank 2 and bijectively corresponding to components of P. Every Seifert piece has a nonpositively curved metric with totally geodesic (flat) boundary by Leeb [120]. In the same paper Leeb gives a gluing procedure for identifying rank 2 cusps and flat boundary components of Seifert pieces. (Leeb assumes that hyperbolic pieces have boundary of zero Euler characteristic but this does not matter since his gluing runs in the cusps, which in our case have rank 2). The result is a complete nonpositively curved metric on the interior of N. $\square$

Remark A variation of the above proof identifies N with a compact locally convex C^1 manifold of $K \leq 0$ with boundary, namely, replace geometrically finite pieces with the ε-neighborhoods of their convex cores, where the cusps are chopped off, and the metric at cusps are modified so that their boundaries are flat and totally geodesic. Thus the group $\pi_1(N)$ is CAT(0), which was first noted in [39, Theorem 4.3] with a slightly different proof.

Remark More information on 3-manifold groups can be found in the survey [10]; in particular, if M is open aspherical 3-manifold with finitely generated fundamental group, then the advances on the virtual Haken conjecture imply that $\pi_1(M)$ virtually embeds into a right angled Artin group, which has a number of group-theoretic consequences, e.g. $\pi_1(M)$ is linear over $\mathbb{Z}$.

By Scott/Shalen compact core theorem any open aspherical 3-manifold M with finitely generated deformation retracts to a compact codimension zero submanifold (to which Theorem 8 applies) but the topology of M is still a mystery.

A open 3-manifold is called *tame* if it is homeomorphic to the interior of a compact manifold. Marden's Tameness Conjecture predicted tameness of every open complete 3-manifold with $K \equiv -1$ and finitely generated fundamental group, and it was proved by Agol [3] and Gabai-Calegari [49]. The following question is still open:

Question 9 Let M be an open 3-manifold with a complete metric of $K \leq 0$ and finitely generated fundamental group. Is M tame?

The same question is open for manifolds of $K \leq -1$ or $-1 \leq K \leq 0$. The answer is yes under any of the following assumptions:

- M admits a complete negatively pinched metric (due to Bowditch [37] who built on proofs of Marden's Tameness Conjecture by Agol, Calegari-Gabai, and Soma).
- M is a cover of the interior of a compact manifold (as proved by Long-Reid in [50] that combines results of Simon with the proof of Tameness Conjecture; here the assumption that M has $K \leq 0$ is not needed).
- M has a complete metric of $-1 \leq K \leq 0$ and $\mathrm{Inj}\,\mathrm{Rad} \to 0$ (as proved by Schroeder [13, Appendix 2]).

By contrast, there exits non-tame, open 3-manifolds with universal cover diffeomorphic to $\mathbb{R}^3$ and fundamental groups isomorphic to $\mathbb{Z}$ [156] and $\mathbb{Z} * \mathbb{Z}$ [80].

Example 10 The non-tame fake open solid torus in [156] does not admit a complete metric of $K \leq 0$ because $\mathbb{R}^2$-bundles over S^1 are the only open, complete 3-manifolds of $K \leq 0$ with infinite cyclic fundamental group. (The isometric $\mathbb{Z}$-action in the universal cover must stabilize a geodesic or a horoball, in the former case the nearest point projection to the geodesic descends to an $\mathbb{R}^2$-bundle over S^1, while in the latter case the quotient is the product of $\mathbb{R}$ and an open surface homotopy equivalent to a circle, which is an $\mathbb{R}$-bundle over S^1).

3 Towards a rough classification of discrete isometry groups

In this section we sketch a higher-dimensional analog of the classification of open complete constant curvature surfaces; details appear in Sections 4–11. We refer to [13, 73, 40] and Section 1 for background on Hadamard manifolds, adopt the following notations:

- X is a Hadamard manifold with ideal boundary $X(\infty)$,
- $\mathrm{Iso}(X)$ is the isometry group of X,
- Γ is a subgroup of $\mathrm{Iso}(X)$,

and consider the following three classes of complete manifolds of $K \leq 0$ of the form X/G where $\Gamma \leq \mathrm{Iso}(X)$ is discrete and torsion-free:

(1) Γ contains no parabolic elements.
(2) Γ contains a rank one element.
(3) Γ fixes a point $\xi \in X(\infty)$, and the associated Γ-action on the space L_ξ of lines asymptotic to ξ is free and properly discontinuous.

There are severe algebraic restrictions on Γ in the cases (1)-(2), and in the case (3) the manifold X/Γ is diffeomorphic to the product of $\mathbb{R}$ and L_ξ/Γ, where L_ξ is diffeomorphic to a Euclidean space.

Example 11 Γ satisfies (3) if it stabilizes a horoball. (Indeed, L_ξ is equivariantly diffeomorphic to a horosphere, and the discreteness of Γ implies that its action on L_ξ is properly discontinuous).

Question 12 Suppose X has rank one. Does every discrete torsion-free isometry group Γ of X satisfy one of the conditions (1), (2), (3) above?

    The answer is yes if X is visibility (i.e. any two points of $X(\infty)$ are endpoints of a geodesic, which happens e.g. if $K \leq -1$), see Corollary 32.

    Without any assumption on X the answer is no, e.g. when X/Γ is the product of two finite volume, complete, open, hyperbolic surfaces.

Remark

(a) The classes (1), (2), (3) are clearly not disjoint, e.g. a the cyclic group generated by a translation in $\mathbb{R}^n$ lies in (1) and (3). On the other hand, any (discrete) subgroup satisfying (2) and (3) is virtually cyclic, see Proposition 24.
(b) A given group may be isomorphic to three different isometry groups satisfying (1), (2), (3) respectively.
(c) If $\Gamma \leq \mathrm{Iso}(X)$ is a discrete subgroup, then the requirement "Γ is isomorphic to a discrete isometry group of a Hadamard manifold that satisfies (3)" does not restrict Γ because the isometric Γ-action on the warped product Hadamard manifold $\mathbb{R} \times_{e^r} X$ stabilizes a horoball. On the other hand, if we fix the dimension, this does become a nontrivial restriction as follows.
(d) Define the *Euclidean action dimension* as the smallest dimension of a Euclidean space on which Γ acts smoothly and properly discontinuously, and if the Euclidean action dimension of Γ equals $\dim(X)$, then Γ-action

on X cannot satisfy (3) because L_ξ is diffeomorphic to the Euclidean space of lower dimension. See [31, 28, 165, 65] for computations of a related invariant called the *action dimension* which usually equals the Euclidean action dimension.

4 Groups of non-parabolic isometries

In this section we discuss groups in the class (1) of Section 3. An isometry of X is *elliptic, axial*, or *parabolic* if the minimum of its displacement function is zero, positive, or not attained, respectively. If γ is a non-parabolic isometry, then the set $\mathrm{Min}(\gamma)$ of points where the displacement functions attains a minimum splits as $C_\gamma \times \mathbb{R}^k$, where C_γ is a closed convex subset with γ acting as the product of the trivial action on C_γ and a translation on $\mathbb{R}^k$ [40, Theorem II.7.1]. If γ is axial, its axes are precisely the lines $\{x\} \times \mathbb{R}$, $x \in C_\gamma$.

Theorems 13, 14, 15(2ab) are part of the flat torus theorem "package" discovered by Gromoll-Wolf [89] and Lawson-Yau [119], and generalized in [40].

Theorem 13 *Let $A \le Iso(X)$ be an abelian discrete subgroup that consists of non-parabolic isometries. Set $\mathrm{Min}(A) := \cap_{a \in A} \mathrm{Min}(a)$. Then*

(1) *$\mathrm{Min}(A)$ is a nonempty, closed, convex A-invariant subset that splits as $C_A \times \mathbb{R}^m$ where A acts trivially on C_A and by translations on $\mathbb{R}^m$;*
(2) *A is finitely generated of rank $\le m \le \dim(X)$;*
(3) *A has finite intersection with each conjugacy class in $Iso(X)$.*

Remark Any periodic abelian discrete subgroup of $Iso(X)$ is finite (because it is countable, hence locally finite, and so is a union of finite subgroups whose fixes points set is a descending family of totally geodesics submanifolds of X, which has to stabilize by dimension reasons). Thus abelian discrete groups of non-parabolic isometries are finitely generated, which is a key feature of the Riemannian setting. By contrast $\mathbb{Q}$ can act properly by non-parabolic isometries on a proper CAT(0) space, which is the product of a simplicial tree and a line [40, Example II.7.13].

Theorem 14 *Let $A \le Iso(X)$ be an abelian discrete subgroup that consists of axial isometries, and let $N \le Iso(X)$ is a subgroup that normalizes A. Then*

(4) *N stabilizes $\mathrm{Min}(A)$ and preserves its product decomposition;*
(5) *A is centralized by a finite index subgroup of N;*
(6) *A is a virtual direct factor of N if N is finitely generated and $A \le N$.*

Proof of Theorems 13, 14   (1) is proved in [13, Lemma 7.1(1)]. Discreteness of A and (1) implies that A-action on $\mathbb{R}^m$ is properly discontinuous, so A has rank $\leq m$, which gives (2). Claim (3) follows from [40, Lemma II.7.17(2)], and (4), (5), (6) are parts of the flat torus theorem [40, Theorem II.7.1]. $\square$

Theorem 15   *Let $\Gamma \leq \mathrm{Iso}(X)$ be a subgroup without parabolic elements.*

(1) *If H is commensurable to Γ, then H is isomorphic to a discrete group of non-parabolic isometries of some Hadamard manifold.*

(2) *The following groups do not embed into Γ:*
   (a) *any solvable group that is not virtually abelian;*
   (b) *the Baumslag-Solitar group $\langle x, y \mid xy^m x^{-1} = y^l \rangle$ with $m \neq \pm l$;*
   (c) *$\pi_1(L)$, where L is a closed aspherical 3-manifold that admits no metric of $K \leq 0$.*

Proof   (1) is immediate via the induced representation construction [113, Theorem 2.3]. To prove (2a) note that G must be polycyclic (combine finite generation of abelian discrete subgroups of non-parabolic isometries with Mal'cev's theorem that a solvable group whose abelian subgroups are finitely generated is polycyclic). Then proceed by induction on Hirsch length, and use Theorem 14(6) to split virtual $\mathbb{Z}$-factors one at a time, see [40, Theorem II.7.16]. Also (2b) follows from Theorem 13(3), see [40, Theorem III.Γ.1.1(iii)]. Finally, to prove (2c) invoke the solution of the virtual Haken conjecture [4]; thus L has a Haken finite cover, so it admits a metric of $K \leq 0$ by the main result in [113, Corollaries 2.6–2.7]. $\square$

Remark

(1) If a closed aspherical 3-manifold admits no metric of $K \leq 0$, then it is Seifert or graph (the manifold is virtually Haken [4] so the claim follows from [120, 113]).

(2) Closed aspherical Seifert manifolds that admit no metric of $K \leq 0$ are precisely those modelled on Nil, Sol, or $\widetilde{SL}_2(\mathbb{R})$ as easily follows from Theorems 13(6), 15(2a), and the observation that the Seifert manifolds modelled on $\mathbb{R}^3$ or $\mathbb{H}^2 \times \mathbb{R}$ are nonpositively curved.

(3) The problem which orientable closed graph manifolds admit metrics of $K \leq 0$ was resolved in [46, 48] who found several combinatorial criteria on the gluing data. (These papers only consider manifolds with no embedded Klein bottles, but the assumption can be easily removed as was explained to the author by Buyalo).

(4) A non-orientable closed 3-manifold admits a metric of $K \leq 0$ if and only if its orientation cover does [113].

(5) A closed aspherical graph manifold admits a metric of $K \leq 0$ if and only if its fundamental group virtually embeds into a right angled Artin group [123].

(6) Kapovich-Leeb used Theorem 15(2c) to give other examples of groups that do not act on Hadamard spaces by non-parabolic isometries [113].

5 Groups with rank one elements: prelude

Sections 5–8 collect what is known on the class (2) of Section 3.

An axial isometry γ has *rank one* if it has an axis that does not bound a flat half plane. This property can be characterized in terms of the splitting $\text{Min}(\gamma) \cong C_\gamma \times \mathbb{R}^k$, namely, γ has rank one if and only if C_γ is compact.

Any discrete subgroup of $\text{Iso}(X)$ that normalizes a rank one element is virtually cyclic (in fact, the normalizer preserves the splitting, and hence fixes a point of C_γ and stabilizing the corresponding axis).

Rank one isometries were introduced by Ballmann [11, Theorem III.3.4], who proved that if X is a rank one and $\Gamma \leq \text{Iso}(X)$ is any subgroup satisfying the duality condition (e.g. a lattice), then Γ contains a rank one element. He then used a ping pong argument to find a copy of non-cyclic free group inside Γ. Various aspects of isometry groups containing rank one elements were further studied in [29, 12, 30, 100, 101, 52]. In particular, the following is due to [30, Proposition 5.11] or [100, Theorem 1.1(4)]:

Theorem 16 *If $\Gamma \leq \text{Iso}(X)$ is a discrete subgroup that contains a rank one element and is not virtually-$\mathbb{Z}$, then Γ contains a non-cyclic free subgroup consisting of rank one elements.*

Sisto proved [160, Theorem 1.4] that if Γ in Theorem 16 is finitely generated, then its generic element has rank one, where "generic" roughly means that the probability that a word written in random finite generating set of G represents a rank one element approaches 1 exponentially with the length of the word.

6 Acylindrically hyperbolic groups and rank one elements

In the last decade it was realized that many groups of geometric origin contain (suitably defined) rank one elements, which allowed for a uniform treatment of such groups and resulted in a host of applications. A crucial notion in these developments is acylindricity which goes back to Sela and Bowditch. In connection with rank one elements different versions of acylindricity were introduced and studied by Bestvina-Fujiwara [29, 30], Hamenstädt [99], Dahmani-Guirardel-Osin [60], Sisto [160], and most recently Osin showed [144] that all these approaches are equivalent.

An isometric action of a group G on a Gromov hyperbolic space (X, d) is called

- *non-elementary* if its limit set consists of > 2 points,
- *acylindrical* if for each $\varepsilon > 0$ there are R, N such that if $d(x, y) \geq R$, then at most N elements $g \in G$ satisfy $d(x, gx) \leq \varepsilon$ and $d(y, gy) \leq \varepsilon$.

A group G is *acylindically hyperbolic* if it admits a non-elementary acylindrical isometric action on a Gromov hyperbolic space.

The class of acylindrically hyperbolic groups includes many groups of geometric origin, e.g. any subgroup of a relatively hyperbolic group that is not virtually-cyclic and does not lie is peripheral subgroup, or all but finitely many mapping class groups; see [144], for other examples.

Of particular importance for this section is the following result of Sisto [160] who actually proves it for any group acting properly and isometrically on a proper CAT(0) space:

Theorem 17 (Sisto) *If $\Gamma \leq Iso(X)$ is a discrete subgroup that contains a rank one element, then Γ is virtually cyclic or acylindrically hyperbolic.*

Dahmani-Guirardel-Osin [60] introduced a notion of a *hyperbolically embedded subgroup of G*, and Osin [144] proved that G is acylindrically hyperbolic if and only if G contains an infinite, proper, hyperbolically embedded subgroup. What Sisto actually showed is that any rank one element $\gamma \in \Gamma$ lies in a virtually cyclic hyperbolically embedded subgroup $E(\gamma)$. We omit the definition of a hyperbolically embedded subgroup, and just note that they are almost malnormal by [60, Proposition 4.33]:

Theorem 18 *If H is a hyperbolically embedded subgroup of a group G, then H is almost malnormal in G, i.e. $H \cap gHg^{-1}$ is finite for all $g \notin H$.*

Here are other applications [60, 144], which hold in particular when G is a discrete, non-virtually-cyclic subgroup of $Iso(X)$ containing a rank one element.

Theorem 19 *If G is acylindrically hyperbolic, then*

(1) *G has a non-cyclic, normal, free subgroup,*
(2) *every countable group embeds into a quotient of G,*
(3) *every infinite subnormal subgroup of G is acylindrically hyperbolic,*

44   *Igor Belegradek*

(4) *G has no nontrivial finite normal subgroups if and only if every conjugacy class in G is infinite,*
(5) *every s-normal subgroup of G is acylindrically hyperbolic,*
(6) *if G equals the product of subgroups $G_1, \ldots, G_k$, then at least one G_i is acylindrically hyperbolic.*
(7) *G is not the direct product of infinite groups.*
(8) *any group commensurable to G is acylindrically hyperbolic.*
(9) *any co-amenable subgroup of G acylindrically hyperbolic.*

Proof Proofs of (1)–(4) are in [60, Theorem 8.6, Lemma 8.11, Theorem 8.12] (1)–(2), while (5)–(7) are proved in [144, Corollary 1.5, Proposition 1.7, Corollary 7.3], and (8) appears in [126, Lemma 3.8].

The claim (9) is due to Osin who communicated to the author the following argument and kindly permitted to include it here. Suppose G is acylindrically hyperbolic and $K \leq G$ is not. In the next paragraph we find a non-cyclic free subgroup $F \leq G$ with $F \cap K = \{1\}$. The group F is not amenable, and its action by left translations on the set G/H is free, so there is no F-invariant finitely additive probability measure on G/K. The contrapositive of (9) now follows because if K were co-amenable, G/K would admit a G-invariant (and hence F-invariant) finitely additive probability measure.

To construct F note that [144, Theorems 1.1-1.2] yield a non-elementary acylindrical G-action on a Gromov hyperbolic space for which the subgroup K is either elliptic or else virtually-$\mathbb{Z}$ and contains a loxodromic element (we refer to [144] for terminology). If K is elliptic, then [144, Theorems 1.1] yields independent loxodromic elements $a, b \in G$. The standard ping-pong argument shows that for some $n \gg 1$, the subgroup $\langle a^n, b^n \rangle$ is free of rank 2 and all its non-trivial elements are loxodromic. In particular, $\langle a^n, b^n \rangle \cap K = \{1\}$. If K is virtually-$\mathbb{Z}$ and contains a loxodromic element c, then [144, Theorems 1.1] gives loxodromic elements $a, b \in G$ such that a, b, c are independent. Again by ping-pong $\langle a^n, b^n, c^n \rangle$ is free of rank 3 so that $\langle a^n, b^n \rangle \cap K = \{1\}$. Thus we get a non-cyclic free subgroup $F = \langle a^n, b^n \rangle$ with $F \cap K = \{1\}$, and in fact all nontrivial elements of F are loxodromic. $\square$

A subgroup $K \leq G$ is *subnormal* if there are subgroups $G_i \leq G$ with $G_0 = G$, $G_k = K$, and such that G_i is a normal in G_{i-1} for all $i = 1, \ldots, k$.

If a group G equals the product of subgroups $G_1, \ldots, G_k$, one says that G *boundedly generated by* $G_1, \ldots, G_k$.

Two groups are *commensurable* if they have isomorphic finite index subgroups.

A subgroup $K \leq G$ is *s-normal* if $K \cap gKg^{-1}$ is infinite for each $g \in G$. Thus the Baumslag-Solitar group $B(m, n) = \langle a, b \mid ab^m = b^n a \rangle$ is not acylindrically hyperbolic, except for $B(0, 0)$, because $\langle b \rangle$ is s-normal and not acylindrically hyperbolic.

A subgroup $K \leq G$ is *co-amenable* if one of the following holds:

(1) every continuous affine G-action on a convex compact subset of a locally convex space with a K-fixed point has a G-fixed point;
(2) $\ell^\infty(G/K)$ has a G-invariant mean;
(3) G/K has a G-invariant finitely additive probability measure;
(4) the inclusion $K \hookrightarrow G$ induces injections in bounded cohomology in all degrees with coefficients in any dual Banach G-module.

The equivalence (1) $\Leftrightarrow$ (2) is proved in [74], while (2) $\Leftrightarrow$ (3) follows from the standard correspondence between means and measures, and (3) $\Leftrightarrow$ (4) can be found in [131]. See also [134] for leisurely discussion of co-amenability.

Here we are mainly interested in examples of non-amenable groups that admit co-amenable subgroups:

- A normal subgroup $N \trianglelefteq G$ is co-amenable if and only if G/N is amenable.
- If K is co-amenable in N, and in turn N is co-amenable in G, then K is co-amenable in G.
- the image of a co-amenable subgroup under an epimorphism $G \to \bar{G}$ is co-amenable.
- If $\theta \colon K \to K$ is a monomorphism, and $G := \langle K, t \mid tkt^{-1} = \theta(k), k \in K \rangle$ is the associated HNN-extension, then K is co-amenable in G.

The first three facts above are straightforward, while the last one is due to Monod-Popa [131].

Example 20 Starting from a group K that is not acylindrically hyperbolic one can use iterated HNN-extensions and extensions with amenable quotient to get many examples of non acylindrically hyperbolic groups. (These constructions preserve finiteness of cohomological dimension if the initial K and every amenable quotient have finite cohomological dimension).

7 Bounded cohomology and rank one elements

Bounded cohomology naturally appear in a variety of contexts, see e.g. [91, 129]. Of particular interest for our purposes is the *comparison map*

$$\iota(G) \colon H_b^2(G; \mathbb{R}) \to H^2(G; \mathbb{R})$$

between the bounded and ordinary cohomology in degree two, which encodes some subtle group-theoretic properties:

- **(Johnson)** If G is amenable, then $H^p(G; \mathbb{R}) = 0$ for $p > 0$ [111, 141].
- **(Burger-Monod)** $\iota(G)$ is injective if G is the fundamental group of an irreducible, finite volume complete manifold of $K \leq 0$, no local Euclidean de Rham factor, and rank ≥ 2 [43].
- **(Bavard)** Injectivity of $\iota(G)$ is equivalent to vanishing of the stable commutator length on $[G, G]$ [17]. Thus if $\iota(G)$ is non-injective, then there is $g \in [G, G]$ such that the minimal number of commutators needed to represent g^n grows linearly with n.

Theorem 21 (Bestvina-Fujiwara, Osin) *If G is acylindrically hyperbolic, then the comparison map $\iota(\Gamma)$ has infinite dimensional kernel.*

This was proved in [29] for a class of groups which according to [144] coincides with the class of acylindrically hyperbolic groups.

Remark For discrete subgroup $\Gamma \leq \text{Iso}(X)$ with rank one elements the above theorem was first established in [30]. Discreteness of Γ in Theorem 21 can be weakened to the weak proper discontinuity [30], but it cannot be dropped, e.g. the projection of any irreducible lattice $\Lambda \leq \text{Iso}(\mathbf{H}^2) \times \text{Iso}(\mathbf{H}^2)$ to either factor acts on the hyperbolic plane isometrically, effectively, and by rank one isometries, but the comparison map $\iota(\Lambda)$ is injective [43].

8 Monod-Shalom's class and rank one elements

In [132, 133], Monod-Shalom introduced and studied the following class of groups, which they thought of as a cohomological manifestation of negative curvature: Let $\mathcal{C}_{\text{reg}}$ be the class of countable groups G such that $H_b^2(G; \ell^2(G)) \neq 0$, which refers to the bounded cohomology of G with coefficients in the regular representation.

A way to prove that $H_b^2(G; \ell^2(G)) \neq 0$ is to show that the corresponding comparison map $H_b^2(G; \ell^2(G)) \to H^2(G; \ell^2(G))$ has infinite dimensional kernel, which was done for many "hyperbolic-like" groups in [128]. The following was proved by Hamenstädt [98], and later from a different perspective by Hull-Osin [105]:

Theorem 22 $\mathcal{C}_{\text{reg}}$ *contains every countable acylindrically hyperbolic group, and hence any discrete subgroup $\Gamma \leq \text{Iso}(X)$ that contains a rank one element and is not virtually-$\mathbb{Z}$.*

For discrete subgroup $\Gamma \leq \text{Iso}(X)$ with rank one elements the above theorem was first established in [101].

As proved in [133, Chapter 7], examples of groups not in C_{reg} include

- amenable groups,
- products of at least two infinite groups,
- lattices in higher-rank simple Lie groups (over any local field),
- irreducible lattices in products of compactly generated non-amenable groups.

and the class C_{reg} is closed under

- passing to an infinite normal subgroup,
- passing to a co-amenable subgroup,
- measure equivalence.

We refer to [85] for a survey on measure equivalence; e.g. commensurable groups are measure equivalent.

Question 23 Is every group in C_{reg} acylindrically hyperbolic?

To transition to our next topic, note that non-virtually-cyclic discrete groups with rank one elements never fix a point at infinity:

Proposition 24 *If Γ is discrete, fixes a point at infinity, and contains a rank one element, then Γ is virtually cyclic.*

Proof This follows from [30, Section 6] provided Γ satisfies the weak proper discontinuity condition, which is implied by discreteness. The idea is that if $g \in \Gamma$ has rank one, then either Γ is virtually-$\mathbb{Z}$, or Γ contains another rank one element h such that their axis A_g, A_h do not have the same sets of endpoints at infinity. A rank one element fixes precisely two points at infinity, the endpoints of its axis. Since Γ has a fixed point, it must be a common endpoint of A_g, A_h, and this contradicts weak proper discontinuity: there is a subsegment I of A_g and an infinite subset Q of Γ such that the distances between the endpoints of I and $g(I)$, $g \in Q$ are uniformly bounded. □

Example 25 The (non-discrete) stabilizer of a boundary point in the hyperbolic plane contains rank one elements without being virtually-$\mathbb{Z}$.

9 Groups that fix a point at infinity: prelude

Basic properties of horoballs, horospheres, and Busemann functions can be found in [13, 73, 11]. A *horoball* in X is the Hausdorff limit of a sequence of metric balls in X with radii going to infinity. A *horosphere* is the boundary of a horoball. Every point at infinity ξ is represented by a Busemann function $b_\xi : X \to \mathbb{R}$, which is determined by ξ up to an additive constant. The fibers of b_ξ are the horospheres centered at ξ, and the sublevel sets of b are horoballs centered at ξ. The function b_ξ is a C^2 Riemannian submersion $X \to \mathbb{R}$, and in particular, each horosphere is diffeomorphic to the Euclidean space of dimension $\dim(X) - 1$.

If Γ fixes a point ξ at infinity of X, then Γ permutes horospheres centered at ξ, and associating to $\gamma \in \Gamma$ the distance by which it moves a horosphere to a concentric one defines a homomorphism $\Gamma \to \mathbb{R}$, which is in general nontrivial (think of the stabilizer of a point at infinity of the hyperbolic plane).

Let L_ξ be the space of lines in X asymptotic to ξ. The geodesic flow towards ξ identifies X with the total space of a principal $\mathbb{R}$-bundle over L_ξ, which is trivial as every horosphere centered at ξ gives a section. In particular, L_ξ has a structure of a smooth manifold diffeomorphic to a horosphere about ξ. If Γ fixes ξ, then it acts smoothly on L_ξ.

Recall the condition (3) of Section 3: Γ fixes a point $\xi \in X(\infty)$, and the associated Γ-action on the space L_ξ of lines asymptotic to ξ is free and properly discontinuous. Under this condition the principal $\mathbb{R}$-bundle $X \to L_\xi$ descends to an orientable (and hence trivial) real line bundle $X/\Gamma \to L_\xi/\Gamma$, so we get:

Lemma 26 *If Γ satisfies the condition* (3) *of Section 3, then X/Γ is diffeomorphic to the product of $\mathbb{R}$ and L_ξ/Γ.*

The prime example of a group satisfying (3) is a discrete torsion-free subgroup Γ that stabilizes a horoball, in which case Γ stabilizes every concentric horoball, so that X is Γ-equivariantly is diffeomorphic to the product of $\mathbb{R}$ with a horosphere, and (3) follows because Γ acts freely and properly discontinuously on X. More examples are needed:

Question 27 Let Γ be any discrete torsion-free isometry group of X whose fixed point set at infinity is nonempty.

- Does Γ satisfies (3) for some ξ?
- If Γ satisfies (3), does Γ stabilize a horoball?
- What is the structure of Γ if it does not stabilize a horoball?

Remark

(1) An axial isometry does not stabilize a horoball centered at an endpoint of one of its axis, but it can stabilize another horoball (e.g. translation in the plane stabilizes any half plane whose boundary is parallel to the translation axis).
(2) Any parabolic isometry stabilizes a horoball by Lemma 33 but different parabolics in Γ can stabilize different horoballs.

(3) An elliptic element fixing a point at infinity stabilizes a horoball centered at the point (because it fixes a ray from a fixes point inside X to the fixed point at infinity).

Question 28 Let $\Gamma \le \mathrm{Iso}(X)$ be discrete, containing a parabolic and no rank one elements. What conditions on X ensure that Γ fixes a point at infinity?

Recall that the limit set $\Lambda(\Gamma)$ is the set of accumulation points of the Γ-orbit of a point of X. Ballmann-Buyalo [12, Proposition 1.10] gave the following characterization of groups containing rank one elements in terms of the Tits radius of the limits set:

Proposition 29 *Γ contains no rank one element if and only if $\Lambda(\Gamma)$ lies in the Tits ball of radius $\le \pi$ centered at a point of $\Lambda(\Gamma)$.*

Combining this characterization with results of Schroeder [13, Appendix 3] one can answer Question 28 when every component of the Tits boundary of X has radius $\le \frac{\pi}{2}$:

Corollary 30 *If every component of $X(\infty)$ equipped with the Tits metric has radius $\le \frac{\pi}{2}$, then a subgroup $\Gamma \le \mathrm{Iso}(X)$ either contains a rank one element or fixes a point at infinity.*

Proof Following [13, page 220], for a subset $Q \subset X(\infty)$, let

$$C_Q = \left\{ z \in Q \mid Q \text{ lies in the Tits ball of radius } \frac{\pi}{2} \text{ centered at } z \right\}.$$

Lower semicontinuity of the Tits distance implies that if Q is closed in the cone topology on $X(\infty)$, then so is C_Q [13, 4.9]. If C_Q is nonempty, then clearly it has Tits diameter $\le \frac{\pi}{2}$.

Apply this to $Q = \Lambda(\Gamma)$, which is a closed Γ-invariant subset. Since Γ has no rank one element, Proposition 29 implies that $\Lambda(\Gamma)$ lies in the Tits ball of radius $\le \pi$ about one of its points, and by our assumption the ball must have radius $\le \frac{\pi}{2}$ (for Tits metric is length so the distance between different components is infinite). Thus $C_{\Lambda(\Gamma)}$ is a closed subset of Tits diameter $\le \frac{\pi}{2}$.

By the main result of [13, Appendix 3] any subset of $X(\infty)$ that is closed in the cone topology and has Tits diameter $\le \frac{\pi}{2}$ has a unique *center*, defined as the center of the closed (Tits) ball of the smallest radius among all balls containing the subset. Let z_G be the unique center of $C_{\Lambda(\Gamma)}$. Since $\Lambda(G)$ is Γ-invariant, so is $C_{\Lambda(\Gamma)}$, and hence Γ fixes z_G. $\square$

Example 31 The components of the Tits boundary are points if (and only if) X is visibility [13, 4.14], so Corollary 30 applies if X is visibility, in which case one can say more:

Corollary 32 *If X is visibility and* Γ *contains a parabolic element but no rank one elements, then* Γ *contains no axial isometries,* $\Lambda(\Gamma)$ *is a point, the fixed point set of* Γ *at infinity equals* $\Lambda(\Gamma)$, *and* Γ *stabilizes every horoball centered at* $\Lambda(\Gamma)$.

Proof Since Γ contains no rank one elements, Proposition 29 implies that $\Lambda(\Gamma)$ is a point (as the Tits distance between any two distinct points is infinite). Being a visibility space, X has no flat half spaces, so Γ contains no axial isometries. The limit set is Γ-invariant, so $\Lambda(\Gamma)$ is a fixed point of Γ. If Γ fixed any other point, it would also be fixed by the cyclic subgroup generated by a parabolic in Γ, but the fixed point set of any abelian subgroup containing a parabolic has Tits radius $\leq \frac{\pi}{2}$, which again is a single point. Thus $\Lambda(\Gamma)$ is a unique fixed point of Γ.   □

Remark By Corollary 32 and Proposition 24 any non-virtually-cyclic, discrete isometry group of a visibility manifold that fixes a point at infinity must stabilize a horoball. There is a sizable class of groups to which this applies, e.g. by Corollary 42 it contains the product of any nontrivial torsion-free groups, see [116] for more examples.

10 Groups whose center contains a parabolic

The following result is implicit in [13, Lemma 7.3, 7.8].

Lemma 33 $\Gamma \leq Iso(X)$ *stabilizes a horoball if it has a finite index subgroup* Γ_0 *whose center* $Z(\Gamma_0)$ *contains a parabolic isometry.*

Proof Fix a parabolic isometry $z \in Z(\Gamma_0)$, and right coset representatives $g_1, \ldots g_k$ of Γ_0 in Γ. Then the function $x \to \sum_i d(g_i^{-1} z g_i x, x)$ is convex and Γ-invariant. Since $x \to d(zx, x)$ does not assume its infimum, neither does the above convex function. Then a limiting process outlined in [13, Lemma 3.9], cf. [40, Lemma II.8.26], gives rise via Arzela-Ascoli theorem to a Γ-invariant Busemann function, whose sublevel sets are G-invariant horoballs.   □

Flat torus theorem [40, Chapter II.7] restricts a discrete isometry group of X whose center consists of axial elements, which can be summarized as follows, see [18].

Theorem 34 *Let G be a group with subgroups H, G_0 such that their centers* $Z(H)$, $Z(G_0)$ *are infinite,* $Z(H) \subseteq Z(G_0)$, *the index of G_0 in G is finite, and one of the following conditions hold:*

(1) $Z(H)$ *is not finitely generated;*
(2) *any homomorphism $H \to \mathbb{R}$ is trivial.*
(3) *H is finitely generated, and $Z(H)$ contains a free abelian subgroup that is not a direct factor of any finite index subgroup of H.*

If a discrete subgroup of Iso(X) is isomorphic to G, then it stabilizes a horoball.

The reader may want to first think through the case when $H = G_0 = G$, and then go on to observe that if Theorem 34 holds for H, G_0, G, then it also does for H, $G_0 \times K$, $G \times K$ for any group K.

Example 35 Theorem 34(1) applies, e.g. to any infinitely generated, torsion-free, countable abelian group of finite rank, such as $(\mathbb{Q}, +)$, where finiteness of rank ensures finiteness of cohomological dimension [32, Theorem 7.10].

Remark It is unknown whether there is a group of type F with infinitely generated center. Such a group cannot be elementary amenable [18], or linear over a field of characteristic zero [5, Corollary 5]. Sanity check: there does exist a finitely presented group with solvable word problem whose center contains every countable abelian group [148, Corollary 3].

Example 36 (groups of type F to which Theorem 34(3) applies, see [18]):

(1) H is the fundamental group of the total space of any principal circle bundle with non-zero rational Euler class and a finite aspherical cell complex as the base.
(2) H is a torsion-free, finitely generated, non-abelian nilpotent group.
(3) H is the fundamental group of any closed orientable Seifert 3-manifold modelled on $\widetilde{SL}_2(\mathbb{R})$.
(4) H is the preimage of any torsion-free lattice in $Sp_{2n}(\mathbb{R})$ under the universal cover $\widetilde{Sp}_{2n}(\mathbb{R}) \to Sp_{2n}(\mathbb{R})$ for $n \geq 2$.
(5) H is the amalgamated product $G_1 *_A G_2$ where G_1, G_2 have type F and are finitely generated, A lies in the center of G_1, G_2 and contains a subgroup that is not a virtual direct factor of G_1.

11 Anchored groups and fixed points at infinity

Let us discuss algebraic conditions that force an isometry group of X to fix a point at infinity.

If a subgroup $\Gamma \leq \text{Iso}(X)$ contains a parabolic element, stabilizes a closed convex noncompact subset $W \subseteq X$, and fixes a point at infinity of W, then we say that Γ *is anchored in* W. (Passing to an invariant closed convex subset is essential in some inductive arguments, e.g. in Theorem 43).

Theorem 37 *Let* $\Gamma \le Iso(X)$ *be a subgroup and* W *be a any closed, convex, noncompact* Γ-*invariant subset of* X. *Then* Γ *is anchored in* W *if one of the following holds:*

(1) Γ *is abelian and contains a parabolic.*

(2) Γ *has a normal subgroup that is anchored in* W.

Proof (1) For $W = X$ this is due to Schroeder [13, Appendix 3], see also cf. [82]. The general case follows from a result of Caprace-Lytchak [53, Corollary 1.5] that the centralizer of a parabolic isometry of a CAT(0) space of finite telescopic dimension has a fixed point at infinity. Closed convex subsets of Hadamard manifolds have finite telescopic dimension, see [53, Section 2.1], and a parabolic isometry of X that stabilizes a closed convex subset acts in that subset as a parabolic isometry.

(2) For $W = X$ this is due to Eberlein [73, Proposition 4.4.4], whose proof generalizes to our setting via [82, Proposition 5.7] or [14, Proposition 1.4]. $\square$

Given a class of Hadamard manifolds $\mathcal{C}$, we say that a group G is *clinging in* $\mathcal{C}$ if for any discrete subgroup $\Gamma \le Iso(Y)$ such that $Y \in \mathcal{C}$ and Γ is isomorphic to G, and for any Γ-invariant closed convex noncompact subset W of Y, the group Γ is anchored in W. If G is clinging in the class of all Hadamard manifolds, we simply call G *clinging*.

In particular, G is clinging in $\mathcal{C}$ if no such Γ exists but we of course are interested in nontrivial examples. Theorem 38 and Corollary 39 below can be found in [18].

Theorem 38 *A group* G *is clinging in* $\mathcal{C}$ *if one of the following is true:*

(1) G *has a clinging in* $\mathcal{C}$ *normal subgroup, or*

(2) G *is the union of a nested sequence of clinging in* $\mathcal{C}$ *subgroups.*

(3) G *is as in* Theorem 34.

(4) G *is virtually solvable and not virtually-$\mathbb{Z}^k$ for any k.*

(5) *a normal abelian subgroup of* G *contains an infinite* G-*conjugacy class.*

Corollary 39 *Let* G *be a finitely generated, torsion-free group that has a nontrivial, normal, elementary amenable subgroup. Then either* G *is clinging, or* G *has a nontrivial, finitely generated, abelian, normal subgroup that is a virtual direct factor of* G.

Splitting results of Schroeder [153] and Monod [130] give another source of groups fixing points at infinity.

Theorem 40 *Let W be a closed, convex, noncompact subset of a Hadamard manifold. If a discrete torsion-free isometry group of W contains two commuting subgroups Γ_1, Γ_2, then one of them fixes a point at infinity of W.*

Proof Suppose neither Γ_1 nor Γ_2 fixes a point at infinity of W, and in particular $\Gamma_1\Gamma_2$ is nontrivial. Since Γ_1 fixes no point at infinity, it follows from [130, Proposition 27, Remark 39 and Subsection 4.6] that W contains a non-empty minimal convex closed Γ_1-invariant subset C_1, and moreover, the union C of such sets splits as $C_1 \times C_2$ for some bounded convex subset C_2 where Γ_1, Γ_2 preserve the splitting and act trivially on C_2, C_1, respectively. (Boundedness of C_2 is a key point, so we explain it here: if C_2 is unbounded, it contains a ray $s \to r(s)$, so given $x \in C_1$ we get a ray $\{x\} \times r$ in X which is mapped by any $\gamma \in \Gamma_1$ to an asymptotic ray as γ maps $(x, r(s))$ to $(\gamma(x), r(s))$; thus Γ_1 fixes a point at infinity contradicting the assumptions). Since C_2 is bounded, Γ_2 fixes the circumcenter of C_2, and hence fixes a point $z_2 \in C$. Repeating the same argument with Γ_2, $\{z_2\}$ in place of Γ_1, C_1 shows that the union Z of minimal convex closed Γ_2-invariant subsets splits as the product of $\{z_2\}$ and a bounded convex subset, and the splitting is invariant under Γ_1, Γ_2. It follows that $\Gamma_1\Gamma_2$ fixes a point of Z. This is where we need that $\Gamma_1\Gamma_2$ lies in a torsion-free discrete subgroup, because it implies that $\Gamma_1\Gamma_2$ is trivial. $\square$

Corollary 41 *If G_1 and G_2 are groups each containing a subgroup as in Theorem 15(2), then $G_1 \times G_2$ is clinging in the class of all Hadamard manifolds.*

Proof Suppose $G_1 \times G_2$ is isomorphic to a discrete subgroup $\Gamma \le \mathrm{Iso}(X)$ stabilizing a closed, convex, noncompact subset W. By Theorem 15(2), each factor contains a parabolic, and one of them fixes a point at infinity by Corollary 40, and hence is anchored in W. So Γ is anchored in W by Theorem 37(2). $\square$

Corollary 42 *If G_1, G_2 are nontrivial torsion-free groups, then $G_1 \times G_1$ is clinging in the class of Hadamard manifolds containing no flat half planes.*

Proof Suppose $G_1 \times G_2$ is realized as a discrete isometry group of a Hadamard manifold with no flat half planes, and suppose W is a closed, convex, noncompact invariant subset. Since $G_1 \times G_2$ is torsion-free, by Theorem 40 one of the factors fixes a point at infinity of W. If say G_1 contains a hyperbolic element h, then h has rank one as X contains no flat half planes, so the centralizer of h in $G_1 G_2$ is cyclic, and also contains h and G_2 violating the assumption that G_2 is nontrivial. Thus G_1 consists of parabolics, and by symmetry so does

G_2. One of the groups G_1, G_2 fixes a point at infinity, hence it is anchored in W, and so is $G_1 G_2$ by Theorem 37(2).   □

Results of Caprace-Monod imply:

Theorem 43   *If H is clinging in a class of Hadamard manifolds C, and G contains H as a co-amenable subgroup, then G is clinging in C.*

Proof   Realize G as a discrete isometry group of a Hadamard manifold in C stabilizing a closed, convex, noncompact subset W. By assumption H contains a parabolic, hence so does G. If G does not fix a point at infinity of W, then by [55, Theorem 4.3] W contains a minimal closed convex G-invariant subspace U. Note that U has no Euclidean de Rham factor (as the other factor would then be a smaller Γ-invariant subset). The co-amenability implies that H fixes no point at infinity of U [54, Proposition 2.1] so it is not clinging in C.   □

Remark   Burger-Schroeder [44] showed that any amenable subgroup $\Gamma \leq$ Iso(X) either fixes a point at infinity or stabilizes a flat, and this generalizes to actions on proper CAT(0) spaces by Adams-Ballmann [2]. Even more generally, Caprace-Monod [54, Corollary 2.2] obtained the same conclusion whenever Γ contains two commuting co-amenable subgroups (and also gave examples with non-amenable Γ). To make this result into a source of groups that fix a point at infinity more examples are needed, and with our focus on manifolds one has to answer the following.

Question 44   Is there a group that contains two commuting co-amenable subgroups, has finite cohomological dimension, and is not virtually solvable?

As mentioned in Section 9, if Γ fixes a point at infinity, then Γ permutes horospheres centered at the point defining a homomorphism $\Gamma \to \mathbb{R}$. Thus *if $\Gamma \leq$ Iso(X) has no nontrivial homomorphism into $\mathbb{R}$, then Γ stabilizes a horoball if and only if Γ fixes a point at infinity.*

Examples of clinging groups with finite abelianization (and hence no nontrivial homomorphisms into $\mathbb{R}$) are abound, see [18], and there are many such groups of finite cohomological dimension, or even of type F (so they may well be the fundamental groups of complete manifolds of $K \leq 0$).

Note that the property of having finite abelianization is inherited by amalgamated products (clearly), and by extensions (due to right exactness of the abelianization functor). Moreover, an extension with a finite quotient often has finite abelianization, e.g. the abelianization of the semidirect product $A \rtimes B$ is $(A^{ab})_B \times B^{ab}$, where $(A^{ab})_B$ is the coinvariants for the B-action on A^{ab}.

12 Homotopy obstructions (after Gromov and Izeki-Nayatani)

If M is a complete manifold of $K \leq 0$, then $\pi_1(M)$ has finite cohomological dimension. A group has finite cohomological dimension if and only if it is the fundamental group of a manifold whose universal cover is diffeomorphic to a Euclidean space.

Gromov asked [93] whether every countable group of finite cohomological dimension is isomorphic to some $\pi_1(M)$ where M is complete of $K \leq 0$. (The question is a good illustration of how little we know about open manifolds of $K \leq 0$.) The answer is no due to groundbreaking works of Gromov [94] and Izeki-Nayatani [107] on groups with strong fixed point properties.

These papers combine certain averaging procedures with ideas of harmonic map superrigidity to produce many a group G such that

(a) any isometric G-action on a Hadamard manifold has a fixed point;
(b) G has type F (i.e. is the fundamental group of a compact aspherical manifold with boundary).

Since the $\pi_1(M)$-action on the universal cover of M is free, it follows that there is no complete manifold M of $K \leq 0$ and $\pi_1(M) \cong G$. The methods actually reach far beyond Hadamard manifolds, and apply to isometric G-actions on a wide variety of spaces, see [108, 137, 109].

Gromov's examples are certain torsion-free hyperbolic groups produced from a sequence of graphs Γ_n whose edges are labeled with words of length j in an alphabet of $d > 1$ letters. The words are chosen randomly, and reversing orientation of the edge corresponds to taking inverse of a word. Given the data let $G(\Gamma_n, d, j)$ be the quotient group of F_d, the free group on d generators, by the relations corresponding to the cycles in Γ_n. The main result is that there is a sequence of expander graphs Γ_n such that for a large enough j the group $G(\Gamma_n, d, j)$ is torsion-free, hyperbolic, and satisfies (a) with probability $\to 1$ as $n \to \infty$. Like any torsion-free hyperbolic group, $G(\Gamma_n, d, j)$ has type F.

Izeki-Nayatani's original example is any uniform torsion-free lattice in $PSL_3(\mathbb{Q}_p)$, which has type F because it acts freely and properly discontinuously on the associated Euclidean building.

Remark The class of groups that satisfy (a) includes any finite group, or more generally any locally finite infinite subgroup such as $\mathbb{Q}/\mathbb{Z}$. There are much deeper examples in [9] of finitely presented, infinite, non-torsion-free groups that fix a point for any action by a homeomorphisms on a contractible manifold. None of these groups satisfies (b) as they have nontrivial finite order elements.

Remark By Davis's reflection group trick any group of type F embeds into the fundamental group of a closed aspherical manifold. Thus there is a closed aspherical manifold that is not homotopy equivalent to a complete manifold of $K \leq 0$.

Question 45 Is there a closed aspherical manifold whose fundamental group satisfies (a)?

13 Homeomorphism obstructions: exploiting $\mathbb{R}$ factors

Call a group G *reductive* if for any epimorphism $G \to H$ such that H is a discrete, non-cocompact, torsion-free isometry group of a Hadamard manifold stabilizing a horoball, or a totally geodesic submanifold where H acts cocompactly.

Example 46 (1) Any quotient of a reductive group is reductive.
(2) Any finitely generated, virtually nilpotent group is reductive (as follows from Sections 4, 10, see [18]).
(3) Any irreducible, uniform lattice in the isometry group of a symmetric space of $K \leq 0$ and real rank > 1 is reductive, by the harmonic map superrigidity [69, Theorem 1.2], see also [18].

A manifold is *covered by* $\mathbb{R}^n$ if its universal cover is diffeomorphic to $\mathbb{R}^n$. Thus any complete n-manifold of $K \leq 0$ is covered by $\mathbb{R}^n$.

Remark It is well-known that an open $K(G, 1)$ manifold covered by $\mathbb{R}^n$ exists if and only if G is a countable group of finite cohomological dimension, see e.g. [18].

A manifold is *covered by* $\mathbb{R} \times \mathbb{R}^{n-1}$ if it is diffeomorphic to the product of $\mathbb{R}$ and a manifold covered by $\mathbb{R}^{n-1}$. For instance, if G is a discrete torsion-free isometry group of a Hadamard manifold that satisfies the condition (3) of Section 3, then M is covered by $\mathbb{R} \times \mathbb{R}^{n-1}$. As we saw above (3) can be forced by purely algebraic assumptions on $\pi_1(M)$:

Example 47 If M is a complete connected manifold of $K \leq 0$ such that $\pi_1(M)$ is either clinging with finite abelianization, or satisfies the assumptions of Theorem 34, then M is covered by $\mathbb{R} \times \mathbb{R}^{n-1}$.

A trivial method of producing manifolds that are covered by $\mathbb{R}^n$ but not covered by $\mathbb{R} \times \mathbb{R}^{n-1}$ is to consider any manifold of minimal dimension among all manifolds in its homotopy type that are covered by a Euclidean space, which yields (see [18]):

Proposition 48 *Any aspherical manifold is homotopy equivalent to a manifold covered by* $\mathbb{R}^n$ *but not covered by* $\mathbb{R} \times \mathbb{R}^{n-1}$.

This method is non-constructive for it is not easy to decide whether a specific open manifold has the minimal dimension in the above sense (see [31, 28, 165, 65] for the manifolds of such minimal dimensions).

Corollary 49 *If G is reductive, clinging with finite abelianization, or as in* Theorem 34, *then any* $K(G, 1)$ *manifold is homotopy equivalent to a manifold that admits no metric of* $K \leq 0$ *and is covered by a Euclidean space.*

Example 50 Corollary 49 applies if $G = \mathbb{Q}$, see Example 35.

An essential tool in understanding manifolds covered by $\mathbb{R} \times \mathbb{R}^{n-1}$ is the recent result of Guilbault [97]: if an open manifold W of dimension ≥ 5 is homotopy equivalent to a finite complex, then $\mathbb{R} \times W$ is diffeomorphic to the interior of a compact manifold. Building on this result, the author [18] proved

Theorem 51 *Let W be an open* $(n-1)$-*manifold with* $n \geq 5$ *that is homotopy equivalent to a finite complex of dimension* $k \leq n - 3$. *Then* $\mathbb{R} \times W$ *is diffeomorphic to the interior of a regular neighborhood of a* k-*dimensional finite subcomplex.*

With more work one gets [18] the following applications:

Theorem 52 *Let L be a finite aspherical CW complex such that* $G = \pi_1(L)$ *is reductive, clinging with finite abelianization, or as in* Theorem 34. *Suppose that L is homotopy equivalent to a complete n-manifold M of* $K \leq 0$ *and* $n \geq 5$, *and set* $l = \dim(L)$.

(1) *If* $l \leq n - 3$, *then M diffeomorphic to the interior of a regular neighborhood of a* k-*dimensional finite subcomplex.*
(2) *If L is a closed manifold of dimension* $< \frac{2n-2}{3}$, *then M is diffeomorphic to the total space of a vector bundle over L.*
(3) *If* $l < \frac{n}{2}$, *then every complete n-manifold of* $K \leq 0$ *in the tangential homotopy type of M is diffeomorphic to M.*
(4) *If* $l \leq n - 3$, *then the tangential homotopy type of M contains countably many open n-manifolds that admit no complete metric of* $K \leq 0$.

Question 53 Can one strengthen the conclusion "countably many" in the part (4) of Theorem 52 to "a continuum of"?

A positive answer is given in [18] under a technical assumption which holds e.g. if L is a closed manifold, or if either $\mathbb{Z}^3$ or $\mathbb{Z} * \mathbb{Z}$ does not embed into G.

Remark Limiting Theorem 52 to certain classes of manifolds of $K \leq 0$ may result in enlarging the class of allowable fundamental groups, e.g. applying the theorem to manifolds with visibility universal cover, we can allow G to be the product of any two nontrivial groups.

As an application of Theorem 51, we get the following characterization of $\mathbb{R}^n$:

Corollary 54 *An open contractible n-manifold W is homeomorphic to $\mathbb{R}^n$ if and only if $W \times S^1$ admits a metric of $K \leq 0$.*

If $n = 4$, then "homeomorphic" in Corollary 54 cannot be upgraded to "diffeomorphic": if W is an exotic $\mathbb{R}^4$, then $W \times S^1$ is diffeomorphic to $\mathbb{R}^4 \times S^1$.

14 Benefits of a lower curvature bound

Complete manifolds of Ric $\geq -(n-1)$ are central to the global Riemannian geometry. For manifolds of $K \leq 0$, a lower Ricci curvature bound at a point is equivalent (by standard tensor algebra considerations) to a lower sectional curvature bound at the same point; by rescaling one can always make the bounds equal the curvature of the hyperbolic n-space.

In the seminal work [91] Gromov uncovered a relation between the simplicial volume and volume growth, which for complete manifolds with Ric $\geq -(n-1)$ is governed by Bishop-Gromov volume comparison. The following can be found in [91, p.13, 37].

Theorem 55 (Gromov) *Let W be an n-manifold such that every component C_i of ∂W is compact. If the interior of W is homeomorphic to a complete manifold of Ric $\geq -(n-1)$, then $\sum_i \|C_i\| \leq \liminf\limits_{r \to \infty} \dfrac{\text{Vol } B_p(r)}{r}$.*

Here $B_p(r)$ is the r-ball in M centered at p, and $\|C_i\|$ is the simplicial volume of C_i.

Example 56 If M in Theorem 55 has finite volume, or more generally sublinear volume growth, then each component of ∂W has zero simplicial volume.

Question 57 Does every complete manifold with $-1 \leq K \leq 0$ and sublinear volume growth admit a finite volume metric?

Example 58 Any infinite cyclic cover a closed connected manifold L of $K \leq 0$, i.e. the cover corresponding to the kernel of a surjective homomorphism $\pi_1(L) \to \mathbb{Z}$, has linear growth.

Problem 59 *Study manifolds of* $-1 \leq K \leq 0$ *with linear volume growth.*

Another consequence of a lower curvature bound is the famous Margulis lemma which appeared in [13] for manifolds of $-1 \leq K \leq 0$ following unpublished ideas of Margulis, and in [84, 115] for manifolds of $K \geq -1$. The following version of the Margulis lemma for manifolds of Ric $\geq -(n-1)$ is due to Kapovitch-Wilking [114] with essential ingredients provided by prior works of Cheeger-Colding.

Theorem 60 (Kapovitch-Wilking) *For each n there are constants m and $\varepsilon \in (0, 1)$ such that if p is a point of a complete n-manifold with Ric $\geq -(n-1)$ on $B_p(1)$, then the image of $\pi_1(B_p(\varepsilon)) \to \pi_1(B_p(1))$ has a nilpotent subgroup generated by n elements and of index $\leq m$.*

In fact, the nilpotent subgroup in Theorem 60 has a generated set $\{s_1, \dots, s_n\}$ such that s_1 is central and the commutator $[s_i, s_j]$ is contained in the subgroup generated by $s_1, \dots, s_{i-1}$ for each $1 < i < j$. Another universal bound on the number of generators of any given r-ball is given by

Theorem 61 (Kapovitch-Wilking) *For each n, r there is a constant k such that if p is a point in a complete Riemannian manifold M such that $\pi_1(B_p(r)) \to \pi_1(M, p)$ is onto and Ric $\geq -(n-1)$ on $B_p(4r)$, then $\pi_1(M, p)$ is generated by $\leq k$ elements.*

An important feature of the two preceding results is that no curvature control is required outside a compact subset.

15 Injectivity radius going to zero at infinity

We say that a subset S of a Riemannian manifold *has $Inj\, Rad \to 0$* if and only if for every $\varepsilon > 0$ the set of points of S with injectivity radius $\geq \varepsilon$ is compact; otherwise, we say S has $Inj\, Rad \nrightarrow 0$.

Remark By volume comparison any finite volume complete manifold of $K \leq 0$ has Inj Rad $\to 0$ [13, 8.4].

Proposition 62 *Any finite volume complete real hyperbolic manifold admits a complete metric with* Inj Rad $\to 0$, *bounded negative curvature, infinite volume, and sublinear volume growth.*

Proof We just do the two dimensional case; the general case is similar. Any end of a finite volume complete hyperbolic surface surface has an annular neighborhood with the metric $dt^2 + e^{-2t}d\phi^2$, $t > 0$. Modify it to the metric

$dt^2 + f^2(t)d\phi^2$ where f is a convex decreasing function such that $f(t) = e^{-t}$ for small t, and $f(t) = t^{-\alpha}$, $\alpha \in (0, 1)$ for large t. Let Σ_f be the resulting complete Riemannian 2-manifold, and let Σ_f^r denote "Σ_f with the portion with $t > r$ chopped off". Now Σ_f has

- Inj Rad $\rightarrow 0$ because f monotonically decreases to zero,
- infinite volume since $\frac{1}{2\pi}\text{Vol}(\Sigma_f^r)$ grows (sublinearly) as $\int_0^r f(s)ds = \frac{r^{1-\alpha}}{1-\alpha}$,
- bounded negative sectional curvature because on the annular neighborhood $K = -\frac{f''}{f} < 0$ which equals $-\frac{\alpha(\alpha+1)}{t^2}$ for $t > r$. $\qquad\square$

Question 63 Is there a complete manifold of $K \leq 0$ and Inj Rad $\rightarrow 0$ that admits no complete finite volume of $K \leq 0$? What is the answer in the presence of a lower curvature bound.

Gromov [90] pioneered the study of ends of negatively curved manifolds via the critical point theory for distance functions, which was extended by Schroeder [13, Appendix 2] as follows:

Theorem 64 (Schroeder) *If M is a complete manifold of* Inj Rad $\rightarrow 0$ *and* $-1 \leq K \leq 0$, *then either M is the interior of a compact manifold, or M contains a sequence of totally geodesic, immersed, flat tori with diameters approaching zero.*

None of the assumptions in the above theorem can be dropped due to examples of Gromov [13, Chapter 11] and Nguyen Phan [139] in which $\pi_1(M)$ is infinitely generated, see Section 20.

Problem 65 *Find a geometrically meaningful compactification of complete manifolds of* Inj Rad $\rightarrow 0$ *and* $-1 \leq K \leq 0$.

In the locally symmetric case this was accomplished in [121, 151, 122], and pinched negatively curved manifolds are naturally compactified by horospheres.

A weak substitute for a geometrically meaningful compactification is given by the following general theorem [58]:

Theorem 66 (Cheeger-Gromov) *For each n there is a constant c such that any complete finite volume n-manifold M of $|K| \leq 1$ admits an exhaustion by compact smooth codimension zero submanifolds M_i with boundary such that $M_i \subset \text{Int}(M_{i+1})$, the norm of the second fundamental form of ∂M_i is $\leq c$, and $\text{Vol}(\partial M_i) \rightarrow 0$ as $i \rightarrow \infty$.*

The proof constructs a controlled exhaustion function on M. For a related work based on different technical tools see [152, Theorem I.4.2] and [59, 163].

If M in Theorem 66 is the interior of a manifold W with compact boundary, then considering the components of ∂M_i that lie in a collar neighborhood of ∂W, we conclude that

Corollary 67 *If W is a manifold with compact connected boundary whose interior admits a complete finite volume metric g of $|K| \leq 1$, then $\mathbb{R} \times \partial W$ contains the sequence of compact separating hypersurfaces H_i which are homologous to $\{0\} \times \partial W$ and satisfy $\mathrm{MinVol}(H_i) \to 0$ as $i \to \infty$. Moreover,*

(1) *∂W has even Euler characteristic and zero simplicial volume;*
(2) *If ∂W is orientable, then its Pontryagin numbers vanish.*
(3) *If g also has $K \leq 0$, then the ℓ^2-Betti numbers of ∂W vanish, and hence ∂W has zero Euler characteristic.*

Proof Projecting onto the ∂W factor yields a degree one map $\partial M_i \to \partial W$ (if ∂W is non-orientable, then so is H_i, and we get a degree one map of their orientation covers). Thus ∂W has zero simplicial volume, which of course we already knew by Theorem 55.

By Chern-Weil theory the Pontryagin numbers $p_I(L)$ of a closed manifold L satisfy $|p_I(L)| \leq c_I \mathrm{MinVol}(L)$ where c_I is a constant depending only on $l = \dim(L)$ [91]. Thus $p_I(H_i) \to 0$ as $i \to \infty$. Since Pontryagin numbers are oriented cobordism invariant, we conclude that if ∂W in Corollary 67 is orientable, then its Pontryagin numbers vanish.

The boundary of a compact manifold has even Euler characteristic [66, Corollary VIII.8.8], so applying this to the cobordism between ∂W and H_i we see that $\chi(\partial W) + \chi(H_i)$ is even, and again Chern-Weil theory implies $\chi(H_i) \to 0$ as $i \to \infty$, and the claim follows.

Finally, vanishing of the ℓ^2-Betti numbers follows from [57, Theorem 1.2], and their alternating sum equals the Euler characteristic. $\qquad\square$

Question 68 Do any of the conclusions of Corollary 67 hold for complete manifolds with $-1 \leq K \leq 0$ and $\mathrm{Inj\,Rad} \to 0$?

Example 69 If Σ_f is as in Proposition 62, then the Riemannian product $\Sigma_f \times \Sigma_f$ has $\mathrm{Inj\,Rad} \to 0$ but superlinear volume growth if $0 < \alpha \leq \frac{1}{2}$ because for large r the subset $\Sigma_f^r \times \Sigma_f^r$ is sandwiched between concentric balls of radii r and $3r$, and its volume grows superlinearly as $\alpha \leq \frac{1}{2}$. Thus proving that the boundary has zero simplicial volume one requires new ideas beyond Theorem 55.

16 Negatively curved manifolds with uniform volume bound

For a connected complete Riemannian manifold M we denote by $\widetilde{M}$ its universal cover with the pullback metric.

Fukaya [83] proved the following result, whose analog for closed manifolds of dimension $\neq 3$ is due to Gromov [90]:

Theorem 70 (Fukaya) *Given V and $n \neq 3, 4$, only finitely many of diffeomorphism classes contain open complete n-manifolds M such that*

(1) $K < 0$ *or $\widetilde{M}$ is visibility,*
(2) $K \geq -1$,
(3) $Vol(M) < V$.

In dimension four Fukaya proved that the class of manifolds satisfying (1)-(3) contains only finitely many homotopy types (the missing ingredient is the weak h-cobordism theorem, which is unknown for h-cobordisms between closed 3-manifolds).

The theorem fails in dimension three as there are infinitely many (both open and closed) hyperbolic 3-manifolds with uniformly bounded volume [161]. Taking products with flat tori demonstrates that (1) cannot be replaced with $K \leq 0$, even though the optimal curvature condition is unclear.

Question 71 Is Theorem 70 true with (1) replaced by "$\widetilde{M}$ has rank one", or "$\widetilde{M}$ contains no flat half planes"?

Remark The proof in [90, 83] established an upper diameter bound in terms of volume, and then applies Cheeger's finiteness theorem (if M is open the diameter bound is for a compact domain D such that $M \setminus D$ is the interior of an h-cobordism). The strategy fails if one merely assumes that $\widetilde{M}$ has rank one by the following example: Chop off a cusp of a finite volume complete real hyperbolic manifold, and modify the metric to have totally geodesic flat boundary and $K \leq 0$. Then double along the boundary, which gives a finite volume complete rank one manifold of $K \leq 0$ and volume bounded roughly by $2\text{Vol}(M)$, but its diameter can be chosen arbitrary large by chopping deeper into the cusp.

Question 72 How does the number of diffeomorphism types of manifolds M in Theorem 70 grow with n and V?

In the locally symmetric case the above question was extensively studied, see [86] and references therein.

17 Non-aspherical ends of nonpositively curved manifolds

If a (not necessarily connected) manifold B is diffeomorphic to the boundary of a connected, smooth (not necessarily compact) manifold W, then we say that B *bounds* W.

Any aspherical manifold B bounds a noncompact aspherical manifold, namely $B \times [0, 1)$, and in fact, the universal cover of $B \times (0, 1)$ is a Euclidean space. Note that $B \times (0, 1)$ admits a complete metric of $K \leq 0$ if B is an infranilmanifold [25], or if B itself admits a complete metric of $K \leq 0$. On the other hand, if $\pi_1(B)$ contains a subgroup with strong fixed point properties as in Section 12, then $B \times (0, 1)$ admits no complete metric of $K \leq 0$. Our ignorance is illustrated by the following

Question 73 Does every closed aspherical manifold bounds a manifold whose interior admits a finite volume complete metric of $K \leq 0$?

In this section we discuss similar matters when B is closed and not aspherical. We focus on easy-to-state results and refer to [27] for a complete account.

Boundaries of compact manifolds with a complete metric of $K \leq 0$ on the interior could be quite diverse:

Example 74

(1) The total space of any vector bundle a closed manifold of $K \leq 0$ admits a complete metric of $K \leq 0$ [6].
(2) Complete finite volume locally symmetric manifold of $K \leq 0$ and $\mathbb{Q}$-rank ≥ 3 are interiors of compact manifolds with non-aspherical boundary.
(3) A complete manifold M of $K \leq 0$ is *convex-cocompact* if it deformation retracts onto a compact locally convex subset; such M is the interior of a compact manifold whose boundary is often non-aspherical.

There seem to be no simple description of closed manifolds that bound aspherical ones, and some obstructions are summarized below. In order for B to bound an aspherical manifold, a certain covering space of B must bound a contractible manifold. In formalizing how this restricts the topology of B, the following definition is helpful: given a class of groups $\mathcal{Q}$, a group is *anti–$\mathcal{Q}$* if it admits no nontrivial homomorphism into a group in $\mathcal{Q}$. Clearly, the class of anti–$\mathcal{Q}$ groups is closed under extensions, quotients, and any group generated by a family of anti–$\mathcal{Q}$ subgroups is anti–$\mathcal{Q}$.

Example 75 Let $\mathcal{A}_n$ denote the class of fundamental groups of aspherical n-manifolds. See [27] for examples of anti–$\mathcal{A}_n$ groups in such as:

(1) Any group generated by a set of finite order elements.
(2) Any irreducible lattice in the isometry group of a symmetric space of rank ≥ 2 and dimension $> n$ [28].

The following summarizes some obstructions that prevent a manifold from bounding an aspherical one.

Theorem 76 *If B bounds an aspherical, non-contractible n-manifold, and $\pi_1(B)$ is anti–$\mathcal{A}_n$, then B is noncompact, parallelizable, its $\mathbb{Z}$-valued intersection form of vanishes, and its $\mathbb{Q}/\mathbb{Z}$-valued torsion linking form vanishes.*

Example 77 The following manifolds do not bound aspherical ones:

(1) The connected sum of lens spaces, because it is a closed manifold whose fundamental group is anti–$\mathcal{A}_n$.
(2) The product of any manifold with CP^k with $k \geq 2$.
(3) The connected sum of any manifold and the product of two closed manifolds whose fundamental groups are anti–$\mathcal{A}_n$.
(4) The product of a punctured 3-dimensional lens space and a closed manifold whose fundamental group is anti–$\mathcal{A}_n$.
(5) Any manifold that contains the manifold in (2)–(4) as an open subset.

Let $\mathcal{NP}_n$ denote the class of the fundamental groups of complete n-manifolds of $K \leq 0$; of course $\mathcal{NP}_n \subseteq \mathcal{A}_n$. Examples of anti-$\mathcal{NP}_n$ groups of type F discussed in Section 12 immediately imply (see [27]):

Theorem 78 *There is a closed non-aspherical manifold that*

 (i) *bounds a manifold whose interior is covered by a Euclidean space;*
(ii) *bounds no manifold whose interior has a complete metric of $K \leq 0$.*

Other obstructions come from results of Section 13. Given groups I, J and a class of groups $\mathcal{Q}$ we say that I *reduces to* J *relative to* $\mathcal{Q}$ if every homomorphism $I \to Q$ with $Q \in \mathcal{Q}$ factors as a composite of an epimorphism $I \to J$ and a homomorphism $J \to Q$. Here we are mainly interested in groups that reduce relative to $\mathcal{NP}_n$ to the groups from the parts (2)-(3) of Example 46, which have finite virtual cohomological dimension.

Theorem 79 *Let $n \geq 6$, let G be a group from Example 46(2)-(3) of virtual cohomological dimension $\leq n - 3$, and let B be a closed $(n - 1)$-manifold such that $\pi_1(B)$ reduces to G relative to $\mathcal{NP}_n$. If B bounds a manifold N such that $\mathrm{Int}(N)$ admits a complete metric of $K \leq 0$, then there is a closed manifold L of dimension $\leq n - 3$ such that*

(1) L is either an infranilmanifold, or an irreducible, locally symmetric manifold of K ≤ 0 and real rank ≥ 2.

(2) N is the regular neighborhood of a PL-embedded copy of L.

(3) If N is not diffeomorphic to the product of a compact manifold and a closed interval, then L admits a metric of K ≤ 0 and N is the total space of a linear disk bundle over L.

In [27] we give examples of manifolds B that cannot bound a manifold N as in the above theorem such as

Example 80 Let B be the total space of a linear S^k bundle over a closed non-flat infranilmanifold such that $k \geq 3$ and the rational Euler class of the bundle is nonzero. Then B does not bound a manifold whose interior admits a complete metric of $K \leq 0$.

A boundary component of a manifold is *incompressible* if its inclusion induces injections on all homotopy groups. Reductive groups were defined in Section 13.

Theorem 81 *Let B be a closed $(n-1)$-manifold such that $\pi_1(B)$ is reductive and any nontrivial quotient of $\pi_1(B)$ in the class $\mathcal{NP}_n$ has cohomological dimension $n-1$. If B bounds a manifold N such that $\mathrm{Int}(N)$ admits a complete metric of $K \leq 0$, then B is incompressible in N.*

Example 82 Theorem 81 applies to whenever $\pi_1(B)$ is isomorphic to $\pi_1(L)$ where L is an $(n-1)$-manifold of from Theorem 79(1) and $\pi_1(L)$ has no proper torsion-free quotients. Examples of such L include any higher rank, irreducible, locally symmetric manifolds of $K \leq 0$ (thanks to the Margulis Normal Subgroup Theorem), as well as certain infranilmanifolds, see [27].

Given a compact boundary component B of a manifold N, an *end E of* $\mathrm{Int}(N)$ *that corresponds to B* is the intersection of $\mathrm{Int}(N)$ with a closed collar neighborhood of B; note that E is diffeomorphic to $[1, \infty) \times B$.

Theorem 83 *Let B be a closed connected manifold that bounds a manifold N, and let E be an end of $\mathrm{Int}(N)$ corresponding to B. If $\pi_1(B)$ is reductive, and $\mathrm{Int}(N)$ admits a complete metric of $K \leq 0$ and $\mathrm{Inj\,Rad} \to 0$ on E, then B is incompressible in N.*

Example 84 If B is the total space of a linear S^k bundle with $k \geq 2$ over a manifold L as in 79(1), then B does not bound a manifold whose interior admits a complete metric of $K \leq 0$ and $\mathrm{Inj\,Rad} \to 0$.

18 Riemannian hyperbolization (after Ontaneda)

A recent work of Ontaneda [142] allows to dramatically expand the list of known finite volume complete manifolds of $-1 \le K < 0$ of dimensions > 3. Unlike earlier examples, Ontaneda's method assembles a manifold of $K \le -1$ in a lego-like fashion from identical blocks according to a combinatorial pattern specified by a cube complex structure on any given manifold. Each block is a compact real hyperbolic manifold with corners, where every boundary face is totally geodesic and the faces's combinatorial pattern is that of a cube. This process results in a singular metric, which Ontaneda is able to smooth into a complete Riemannian metric of $K \le -1$, provided the block's faces have a sufficiently large normal injectivity radii; such blocks exist.

The idea of building locally CAT(0) manifold out of identical blocks is due to Gromov [92] who came up with several hyperbolization procedures turning a simplicial complex into a locally CAT(0) cubical complex. Gromov's ideas were developed and made precise in [62, 63], cf. [61], and furthermore Charney-Davis [56] developed the *strict hyperbolization* that turns a cubical complex into a piecewise polyhedral complex whose faces are the blocks of the previous paragraph. A key feature of the procedures is that the link of every vertex of the hyperbolized polyhedral complex is a subdivision of the corresponding link in the original complex, which has two consequences:

- The hyperbolization(s) turns a manifold triangulation into a locally CAT(0) manifold with a cubical complex structure;
- The strict hyperbolization turns a locally CAT(0) cubical complex (or manifold) into a locally CAT(-1) polyhedral complex (manifold, respectively).

Trivial exceptions aside, the resulting piecewise Euclidean (or piecewise hyperbolic) metric is non-Riemannian. Smoothing the piecewise hyperbolic metric into a Riemannian metric of $K \le -1$ in [142] is a technological tour de force.

Remark

(1) The cube complex fed into Ontaneda's construction need not be locally CAT(0), so the resulting manifolds of $K \le -1$ in [142] are a priori not homeomorphic to locally CAT(-1) manifolds of [56].

(2) Charney-Davis [56] describe a canonical smoothing of their manifolds (but not of the metrics), yet this smoothing is not necessarily equal to the smooth structure in [142] even when there is a face preserving homeomorphism between the two manifolds.

(3) Taking the normal injectivity radius of the block's faces large enough, one can make the sectional curvature of the resulting manifold arbitrary close to -1.

(4) If one starts with a closed manifold, then Ontaneda's procedure yields a closed manifold of $K \leq -1$, but if the initial manifold is open, its triangulation contains infinitely many simplices, so the resulting Riemannian manifold of $K \leq -1$ has infinite volume.

In order to produce finite volume examples, Ontaneda relativizes the construction as follows: Start with a closed manifold B that bounds a compact manifold W, cone off the boundary, apply the strict hyperbolization, and remove the cone point. The result is a compact manifold with boundary, and Ontaneda gives it a certain smooth structure in which the boundary becomes diffeomorphic to B. Topological properties of the resulting smooth manifold $R_{W,B}$ mirror those of W, see a summary in [20], and in particular by varying W, one finds $R_{W,B}$'s such that

- $R_{W,B}$ has a nontrivial rational Pontryagin class if $\dim(W) \geq 4$,
- $H^*(R_{W,B})$ contains a subring isomorphic to the cohomology ring of a given finite CW complex of dimension $< \dim(W)/2$, see [142],
- $\pi_1(R_{W,B})$ surjects on a given finitely presented group if $\dim(W) \geq 4$.

Moreover $R_{W,B}$ enjoys the following properties:

- $R_{W,B}$ is orientable if so is W,
- the inclusion of each component of B into $R_{W,B}$ is π_1-injective [63],
- $R_{W,B}$ is aspherical if and only if each component of B is aspherical [63],
- $\pi_1(R_{W,B})$ is hyperbolic relative to the images of the fundamental groups of components of B [20].

The piecewise hyperbolic metric on $R_{W,B}$ is singular and incomplete near the removed cone point, but again when the normal injectivity radius of the block's faces large enough, Ontaneda is able to smooth the metric away from a punctured neighborhood of the cone point, while on that neighborhood the metric has to be constructed by an ad hoc method depending on B. The following result is implicit in [142].

Theorem 85 (Ontaneda) *Let B be the boundary of a compact n-manifold and suppose that*

(i) *if $n \geq 6$, then any h-cobordism from B to another manifold is a product;*
(ii) *$\mathbb{R} \times B$ admits a complete metric g of sectional curvature within $[-1, 0)$ such that $(-\infty, 0] \times B$ has finite g-volume, and $g = dr^2 + e^{2r}g_B$ on $[c, \infty) \times B$ for some $c > 0$ and a metric g_B on B.*

Then B bounds a compact smooth manifold whose interior admits a complete metric of finite volume and sectional curvature in $[-1, 0)$.

Condition (ii) implies that each component of B is aspherical, and hence has torsion-free fundamental group. The Whitehead Torsion Conjecture, which is true for many groups of geometric origin [15, 16], predicts that all torsion-free groups have zero Whitehead torsion. If the conjecture is true for the fundamental group of each component of B, then (i) holds. Condition (ii) has been checked in a number of cases in [25, 138, 23] implying:

Theorem 86 *A manifold B bounds a compact manifold whose interior admits a complete metric of finite volume and sectional curvature in $[-1, 0)$ if*

(1) *B is a closed 3-dimensional Sol manifolds [138], or*
(2) *B bounds a compact manifold, and belongs to the class $\mathcal{B}$ [23].*

Here $\mathcal{B}$ is the smallest class of closed manifolds of positive dimension such that

- $\mathcal{B}$ contains each infranilmanifold [25], each closed manifold of $K \leq 0$ with a local Euclidean de Rham factor of positive dimension, and every circle bundle of type (K);
- $\mathcal{B}$ is closed under products, disjoint unions, and products with any compact manifold of $K \leq 0$.

An orientable circle bundle *has type* (K) if its base is a closed complex hyperbolic n-manifold whose holonomy representation lifts from $PU(n, 1)$ to $U(n, 1)$, and if the Euler class of the bundle equals $-m\frac{\omega}{4\pi}$ for some nonzero integer m, where ω is the Kähler form of the base. For example, every nontrivial orientable circle bundle over a genus two orientable closed surface has type (K), see [23].

Question 87 Consider the class of closed manifolds B to which Theorem 85 applies. Does the class contain every circle bundle over a closed manifold of $K < 0$? Every closed infrasolvmanifold? Is the class closed under products?

19 Topology of known manifolds of bounded negative curvature

In dimensions > 3 most (if not all) known examples of complete manifolds of $K \leq 0$ come from combining

- locally symmetric metrics of $K \leq 0$ arising from arithmetic or reflection groups,

- iterated and multiple warped products, building on the seminar work of Bishop-O'Neill on singly warped products of $K \leq 0$ [33].

Ontaneda's work [142] illustrates how the two methods combine: the block is an arithmetic real hyperbolic manifold with corners, and smoothing the metric involves sophisticated warped product considerations.

Prior examples of finite volume complete manifold of $K \leq 0$ that are not locally symmetric were typically produced by starting with a locally symmetric complete manifold of $K \leq 0$ and performing one of the following operations, which can be combined or iterated:

- doubles (Heintze, see [155]) and twisted doubles [143],
- branched covers [136, 95, 79, 166, 8, 64, 77],
- cusp closing [154, 47, 106, 7],
- cut and paste a tubular neighborhood of a totally geodesic submanifold of codimension 1, or dimensions 0 or 1, see [78] for a survey,
- remove a family of totally geodesic submanifolds of codimension two, which results in an incomplete metric that in some cases can be modified to a complete metric of $K \leq 0$ [81, 1, 45, 21, 22].

With few general results available, it makes sense to study topology of known examples in more detail. For the rest of the section let M be an open connected complete finite volume manifold of $-1 \leq K < 0$. The prior discussions gives

- (Section 6) $\pi_1(M)$ is acylindrically hyperbolic because M has finite volume and $K < 0$,
- (Theorem 64) M is the interior of a compact manifold N, which is uniquely determined up to attaching an h-cobordism to the boundary.
- (Corollary 67) ∂N has zero simplicial volume and Euler characteristic, and if ∂N is orientable, ∂N has zero Pontryagin numbers.

We would get a lot more information if $\pi_1(M)$ were hyperbolic relative to a collection of easy-to-understand peripheral subgroups (mainly because relatively hyperbolic groups inherit many properties from their peripheral subgroups, see below). Here is a prototypical example:

Example 88 If M is negatively pinched (i.e K is bounded between two negative constants), then $\pi_1(N)$ is hyperbolic relatively the fundamental groups of the components of ∂N [75, 38] which are virtually nilpotent [13].

One might expect that $\pi_1(M) \cong \pi_1(N)$ is always hyperbolic relative to the fundamental groups of components of ∂N. This idea runs into difficulties because ∂N need not be π_1-incompressible [45], but when it works it does so to great effect, and become a major source of information about $\pi_1(M)$.

We refer to [146] for background on relatively hyperbolic groups; as a part of the definition we require that relatively hyperbolic groups are finitely generated and not virtually cyclic, and their peripheral subgroups are infinite and proper.

Theorem 89 (Belegradek) *If W is a compact aspherical manifold of dimension ≥ 3 such that the components of ∂W are aspherical and π_1-injectively embedded, and $\pi_1(W)$ is hyperbolic relatively to the fundamental groups of the components of ∂W, then*

(a) *if $\pi_1(W)$ splits nontrivially as amalgamated product or HNN extension over a subgroup K, then K is acylindrically hyperbolic.*
(b) *$\pi_1(W)$ is co-Hopf, i.e. its injective endomorphisms are surjective.*
(c) *$\mathrm{Out}(\pi_1(W))$ is finite if for every component B of ∂W the group $\pi_1(B)$ is either not relatively hyperbolic or has a relatively hyperbolic structure whose peripheral subgroups are not relatively hyperbolic.*

Sketch of the proof Mayer-Vietoris sequence in the group cohomology applied to the fundamental group of the double of W along ∂W can be used to show that $\pi_1(W)$ can only split over a non-elementary subgroup proving (a), which in turn implies (b)–(c) by results of Druţu-Sapir [68]. An alternative proof of (b) is immediate from the fact that $(W, \partial W)$ has positive simplicial norm [127]. Details can be found in [20]. ☐

Example 90 The fundamental group of a closed aspherical manifold with zero simplicial volume is not relatively hyperbolic (as noted in [24] this follows from [127]). Thus if W is as in Theorem 89 and $\mathrm{Int}(W)$ admits a complete metric finite volume of $\mathrm{Ric} \geq -(n-1)$, then ∂W has zero simplicial volume by Theorem 55, so $\mathrm{Out}(\pi_1(W))$ is finite.

Proving relative hyperbolicity was the main goal of the author's work in [20, 21, 22, 24] where it was accomplished in the following cases:

Example 91

(1) If $R_{W,B}$ is from Section 18, then $\pi_1(R_{W,B})$ is hyperbolic relative to the fundamental groups of components of B [20]. If B is as in Theorem 86, then $R_{W,B}$ admits a complete finite volume metric of $-1 \leq K < 0$.
(2) If V is a closed manifold of $K < 0$, and S is an embedded, codimension two, compact, totally geodesic submanifold, then $\pi_1(V \setminus S)$ is hyperbolic relative to the fundamental groups of boundary components of a tubular neighborhood of S in V [24]. A finite volume complete metric of

$-1 \leq K < 0$ on $V \setminus S$ was constructed in [21, 22] when either V is real hyperbolic, or V and S are complex hyperbolic.

(3) Let V be a closed manifold of $K < 0$, and S be an immersed, codimension two, compact, totally geodesic submanifold whose preimage to the universal cover $\tilde{V}$ of V is *normal* in the sense of Allcock and is "sparse" in the sense that and two disjoint components are sufficiently separated. With these assumptions $\pi_1(V \setminus S)$ is hyperbolic relative to the fundamental group of the boundary components of a regular neighborhood of S in V [24]. If V is real hyperbolic, then $V \setminus S$ admits a complete finite volume metric of $-1 \leq K < 0$ by [1].

Remark A version of the claim in Example 91(2)-(3) holds when V is an open complete finitely volume negatively pinched manifold.

Remark Without the assumptions that the preimage of S to $\tilde{V}$ is normal and "sparse", it seems unlikely that $\pi_1(V \setminus S)$ in Example 91(3) is hyperbolic relative to some easy-to-understand peripheral subgroups, so we ask:

Question 92 Let V be a finite volume complete negatively pinched manifold, and S be an immersed, codimension two, compact, totally geodesic submanifold. Is $\pi_1(V \setminus S)$ acylindrically hyperbolic?

If G is a finitely generated group that is hyperbolic relatively to a finite family of peripheral subgroups, then G inherits the following properties of its peripheral subgroups:

(1) solvability of the word problem [75, 146] (which is a litmus test for decency of a group).
(2) solvability of conjugacy problem [42].
(3) being fully residually hyperbolic [147, 96]; here given a class of groups $\mathcal{C}$, a group G is *fully residually* $\mathcal{C}$ if any finite subset of G can be mapped injectively by a homomorphism of G onto a group in $\mathcal{C}$.
(4) being biautomatic [149].
(5) finiteness of asymptotic dimension [145]
(6) rapid decay property [67]
(7) Tits alternative: a subgroup of a relatively hyperbolic group that does not contain a non-abelian free subgroup is *elementary*, i.e. virtually-$\mathbb{Z}$, finite, or contained in a peripheral subgroup [162].

Example 93 If V and S are as in Example 91(2), then the peripheral subgroups in the relatively hyperbolic groups structure on $\pi_1(V \setminus S)$ are the fundamental groups of circle bundles over components of S. The peripheral subgroups have solvable word and conjugacy problems, have finite asymptotic dimension and rapid decay property, are biautomatic, residually hyperbolic, and their non-virtually-abelian groups contain free nonabelian subgroups, see [21, 22]. Hence all these properties are inherited by $\pi_1(V \setminus S)$.

The fact that most real hyperbolic manifolds are a-Kähler [93] was used in [20] to show

Theorem 94 $R_{W,B}$ *is not homeomorphic to an open subset of a Kähler manifold of real dimension* ≥ 4.

In fact, "not homeomorphic" can be replaced with "not proper homotopy equivalent" under a mild assumption on the strict hyperbolization block [20].

20 Zoo of finite volume rank one manifolds

In this section we discuss examples and structure of connected open complete finite volume manifolds of $K \leq 0$ and rank one, with a particular focus on manifolds of $K < 0$.

Theorem 95 *If V is a complete finite volume manifold of rank one, then $\pi_1(V)$ is acylindically hyperbolic.*

Proof The universal cover $\tilde{V}$ has rank one, and Ballmann [11] proved that a lattice in the isometry group of a rank one Hadamard manifold contains a rank one element that lies in a noncyclic free subgroup, so Sisto's Theorem 17 applies. $\square$

Question 96 Does every complete rank one manifold with $K \leq 0$ and Inj Rad $\to 0$ have acylindically hyperbolic fundamental group?

Kapovitch-Wilking's version of the Margulis lemma only calls for $K \geq -1$ on a ball of radius 1, and that curvature bound can always be achieved by rescaling. Since acylindically hyperbolic groups are not virtually nilpotent, Theorem 60 immediately implies:

Corollary 97 *If ε is the constant in* Theorem 60, *then a finite volume complete rank one manifold M contains no point p such that $K \geq -1$ on $B_p(1)$ and the inclusion $B_p(\varepsilon) \hookrightarrow M$ is π_1-surjective.*

Thus if $\pi_1(M)$ is "concentrated" on an ε-ball, then K blows up near that ball.

Known examples of finite volume complete manifolds of $K < 0$ that admit no finite volume metric of $-1 \leq K \leq 0$ are based on Theorem 55 that a compact boundary component of a manifold whose interior has a complete metric of Ric $\geq -(n-1)$ has zero simplicial volume. Indeed, Nguyen Phan [139] proved

Theorem 98 (Nguyen Phan) *In each dimension* ≥ 3 *there exists a finite volume complete manifold of $K < 0$ that is the interior of a compact manifold whose boundary admits a real hyperbolic metric.*

Recall that a closed real hyperbolic manifold has a positive simplicial volume. The proof of the above theorem is by explicit construction; alternatively, it follows from Ontaneda's Theorem 85 by inserting in the cusp the warped product $\mathbb{R} \times_{e^r} B$ where B is any closed manifold of $K \leq 0$ with nonzero simplicial volume. In fact, this argument proves:

Theorem 99 (Ontaneda) *If a closed manifold of $K \leq 0$ is diffeomorphic to the boundary of a compact manifold, then it is diffeomorphic to the boundary of a compact manifold whose interior admits a complete finite volume metric of $K \leq -1$.*

Example 100 There are closed manifolds of $K \leq 0$ that are not homeomorphic to the boundaries of compact manifolds. One source of such examples is evenness of the Euler characteristic of the boundary of a compact manifold [66, Corollary VIII.8.8]. Examples of closed manifold of $K \leq 0$ with odd Euler characteristic include suitable closed non-orientable hyperbolic surfaces, Mumford's complex hyperbolic surface of Euler characteristic 3 [104, Proposition2.2], or their products.

Iterated gluing along totally geodesic boundaries sometimes yields finite volume M with infinitely generated fundamental group. This phenomenon was discovered by Gromov [13, Chapter 11] who produced such 3-dimensional graph manifolds with $K \in [-1, 0]$ and no local Euclidean de Rham factor. Other examples with infinitely generated fundamental groups, due to Nguyen Phan [139], are finite volume manifolds with $K \leq -1$ and infinitely many ends, appearing in all dimensions ≥ 2.

A related idea of Nguyen Phan [139] uses infinite cyclic covers of closed manifolds of $K < 0$ to produce two-ended finite volume manifolds of $K < 0$:

Theorem 101 (Nguyen Phan) *If $\widehat{L}$ is an infinite cyclic cover of a closed manifold L of $K < 0$ of dimension $l \geq 3$, then $\widehat{L}$ admits a complete finite volume metric metric of $K < 0$.*

Here $\widehat{L} \to L$ is the covering that corresponds to the kernel of any epimorphism $\pi_1(L) \to \mathbb{Z}$. A most famous example is when $\widehat{L}$ corresponds to the fiber group in a closed hyperbolic 3-manifolds that fibers over a circle.

Example 102 (of $\widehat{L}$ with infinitely generated fundamental group) Suppose that $\pi_1(L)$ surjects onto a noncyclic free group F_r. The kernel of any epimorphism $F_r \to \mathbb{Z}$ is infinitely generated, so hence so is the kernel of the composite $\pi_1(L) \to F_r \to \mathbb{Z}$.

Question 103 Is $\pi_1(\widehat{L})$ always infinitely generated when $\dim(\widehat{L}) > 3$?

Remark If $\dim(\widehat{L}) \geq 6$ and $\widehat{L}$ is homotopy equivalent to a finite cell complex (or more generally is finitely dominated), then L (smoothly) fibers over a circle, and the fiber is a closed aspherical manifold whose inclusion into L corresponds homotopically to the covering $\widehat{L} \to L$. Indeed, Siebenmann's version [159] of Farrell's fibering obstruction lies in the group $\mathrm{Wh}(\pi_1(L))$, which is zero by [76].

Remark Another restriction on $\pi_1(\widehat{L})$ is that its outer automorphism group is infinite (as easily follows from the fact that $\pi_1(\widehat{L})$ has trivial centralizer in $\pi_1(L)$). Combining with the previous remark we see that if $\dim(\widehat{L}) \geq 6$ and $\widehat{L}$ is finitely dominated, then $\widehat{L}$ is homotopy equivalent to a closed aspherical manifold of dimension ≥ 5 whose fundamental group has infinite outer automorphism group and embeds into the hyperbolic group $\pi_1(L)$; it seems such a closed aspherical manifold cannot exist, so we ask:

Question 104 Can $\widehat{L}$ ever be finitely dominated when $\dim(\widehat{L}) > 3$?

Example 105 In dimension three Theorem 101 gives a finite volume metric of $K < 0$ on the product of a closed hyperbolic surface with $\mathbb{R}$, and interestingly, the metric can be chosen so that the corresponding nonuniform lattice contains no parabolics, see [139].

Question 106 Is there a nonuniform lattice in the isometry group of a Hadamard manifold of dimension > 3 that contains no parabolics?

Borel-Serre [36] compute the $\mathbb{Q}$-rank of a finite volume complete locally symmetric n-manifold V as $n - \mathrm{cd}(\pi_1(V))$, where cd denotes the cohomological dimension; alternatively, $\mathbb{Q}$-rank equals the dimension of an asymptotic cone of V [103], cf. [110], and $\mathbb{Q}$-rank can also be defined in terms of flats in V [135]. If V has rank one (i.e. contains a rank one geodesic), then the $\mathbb{Q}$-rank of V equals 1. One wonders whether any of these relations between cd, asymptotic cone, and absence of 2-dimensional flats extend to finite volume complete manifolds of rank one.

Question 107 Is the asymptotic cone of a complete finite volume rank one manifold of $K \leq 0$ always a tree?

Question 108 What values does cd take on the fundamental groups of open complete finite volume n-manifolds of $K \leq -1$, rank one, or $K < 0$? Is cd always equal $n - 1$?

To better understand the above question let us relate bounds on cd with the fundamental group at infinity of an open aspherical n-manifold M. First note that if $\mathrm{cd}(\pi_1(M)) \leq n - 2$, then M is one-ended [158, Proposition 1.2] but the converse fails (think of the open Möbius band). In the simplest case when M is the interior of a compact manifold, we get the following clean statement:

Proposition 109 *Let N be a compact aspherical n-manifold with boundary. Then $\mathrm{cd}(\pi_1(M)) \leq n - 2$ if and only if ∂N is connected and the inclusion $\partial N \hookrightarrow N$ is π_1-surjective.*

Proof Clearly $\mathrm{cd}(\pi_1(M)) \leq n - 1$. Now Poincaré-Lefschetz duality in the universal cover [41, Corollary VIII.8.3] implies that $\mathrm{cd}(\pi_1(M)) = n - 1$ if and only if the boundary of the universal cover of N is not connected. The latter is equivalent to "either ∂N is not connected or $\partial N \hookrightarrow N$ is not π_1-surjective" by elementary covering space considerations. □

Question 108 should be compared with Nguyen Phan's Example 105 of a finite volume manifold of $K < 0$ that is quite small homologically, and perhaps there are even smaller examples. We finish with the following tantalizing

Question 110 Given $n > 2$, does the interior of the n-dimensional handlebody admit a complete finite volume metric of $K \leq 0$?

Example 111 The interior of an odd-dimensional handlebody admits no complete finite volume metric of $-1 \leq K \leq 0$. (The boundary of an odd-dimensional handlebody with g handles has Euler characteristic $2 - 2g$ while under our geometric assumptions the Euler characteristic vanishes by Theorem 67 and $g \neq 1$ because $\mathbb{Z}$ is not acylindrically hyperbolic).

Igor Belegradek
School of Mathematics, Georgia Institute of Technology,
Atlanta, GA 30332-0160 U.S.A.
E-mail: ib@math.gatech.edu

References

[1] Abresch, U., and Schroeder, V. 1992. Graph manifolds, ends of negatively curved spaces and the hyperbolic 120-cell space. *J. Differential Geom.*, **35**(2), 299–336.

[2] Adams, S., and Ballmann, W. 1998. Amenable isometry groups of Hadamard spaces. *Math. Ann.*, **312**(1), 183–195.

[3] Agol, I. *Tameness of hyperbolic 3-manifolds*. arXiv:math/0405568.

[4] Agol, I. 2013. The virtual Haken conjecture. *Doc. Math.*, **18**, 1045–1087. With an appendix by Agol, D. Groves, and J. Manning.

[5] Alperin, R. C., and Shalen, P. B. 1982. Linear groups of finite cohomological dimension. *Invent. Math.*, **66**(1), 89–98.

[6] Anderson, M. T. 1987. Metrics of negative curvature on vector bundles. *Proc. Amer. Math. Soc.*, **99**(2), 357–363.

[7] Anderson, M. T. 2006. Dehn filling and Einstein metrics in higher dimensions. *J. Differential Geom.*, **73**(2), 219–261.

[8] Ardanza-Trevijano Moras, S. I. 2000. *Exotic smooth structures on negatively curved manifolds that are not of the homotopy type of a locally symmetric space.* ProQuest LLC, Ann Arbor, MI. Thesis (Ph.D.)–SUNY at Binghamton.

[9] Arzhantseva, G., Bridson, M. R., Januszkiewicz, T., Leary, I. J., Minasyan, A., and Świątkowski, J. 2009. Infinite groups with fixed point properties. *Geom. Topol.*, **13**(3), 1229–1263.

[10] Aschenbrenner, M., Friedl, S., and Wilton, H. *3-manifold groups*. arXiv:1205.0202.

[11] Ballmann, W. 1995. *Lectures on spaces of nonpositive curvature.* DMV Seminar, vol. 25. Basel: Birkhäuser Verlag. With an appendix by Misha Brin.

[12] Ballmann, W., and Buyalo, S. 2008. Periodic rank one geodesics in Hadamard spaces. Pages 19–27 of: *Geometric and probabilistic structures in dynamics.* Contemp. Math., vol. 469. Providence, RI: Amer. Math. Soc.

[13] Ballmann, W., Gromov, M., and Schroeder, V. 1985. *Manifolds of nonpositive curvature.* Progress in Mathematics, vol. 61. Boston, MA: Birkhäuser Boston Inc.

[14] Balser, A., and Lytchak, A. 2005. Centers of convex subsets of buildings. *Ann. Global Anal. Geom.*, **28**(2), 201–209.

[15] Bartels, A., and Lück, W. 2012. The Borel conjecture for hyperbolic and CAT(0)-groups. *Ann. of Math. (2)*, **175**(2), 631–689.

[16] Bartels, A., Farrell, F. T., and Lück, W. 2014. The Farrell-Jones Conjecture for cocompact lattices in virtually connected Lie groups. *J. Amer. Math. Soc.*, **27**(2), 339–388.

[17] Bavard, C. 1991. Longueur stable des commutateurs. *Enseign. Math. (2)*, **37**(1-2), 109–150.

[18] Belegradek, I. *Obstructions to nonpositive curvature for open manifolds.* arXiv:1208.5220v1, to appear in Proc. Lond. Math. Soc.

[19] Belegradek, I. 2001. Pinching, Pontrjagin classes, and negatively curved vector bundles. *Invent. Math.*, **144**(2), 353–379.

[20] Belegradek, I. 2007. Aspherical manifolds with relatively hyperbolic fundamental groups. *Geom. Dedicata*, **129**, 119–144.

[21] Belegradek, I. 2012a. Complex hyperbolic hyperplane complements. *Math. Ann.*, **353**(2), 545–579.

[22] Belegradek, I. 2012b. Rigidity and relative hyperbolicity of real hyperbolic hyperplane complements. *Pure Appl. Math. Q.*, **8**(1), 15–51.

[23] Belegradek, I. 2013. An assortment of negatively curved ends. *J. Topol. Anal.*, **5**(4), 439–449.

[24] Belegradek, I., and Hruska, G. C. 2013. Hyperplane arrangements in negatively curved manifolds and relative hyperbolicity. *Groups Geom. Dyn.*, **7**(1), 13–38.

[25] Belegradek, I., and Kapovitch, V. 2005. Pinching estimates for negatively curved manifolds with nilpotent fundamental groups. *Geom. Funct. Anal.*, **15**(5), 929–938.

[26] Belegradek, I., and Kapovitch, V. 2006. Classification of negatively pinched manifolds with amenable fundamental groups. *Acta Math.*, **196**(2), 229–260.

[27] Belegradek, I., and Nguyễn Phan, T. T. *Non-aspherical ends and nonpositive curvature.* arXiv:1212.3303.

[28] Bestvina, M., and Feighn, M. 2002. Proper actions of lattices on contractible manifolds. *Invent. Math.*, **150**(2), 237–256.

[29] Bestvina, M., and Fujiwara, K. 2002. Bounded cohomology of subgroups of mapping class groups. *Geom. Topol.*, **6**, 69–89.

[30] Bestvina, M., and Fujiwara, K. 2009. A characterization of higher rank symmetric spaces via bounded cohomology. *Geom. Funct. Anal.*, **19**(1), 11–40.

[31] Bestvina, M., Kapovich, M., and Kleiner, B. 2002. Van Kampen's embedding obstruction for discrete groups. *Invent. Math.*, **150**(2), 219–235.

[32] Bieri, R. 1981. *Homological dimension of discrete groups.* Second edn. Queen Mary College Mathematical Notes. London: Queen Mary College Department of Pure Mathematics.

[33] Bishop, R. L., and O'Neill, B. 1969. Manifolds of negative curvature. *Trans. Amer. Math. Soc.*, **145**, 1–49.

[34] Bonahon, F., and Siebenmann, L. C. 1987. The characteristic toric splitting of irreducible compact 3-orbifolds. *Math. Ann.*, **278**(1-4), 441–479.

[35] Borel, A., and Ji, L. 2005. Compactifications of symmetric and locally symmetric spaces. Pages 69–137 of: *Lie theory*. Progr. Math., vol. 229. Boston, MA: Birkhäuser Boston.

[36] Borel, A., and Serre, J.-P. 1973. Corners and arithmetic groups. *Comment. Math. Helv.*, **48**, 436–491. Avec un appendice: Arrondissement des variétés à coins, par A. Douady et L. Hérault.

[37] Bowditch, B. H. 2010. Notes on tameness. *Enseign. Math. (2)*, **56**(3-4), 229–285.

[38] Bowditch, B. H. 2012. Relatively hyperbolic groups. *Internat. J. Algebra Comput.*, **22**(3), 1250016, 66.

[39] Bridson, M. R. 2001. On the subgroups of semihyperbolic groups. Pages 85–111 of: *Essays on geometry and related topics, Vol. 1, 2*. Monogr. Enseign. Math., vol. 38. Geneva: Enseignement Math.

[40] Bridson, M. R., and Haefliger, A. 1999. *Metric spaces of non-positive curvature*. Grundlehren der Mathematischen Wissenschaften [Fundamental Principles of Mathematical Sciences], vol. 319. Berlin: Springer-Verlag.

[41] Brown, K. S. 1994. *Cohomology of groups*. Graduate Texts in Mathematics, vol. 87. New York: Springer-Verlag. Corrected reprint of the 1982 original.

[42] Bumagin, I. 2004. The conjugacy problem for relatively hyperbolic groups. *Algebr. Geom. Topol.*, **4**, 1013–1040.

[43] Burger, M., and Monod, N. 2002. Continuous bounded cohomology and applications to rigidity theory. *Geom. Funct. Anal.*, **12**(2), 219–280.

[44] Burger, M., and Schroeder, V. 1987. Amenable groups and stabilizers of measures on the boundary of a Hadamard manifold. *Math. Ann.*, **276**(3), 505–514.

[45] Buyalo, S. V. 1993. An example of a four-dimensional manifold of negative curvature. *Algebra i Analiz*, **5**(1), 193–199.

[46] Buyalo, S. V., and Kobel'skii, V. L. 1995. Geometrization of graph-manifolds. II. Isometric geometrization. *Algebra i Analiz*, **7**(3), 96–117.

[47] Buyalo, S. V., and Kobel'skii, V. L. 1996. Cusp closing of hyperbolic manifolds. *Geom. Dedicata*, **59**(2), 147–156.

[48] Buyalo, S. V., and Svetlov, P. V. 2004. Topological and geometric properties of graph manifolds. *Algebra i Analiz*, **16**(2), 3–68.

[49] Calegari, D., and Gabai, D. 2006. Shrinkwrapping and the taming of hyperbolic 3-manifolds. *J. Amer. Math. Soc.*, **19**(2), 385–446.

[50] Canary, R. D. 2008. Marden's tameness conjecture: history and applications. Pages 137–162 of: *Geometry, analysis and topology of discrete groups*. Adv. Lect. Math. (ALM), vol. 6. Int. Press, Somerville, MA.

[51] Canary, R. D., and McCullough, D. 2004. Homotopy equivalences of 3-manifolds and deformation theory of Kleinian groups. *Mem. Amer. Math. Soc.*, **172**(812), xii+218.

[52] Caprace, P.-E., and Fujiwara, K. 2010. Rank-one isometries of buildings and quasi-morphisms of Kac-Moody groups. *Geom. Funct. Anal.*, **19**(5), 1296–1319.

[53] Caprace, P.-E., and Lytchak, A. 2010. At infinity of finite-dimensional CAT(0) spaces. *Math. Ann.*, **346**(1), 1–21.

[54] Caprace, P.-E., and Monod, N. 2009a. Isometry groups of non-positively curved spaces: discrete subgroups. *J. Topol.*, **2**(4), 701–746.

[55] Caprace, P.-E., and Monod, N. 2009b. Isometry groups of non-positively curved spaces: structure theory. *J. Topol.*, **2**(4), 661–700.

[56] Charney, R. M., and Davis, M. W. 1995. Strict hyperbolization. *Topology*, **34**(2), 329–350.

[57] Cheeger, J., and Gromov, M. 1985. Bounds on the von Neumann dimension of L^2-cohomology and the Gauss-Bonnet theorem for open manifolds. *J. Differential Geom.*, **21**(1), 1–34.

[58] Cheeger, J., and Gromov, M. 1991. Chopping Riemannian manifolds. Pages 85–94 of: *Differential geometry*. Pitman Monogr. Surveys Pure Appl. Math., vol. 52. Harlow: Longman Sci. Tech.

[59] Dafermos, M. 1997. Exhaustions of complete manifolds of bounded curvature. Pages 319–322 of: *Tsing Hua lectures on geometry & analysis (Hsinchu, 1990–1991)*. Int. Press, Cambridge, MA.

[60] Dahmani, F., Guirardel, V., and Osin, D. *Hyperbolically embedded subgroups and rotating families in groups acting on hyperbolic spaces.* http://arXiv:1111.7048v3.

[61] Davis, M. W. 2008. *The geometry and topology of Coxeter groups.* London Mathematical Society Monographs Series, vol. 32. Princeton, NJ: Princeton University Press.

[62] Davis, M. W., and Januszkiewicz, T. 1991. Hyperbolization of polyhedra. *J. Differential Geom.*, **34**(2), 347–388.

[63] Davis, M. W., Januszkiewicz, T., and Weinberger, S. 2001. Relative hyperbolization and aspherical bordisms: an addendum to "Hyperbolization of polyhedra". *J. Differential Geom.*, **58**(3), 535–541.

[64] Deraux, M. 2005. A negatively curved Kähler threefold not covered by the ball. *Invent. Math.*, **160**(3), 501–525.

[65] Despotovic, Z. 2006. *Action dimension of mapping class groups.* ProQuest LLC, Ann Arbor, MI. Thesis (Ph.D.)–The University of Utah.

[66] Dold, A. 1995. *Lectures on algebraic topology.* Classics in Mathematics. Berlin: Springer-Verlag. Reprint of the 1972 edition.

[67] Druţu, C., and Sapir, M. 2005. Relatively hyperbolic groups with rapid decay property. *Int. Math. Res. Not.*, 1181–1194.

[68] Druţu, C., and Sapir, M. V. 2008. Groups acting on tree-graded spaces and splittings of relatively hyperbolic groups. *Adv. Math.*, **217**(3), 1313–1367.

[69] Duchesne, B. *Superrigidity In Infinite Dimension And Finite Rank Via Harmonic Maps.* arXiv:1206.1964v1.

[70] Eberlein, P. 1979. Surfaces of nonpositive curvature. *Mem. Amer. Math. Soc.*, **20**(218), x+90.

[71] Eberlein, P. 1980. Lattices in spaces of nonpositive curvature. *Ann. of Math. (2)*, **111**(3), 435–476.

[72] Eberlein, P., and O'Neill, B. 1973. Visibility manifolds. *Pacific J. Math.*, **46**, 45–109.

[73] Eberlein, P. B. 1996. *Geometry of nonpositively curved manifolds.* Chicago Lectures in Mathematics. Chicago, IL: University of Chicago Press.

[74] Eymard, P. 1972. *Moyennes invariantes et représentations unitaires.* Lecture Notes in Mathematics, Vol. 300. Berlin: Springer-Verlag.

[75] Farb, B. 1998. Relatively hyperbolic groups. *Geom. Funct. Anal.*, **8**(5), 810–840.

[76] Farrell, F. T., and Jones, L. E. 1986. *K*-theory and dynamics. I. *Ann. of Math. (2)*, **124**(3), 531–569.

[77] Farrell, F. T., and Ontaneda, P. 2006. Branched covers of hyperbolic manifolds and harmonic maps. *Comm. Anal. Geom.*, **14**(2), 249–268.

[78] Farrell, F. T., Jones, L. E., and Ontaneda, P. 2007. Negative curvature and exotic topology. Pages 329–347 of: *Surveys in differential geometry. Vol. XI.* Surv. Differ. Geom., vol. 11. Int. Press, Somerville, MA.

[79] Fornari, S., and Schroeder, V. 1990. Ramified coverings with nonpositive curvature. *Math. Z.*, **203**(1), 123–128.

[80] Freedman, M. H., and Gabai, D. 2007. Covering a nontaming knot by the unlink. *Algebr. Geom. Topol.*, **7**, 1561–1578.

[81] Fujiwara, K. 1988. A construction of negatively curved manifolds. *Proc. Japan Acad. Ser. A Math. Sci.*, **64**(9), 352–355.

[82] Fujiwara, K., Nagano, K., and Shioya, T. 2006. Fixed point sets of parabolic isometries of CAT(0)-spaces. *Comment. Math. Helv.*, **81**(2), 305–335.

[83] Fukaya, K. 1984. A finiteness theorem for negatively curved manifolds. *J. Differential Geom.*, **20**(2), 497–521.

[84] Fukaya, K., and Yamaguchi, T. 1992. The fundamental groups of almost non-negatively curved manifolds. *Ann. of Math. (2)*, **136**(2), 253–333.

[85] Furman, A. 2011. A survey of measured group theory. Pages 296–374 of: *Geometry, rigidity, and group actions*. Chicago Lectures in Math. Chicago, IL: Univ. Chicago Press.

[86] Gelander, T. 2004. Homotopy type and volume of locally symmetric manifolds. *Duke Math. J.*, **124**(3), 459–515.

[87] Gelander, T. 2011. Volume versus rank of lattices. *J. Reine Angew. Math.*, **661**, 237–248.

[88] Goldman, M. E. 1971. An algebraic classification of noncompact 2-manifolds. *Trans. Amer. Math. Soc.*, **156**, 241–258.

[89] Gromoll, D., and Wolf, J. A. 1971. Some relations between the metric structure and the algebraic structure of the fundamental group in manifolds of nonpositive curvature. *Bull. Amer. Math. Soc.*, **77**, 545–552.

[90] Gromov, M. 1978. Manifolds of negative curvature. *J. Differential Geom.*, **13**(2), 223–230.

[91] Gromov, M. 1982. Volume and bounded cohomology. *Inst. Hautes Études Sci. Publ. Math.*, 5–99 (1983).

[92] Gromov, M. 1987. Hyperbolic groups. Pages 75–263 of: *Essays in group theory*. Math. Sci. Res. Inst. Publ., vol. 8. New York: Springer.

[93] Gromov, M. 1993. Asymptotic invariants of infinite groups. Pages 1–295 of: *Geometric group theory, Vol. 2 (Sussex, 1991)*. London Math. Soc. Lecture Note Ser., vol. 182. Cambridge: Cambridge Univ. Press.

[94] Gromov, M. 2003. Random walk in random groups. *Geom. Funct. Anal.*, **13**(1), 73–146.

[95] Gromov, M., and Thurston, W. 1987. Pinching constants for hyperbolic manifolds. *Invent. Math.*, **89**(1), 1–12.

[96] Groves, D., and Manning, J. F. 2008. Dehn filling in relatively hyperbolic groups. *Israel J. Math.*, **168**, 317–429.

[97] Guilbault, C. R. 2007. Products of open manifolds with $\mathbb{R}$. *Fund. Math.*, **197**, 197–214.

[98] Hamenstädt, U. 2008. Bounded cohomology and isometry groups of hyperbolic spaces. *J. Eur. Math. Soc. (JEMS)*, **10**(2), 315–349.

[99] Hamenstädt, U. 2009a. Isometry groups of proper hyperbolic spaces. *Geom. Funct. Anal.*, **19**(1), 170–205.

[100] Hamenstädt, U. 2009b. Rank-one isometries of proper CAT(0)-spaces. Pages 43–59 of: *Discrete groups and geometric structures*. Contemp. Math., vol. 501. Providence, RI: Amer. Math. Soc.

[101] Hamenstädt, U. 2012. Isometry groups of proper CAT(0)-spaces of rank one. *Groups Geom. Dyn.*, **6**(3), 579–618.

[102] Hatcher, A. *Basic 3-Manifold Topology*. www.math.cornell.edu/ hatcher/.

[103] Hattori, T. 1996. Asymptotic geometry of arithmetic quotients of symmetric spaces. *Math. Z.*, **222**(2), 247–277.

[104] Hersonsky, S., and Paulin, F. 1996. On the volumes of complex hyperbolic manifolds. *Duke Math. J.*, **84**(3), 719–737.

[105] Hull, M., and Osin, D. 2013. Induced quasicocycles on groups with hyperbolically embedded subgroups. *Algebr. Geom. Topol.*, **13**(5), 2635–2665.

[106] Hummel, C., and Schroeder, V. 1996. Cusp closing in rank one symmetric spaces. *Invent. Math.*, **123**(2), 283–307.

[107] Izeki, H., and Nayatani, S. 2005. Combinatorial harmonic maps and discrete-group actions on Hadamard spaces. *Geom. Dedicata*, **114**, 147–188.

[108] Izeki, H., and Nayatani, S. 2010. An approach to superrigidity and fixed-point theorems via harmonic maps. Pages 135–160 of: *Selected papers on analysis and differential equations*. Amer. Math. Soc. Transl. Ser. 2, vol. 230. Providence, RI: Amer. Math. Soc.

[109] Izeki, H., Kondo, T., and Nayatani, S. 2012. *N*-step energy of maps and the fixed-point property of random groups. *Groups Geom. Dyn.*, **6**(4), 701–736.

[110] Ji, L., and MacPherson, R. 2002. Geometry of compactifications of locally symmetric spaces. *Ann. Inst. Fourier (Grenoble)*, **52**(2), 457–559.

[111] Johnson, B. E. 1972. *Cohomology in Banach algebras*. Providence, R.I.: American Mathematical Society. Memoirs of the American Mathematical Society, No. 127.

[112] Kapovich, M. 2001. *Hyperbolic manifolds and discrete groups*. Progress in Mathematics, vol. 183. Boston, MA: Birkhäuser Boston Inc.

[113] Kapovich, M., and Leeb, B. 1996. Actions of discrete groups on nonpositively curved spaces. *Math. Ann.*, **306**(2), 341–352.

[114] Kapovitch, V., and Wilking, B. *Structure of fundamental groups of manifolds with Ricci curvature bounded below*. arXiv:1105.5955.

[115] Kapovitch, V., Petrunin, A., and Tuschmann, W. 2010. Nilpotency, almost nonnegative curvature, and the gradient flow on Alexandrov spaces. *Ann. of Math. (2)*, **171**(1), 343–373.

[116] Karlsson, A., and Noskov, G. A. 2004. Some groups having only elementary actions on metric spaces with hyperbolic boundaries. *Geom. Dedicata*, **104**, 119–137.

[117] Kerékjártó, B. 1923. *Vorlesungen über Topologie, I.* Berlin: Springer.

[118] Kleiner, B. 1999. *Integrating infinitesimal flats in Hadamard manifolds*. preprint.

[119] Lawson, Jr., H. B., and Yau, S. T. 1972. Compact manifolds of nonpositive curvature. *J. Differential Geometry*, **7**, 211–228.

[120] Leeb, B. 1995. 3-manifolds with(out) metrics of nonpositive curvature. *Invent. Math.*, **122**(2), 277–289.

[121] Leuzinger, E. 1995. An exhaustion of locally symmetric spaces by compact submanifolds with corners. *Invent. Math.*, **121**(2), 389–410.

[122] Leuzinger, E. 2004. On polyhedral retracts and compactifications of locally symmetric spaces. *Differential Geom. Appl.*, **20**(3), 293–318.

[123] Liu, Yi. 2013. Virtual cubulation of nonpositively curved graph manifolds. *J. Topol.*, **6**(4), 793–822.

[124] Lück, W. 2010. Survey on aspherical manifolds. Pages 53–82 of: *European Congress of Mathematics*. Eur. Math. Soc., Zürich.

[125] Margulis, G. A. 1991. *Discrete subgroups of semisimple Lie groups*. Ergebnisse der Mathematik und ihrer Grenzgebiete (3) [Results in Mathematics and Related Areas (3)], vol. 17. Berlin: Springer-Verlag.

[126] Minasyan, A., and Osin, D. *Acylindrically hyperbolic groups acting on trees*. in preparation.

[127] Mineyev, I., and Yaman, A. *Relative hyperbolicity and bounded cohomology*. www.math.uiuc.edu/~mineyev/math/.

[128] Mineyev, I., Monod, N., and Shalom, Y. 2004. Ideal bicombings for hyperbolic groups and applications. *Topology*, **43**(6), 1319–1344.

[129] Monod, N. 2006a. An invitation to bounded cohomology. Pages 1183–1211 of: *International Congress of Mathematicians. Vol. II*. Eur. Math. Soc., Zürich.

[130] Monod, N. 2006b. Superrigidity for irreducible lattices and geometric splitting. *J. Amer. Math. Soc.*, **19**(4), 781–814.

[131] Monod, N., and Popa, S. 2003. On co-amenability for groups and von Neumann algebras. *C. R. Math. Acad. Sci. Soc. R. Can.*, **25**(3), 82–87.

[132] Monod, N., and Shalom, Y. 2004. Cocycle superrigidity and bounded cohomology for negatively curved spaces. *J. Differential Geom.*, **67**(3), 395–455.

[133] Monod, N., and Shalom, Y. 2006. Orbit equivalence rigidity and bounded cohomology. *Ann. of Math. (2)*, **164**(3), 825–878.

[134] Moon, S. 2009. *Amenable actions of discrete groups*. Ph.D. thesis, Universite de Neuchatel, http://doc.rero.ch/record/18088.

[135] Morris, D. *Introduction to arithmetic groups*. arXiv:math/0106063.

[136] Mostow, G. D., and Siu, Y. T. 1980. A compact Kähler surface of negative curvature not covered by the ball. *Ann. of Math. (2)*, **112**(2), 321–360.

[137] Naor, A., and Silberman, L. 2011. Poincaré inequalities, embeddings, and wild groups. *Compos. Math.*, **147**(5), 1546–1572.

[138] Nguyễn Phan, T. T. *Nil happens. What about Sol?* arXiv:1207.1734.

[139] Nguyễn Phan, T. T. *On finite volume, negatively curved manifolds*. arXiv:1110.4087.

[140] Nguyễn Phan, T. T. 2013. Nonpositively curved manifolds containing a prescribed nonpositively curved hypersurface. *Topology Proc.*, **42**, 39–41.

[141] Noskov, G. A. 1990. Bounded cohomology of discrete groups with coefficients. *Algebra i Analiz*, **2**(5), 146–164.

[142] Ontaneda, P. *Pinched smooth hyperbolization*. arXiv:1110.6374v1.

[143] Ontaneda, P. 2003. The double of a hyperbolic manifold and non-positively curved exotic *PL* structures. *Trans. Amer. Math. Soc.*, **355**(3), 935–965 (electronic).

[144] Osin, D. V. *Acylindrically hyperbolic groups*. arXiv:1304.1246v1.

[145] Osin, D. V. 2005. Asymptotic dimension of relatively hyperbolic groups. *Int. Math. Res. Not.*, 2143–2161.

[146] Osin, D. V. 2006. Relatively hyperbolic groups: intrinsic geometry, algebraic properties, and algorithmic problems. *Mem. Amer. Math. Soc.*, **179**(843), vi+100.

[147] Osin, D. V. 2007. Peripheral fillings of relatively hyperbolic groups. *Invent. Math.*, **167**(2), 295–326.

[148] Ould Houcine, A. 2007. Embeddings in finitely presented groups which preserve the center. *J. Algebra*, **307**(1), 1–23.

[149] Rebbechi, D. Y. 2001. *Algorithmic properties of relatively hyperbolic groups.* Ph.D. thesis, Rutgers Newark arXiv:math/0302245v1.

[150] Richards, I. 1963. On the classification of noncompact surfaces. *Trans. Amer. Math. Soc.*, **106**, 259–269.

[151] Saper, L. 1997. Tilings and finite energy retractions of locally symmetric spaces. *Comment. Math. Helv.*, **72**(2), 167–202.

[152] Schoen, R., and Yau, S.-T. 1994. *Lectures on differential geometry.* Conference Proceedings and Lecture Notes in Geometry and Topology, I. Cambridge, MA: International Press.

[153] Schroeder, V. 1985. A splitting theorem for spaces of nonpositive curvature. *Invent. Math.*, **79**(2), 323–327.

[154] Schroeder, V. 1989. A cusp closing theorem. *Proc. Amer. Math. Soc.*, **106**(3), 797–802.

[155] Schroeder, V. 1991. Analytic manifolds of nonpositive curvature with higher rank subspaces. *Arch. Math. (Basel)*, **56**(1), 81–85.

[156] Scott, P., and Tucker, T. 1989. Some examples of exotic noncompact 3-manifolds. *Quart. J. Math. Oxford Ser. (2)*, **40**(160), 481–499.

[157] Siebenmann, L. C. *Classifcation des 2-variéetés ouvertes.* preprint, 2008.

[158] Siebenmann, L. C. 1969. On detecting open collars. *Trans. Amer. Math. Soc.*, **142**, 201–227.

[159] Siebenmann, L. C. 1970. A total Whitehead torsion obstruction to fibering over the circle. *Comment. Math. Helv.*, **45**, 1–48.

[160] Sisto, A. *Contracting elements and random walks.* arXiv:1112.2666v1.

[161] Thurston, W. P. *The geometry and topology of three-manifolds.* Princeton lecture notes (1978-1981).

[162] Tukia, P. 1994. Convergence groups and Gromov's metric hyperbolic spaces. *New Zealand J. Math.*, **23**(2), 157–187.

[163] Wang, M.-T., and Lin, C.-H. 1997. A note on the exhaustion function for complete manifolds. Pages 269–277 of: *Tsing Hua lectures on geometry & analysis (Hsinchu, 1990–1991).* Int. Press, Cambridge, MA.

[164] Wu, Y. *Translation lengths of parabolic isometries of CAT(0) spaces and their applications.* http://math.rice.edu/ yw22/.

[165] Yoon, S. Y. 2004. A lower bound to the action dimension of a group. *Algebr. Geom. Topol.*, **4**, 273–296 (electronic).

[166] Zheng, F. 1996. Examples of non-positively curved Kähler manifolds. *Comm. Anal. Geom.*, **4**(1-2), 129–160.

3

Cohomologie et actions isométriques propres sur les espaces L_p

MARC BOURDON

Abstract

These notes present and discuss some of the basic properties and results of the ℓ_p-cohomology of groups. We also use the ℓ_p-cohomology to study the proper isometric actions of word hyperbolic groups on L_p-spaces.[1]

1 Introduction

Ce texte traite de certains aspects de la cohomologie ℓ_p des groupes, et des actions isométriques sur les espaces L_p. Pratiquement aucun résultat qu'il contient n'est original. Il comporte deux parties. La première est une relecture détaillée et commentée de certains passages du livre [32] de M. Gromov qui portent sur la cohomologie ℓ_p. Cette partie a également bénéficié des nombreuses discussions que j'ai pu avoir avec P. Pansu. Dans la seconde partie, la cohomologie ℓ_p est utilisée pour étudier les actions isométriques propres des groupes (Gromov) hyperboliques sur les espaces L_p. Chaque partie se termine par un survol de quelques résultats complémentaires et par des questions.

1.1 Cohomologie ℓ_p

Le premier thème abordé est celui de la cohomologie ℓ_p des groupes de type fini Γ. Nous en présentons des définitions et résultats de base. Sont discutés en particulier:

[1] **Keywords:** ℓ_p-cohomology, proper isometric actions on Banach spaces.
AMS codes: 20F67, 20J06, 43A07, 43A15, 58E40.

- L'invariance par quasi-isométrie de la cohomologie ℓ_p de Γ (Th.3 et Def.4),
- Plusieurs caractérisations de la moyennabilité en termes d'homologie et de cohomologie ℓ_p (Th.6),
- L'annulation de la cohomologie ℓ_p réduite de Γ lorsque le centre Γ est infini (Prop.10),
- Des énoncés d'annulation de la 1-cohomologie ℓ_p en présence d'un sous-groupe distingué ayant des propriétés spéciales (Prop.13 et Th.14),
- Un résultat de représentation harmonique de la 1-cohomologie ℓ_p des groupes non moyennables (Cor.7).

1.2 Actions isométriques propres

La 1-cohomologie ℓ_p de Γ décrit les actions isométriques de Γ sur $\ell_p(\Gamma)$ associées à la représentation régulière droite. Elle participe donc au second thème abordé dans ces notes, qui est celui des actions isométriques sur les espaces L_p.

Rappelons qu'une action de Γ sur un Banach V est dite *propre* si pour toute partie bornée $P \subset V$ le cardinal des $g \in \Gamma$ tels que $gP \cap P \neq \emptyset$ est fini.

Un groupe est dit *a-T-menable* s'il possède une action isométrique propre sur un Hilbert. Cette notion apparait dans [32] p.177. Elle joue un rôle de premier plan dans l'étude des groupes via leurs actions sur les espaces de Hilbert (voir notamment [16]).

On s'intéresse ici à sa généralisation naturelle aux actions sur les espaces L_p. L'un des objectifs de ces notes est d'établir le résultat suivant qui est une version quantitative d'un théorème de G. Yu [57]. Dans l'énoncé Confdim($\partial\Gamma$) désigne la dimension conforme Ahlfors régulière du bord de Γ, il s'agit d'un invariant de quasi-isométrie de Γ dont la définition est rappelée au paragraphe 3.1.

Théorème 1 *Soit Γ un groupe hyperbolique (au sens de Gromov), alors pour tout $p > \mathrm{Confdim}(\partial\Gamma)$ il possède une action isométrique propre sur ℓ_p.*

G. Yu démontre l'existence d'une telle action pour p assez grand, en utilisant la construction d'I. Mineyev du' "fibré unitaire tangent" à un groupe hyperbolique. Une de ses motivations provient de la K-théorie des C^*-algèbres [35]. Nous montrons le théorème en utilisant la 1-cohomologie ℓ_p de Γ et le bord à l'infini de Γ. Récemment B. Nica [40] a donné une autre preuve du théorème de G. Yu qui repose sur des structures conformes au bord construites par I. Mineyev.

Dans le cas où Γ est un réseau cocompact d'un groupe de Lie simple de rang 1, le théorème 1 est dû à Y. De Cornulier, R. Tessera, A. Valette [19] par des méthodes différentes.

Il est intéressant de noter que le théorème entraîne que les groupes hyperboliques de Kazhdan agissent proprement sur les espaces ℓ_p lorsque p est assez grand, alors que leurs actions isométriques sur ℓ_2 fixent toujours un point.

1.3 Remerciements

Je remercie les éditeurs C.S. Aravinda, F.T. Farrell et J.F. Lafont pour avoir considéré ce texte pour les actes de la conférence "Geometry, Topology and Dynamics in Negative Curvature" de Bangalore d'aout 2010. Ces notes reprennent et complètent un mini-cours donné lors de la rencontre "Cohomologie L^p et actions affines sur les espaces L^p" en juin 2009 à Orléans. Je remercie ses organisateurs I. Chatterji, Y. de Cornulier, V. Lafforgue et Y. Stalder. Je suis reconnaissant à G. Yu qui m'a posé le problème d'utiliser la cohomologie ℓ_p pour donner une autre démonstration de son théorème. Je salue D. Gaboriau en souvenir de nos discussions et tentatives communes. Je remercie P. Pansu pour les discussions éclairantes. Je suis reconnaissant au rapporteur pour avoir suggéré plusieurs ajouts aux paragraphes 2.6.1, 2.6.3, 3.4.1 et 3.4.2. Ce travail a bénéficié d'une aide du projet ANR GdSous et du Labex Cempi.

1.4 Conventions et notations

Etant donné $p \in [1, +\infty)$ et un ensemble dénombrable E, on désignera indifféramment par $\ell_p(E)$ l'espace de Banach des suites réelles $(a_i)_{i \in E}$ ou des fonctions $f : E \to \mathbb{R}$, qui sont p-sommables. On notera simplement ℓ_p tout espace de Banach $\ell_p(E)$ avec E dénombrable infini (et muni de la mesure de comptage).

On appelle *espace L_p* tout espace de Banach de la forme $L_p(\mathbb{R}, \mu)$ où μ est une mesure borélienne quelconque sur $\mathbb{R}$.

2 Cohomologie ℓ_p

Dans ce chapitre sont présentés des définitions et résultats de base de la cohomologie ℓ_p des groupes. Au premier paragraphe nous définissons l'homologie et cohomologie ℓ_p des complexes simpliciaux et étudions leur dualité. Au second paragraphe l'invariance par quasi-isométrie est discutée. Celle-ci nous permet au paragraphe 3 de définir la cohomologie ℓ_p des groupes qui agissent par isométries de manière proprement discontinue et compacte sur des complexes simpliciaux contractiles. Des liens entre cohomologie ℓ_p et cohomologie à valeurs dans une représentation sont établis. Le paragraphe 4 porte sur la cohomologie ℓ_p en degré 1 des groupes de type fini généraux. On y donne, entre autres choses, plusieurs caractérisations de la moyennabilité. Au paragraphe 5 des exemples de phénomènes d'annulation et de non-annulation de cohomologie sont présentés. On termine au paragraphe 6 par un survol de quelques résultats complémentaires et par des questions.

2.1 Cohomologie ℓ_p des complexes simpliciaux

On ne considérera ici que des complexes simpliciaux X munis d'une métrique de longueur notée $|\cdot - \cdot|$, telle que

(i) il existe une constante $C \geq 0$ telle que tout simplexe de X soit de diamètre inférieur à C,
(ii) il existe une fonction $N : [0, +\infty[\to \mathbb{N}$ telle que toute boule de rayon r contienne au plus $N(r)$ simplexes de X.

Un tel complexe sera dit *géométrique*. Suivant [32], on définit leur homologie et cohomologie ℓ_p pour $p \in [1, +\infty)$, et on discute quelques-unes de leurs propriétés.

Pour $k \in \mathbb{N}$ soit X_k l'ensemble des k-simplexes de X. Pour $p \in [1, +\infty)$ on pose:

$$\ell_p C_k(X) := \{\Sigma_{\sigma \in X_k} a_\sigma \sigma \; ; \; (a_\sigma)_{\sigma \in X_k} \in \ell_p(X_k)\},$$
$$\ell_p C^k(X) := \{\omega : X_k \to \mathbb{R} \; ; \; \omega \in \ell_p(X_k)\}.$$

L'opérateur bord standard induit des opérateurs bornés (grâce à la propriété (ii) ci-dessus):

$$\partial_k : \ell_p C_k(X) \to \ell_p C_{k-1}(X) \, , \; d_k : \ell_p C^k(X) \to \ell_p C^{k+1}(X),$$

qui satisfont pour tout $\omega \in \ell_p C^k(X)$ et tout $\sigma \in X^{k+1}$: $(d_k\omega)(\sigma) = \omega(\partial_{k+1}\sigma)$.

Le k-ième groupe d' homologie ℓ_p de X (resp. d'homologie ℓ_p réduite) est

$$\ell_p H_k(X) = \ker \partial_k / \mathrm{Im} \, \partial_{k+1},$$

(resp. $\ell_p \overline{H_k}(X) = \ker \partial_k / \overline{\mathrm{Im} \, \partial_{k+1}}$, où $\overline{\mathrm{Im} \, \partial_{k+1}}$ désigne l'adhérence pour la topologie ℓ_p).

De même le k-ième groupe de cohomologie ℓ_p de X (resp. de cohomologie ℓ_p réduite) est

$$\ell_p H^k(X) = \ker d_k / \operatorname{Im} d_{k-1},$$

(resp. $\ell_p \overline{H^k}(X) = \ker d_k / \overline{\operatorname{Im} d_{k-1}}$).

On munit ces espaces vectoriels de la topologie quotient provenant de la topologie ℓ_p. Les groupes de cohomologie réduite sont alors des espaces de Banach.

Proposition 2 *Pour $p, q \in (1, +\infty)$ avec $p^{-1} + q^{-1} = 1$ et $k \in \mathbb{N}$, l'espace $\ell_p \overline{H^k}(X)$ est canoniquement isomorphe au dual de $\ell_q \overline{H_k}(X)$. De même $\ell_q \overline{H_k}(X)$ est canoniquement isomorphe au dual de $\ell_p \overline{H^k}(X)$.*

En général il est plus difficile de comparer les cohomologies et homologies non réduites. Par exemple pour $p \in (1, +\infty)$ la cohomologie ℓ_p d'un groupe infini de type fini est toujours nulle en degré 0, par contre son homologie ℓ_p en degré 0 est nulle si et seulement si il est non moyennable (voir Th. 6).

Preuve de la proposition: Pour $p, q \in (1, +\infty)$ avec $p^{-1} + q^{-1} = 1$ et $k \in \mathbb{N}$, considérons l'accouplement

$$\langle \cdot, \cdot \rangle : \ell_p C^k(X) \times \ell_q C_k(X) \to \mathbb{R},$$

défini par:

$$\forall \omega \in \ell_p C^k(X), \ \forall \tau = \Sigma_{\sigma \in X_k} a_\sigma \sigma \in \ell_q C_k(X) : \langle \omega, \tau \rangle = \sum_{\sigma \in X_k} a_\sigma \omega(\sigma).$$

Il satisfait pour $\omega \in \ell_p C^k(X)$ et $\tau \in \ell_{q+1} C_k(X)$: $\langle \omega, \partial \tau \rangle = \langle d\omega, \tau \rangle$. Désignons respectivement les espaces $\ker \partial_k$, $\ker d_k$, $\operatorname{Im} \partial_{k+1}$, $\operatorname{Im} d_{k-1}$ par Z_k, Z^k, B_k, B^k. D'après la relation ci-dessus on a $\langle Z^k, \overline{B_k} \rangle = \langle \overline{B^k}, Z_k \rangle = 0$. Donc $\langle \cdot, \cdot \rangle$ induit un accouplement entre $\ell_p \overline{H^k}(X)$ et $\ell_q \overline{H_k}(X)$.

Montrons que l'application

$$\ell_p \overline{H^k}(X) \to \ell_q \overline{H_k}(X)^*, \ [\omega] \mapsto \langle \omega, \cdot \rangle,$$

est surjective. Soit $f \in \ell_q \overline{H_k}(X)^*$, elle se relève en une forme linéaire continue de Z_k puis se prolonge en une forme linéaire continue de $\ell_q C_k(X)$ grâce au théorème de Hahn-Banach. Notons la $\tilde{f}$. Puisque $\ell_p C^k(X)$ est canoniquement isomorphe au dual de $\ell_q C_k(X)$, il existe $\omega \in \ell_p C^k(X)$ avec $\tilde{f} = \langle \omega, \cdot \rangle$. Par définition de $\tilde{f}$ on a $\tilde{f}(B_k) = 0$, donc pour tout $\tau \in \ell_q C_{k+1}(X)$ on obtient:

$$\langle d\omega, \tau \rangle = \langle \omega, \partial \tau \rangle = \tilde{f}(\partial \tau) = 0.$$

Par suite $d\omega = 0$ et donc $\omega \in Z^k$, ce qui montre la surjectivité.

Etablissons à présent l'injectivité. On sait d'après ci-dessus (en échangeant les rôles de $\ell_p\overline{H^k}(X)$ et $\ell_q\overline{H_k}(X)$) que toute forme linéaire de $\ell_p\overline{H^k}(X)$ s'écrit $\langle \cdot, \tau \rangle$ pour un certain $[\tau] \in \ell_q\overline{H_k}(X)$. D'après Hahn-Banach, pour tout élément non nul $[\omega] \in \ell_p\overline{H^k}(X)$ il existe une forme linéaire de $\ell_p\overline{H^k}(X)$ qui ne s'annule pas en $[\omega]$. Ainsi il existe $[\tau] \in \ell_q\overline{H_k}(X)$ tel que $\langle \omega, \tau \rangle \neq 0$. D'où l'injectivité. $\qquad\square$

2.2 Invariance par quasi-isométrie

Lorsque X est un complexe simplicial fini, ses homologies et cohomologies ℓ_p se confondent avec les homologies et cohomologies ordinaires à coefficients réels. A l'opposé, lorsque X est un complexe simplicial géométrique uniformément contractile, ses homologies et cohomologies ℓ_p sont des invariants de quasi-isométrie. Ce résultat, dû à M. Gromov, est l'objet du théorème 3 ci-dessous (voir aussi [43] et [24] pour des résultats voisins). Rappelons d'abord les définitions nécessaires à son énoncé.

Un complexe géométrique X est dit *uniformément contractile* s'il est contractile et s'il existe une fonction $\phi : \mathbb{R}^+ \to \mathbb{R}^+$ telle que toute boule $B(x, r)$ soit contractile dans $B(x, \phi(r))$.

Une application quelconque $F : X \to Y$ entre deux espaces métriques est dite *quasi-Lipschitz* s'il existe des constantes C, D positives telles que

$$\forall x, x' \in X, \ |F(x) - F(x')| \leq C|x - x'| + D.$$

C'est un *plongement uniforme* si elle est quasi-Lipschitz et s'il existe une fonction $\Psi : \mathbb{R}^+ \to \mathbb{R}^+$ avec $\lim_{t \to +\infty} \Psi(t) = +\infty$ telle que

$$\forall x, x' \in X, \ \Psi(|x - x'|) \leq |F(x) - F(x')|.$$

C'est une *quasi-isométrie* si elle est quasi-Lipschitz, et s'il existe une application quasi-Lipschitz $G : Y \to X$, telle que $F \circ G$ et $G \circ F$ soient à distance bornée de l'identité.

Voici des exemples de telles situations. Un complexe simplicial géométrique et contractile qui possède un groupe d'isométries cocompact est uniformément contractile. Supposons qu'un groupe Γ agisse par isométries de manière proprement discontinue sur deux complexes simpliciaux géométriques X et Y, soit $x_0 \in X$ et $y_0 \in Y$. Si l'action de Γ sur X est cocompacte alors l'application de Γx_0 dans Y, $gx_0 \mapsto gy_0$, se prolonge en un plongement uniforme F de X dans Y. Si de plus Γ agit de manière cocompacte sur Y, alors F est une quasi-isométrie de X sur Y.

Théorème 3 *([32] p. 219). Soit X et Y deux complexes simpliciaux géométriques uniformément contractiles, et soit $p \in [1, +\infty)$.*

a) *Tout plongement uniforme $F : X \to Y$ induit canoniquement des morphismes continus $F^\bullet : \ell_p H^\bullet(Y) \to \ell_p H^\bullet(X)$.*

b) *Si F_i, $i = 1, 2$, sont comme en a) et sont à distance bornée, alors $F_1^\bullet = F_2^\bullet$.*

c) *Si $F : X \to Y$ est une quasi-isométrie alors les $F^\bullet$ sont des isomorphismes d'espaces vectoriels topologiques.*

De plus, ces résultats persistent en cohomologie ℓ_p réduite.

Sa preuve est seulement esquissée dans [32] p. 219, nous renvoyons à [8], [27] pour une preuve détaillée.

2.3 Cohomologie ℓ_p des groupes

Soit Γ un groupe agissant par isométries, de manière proprement discontinue et cocompacte, sur un complexe simplicial géométrique et uniformément contractile X. Soit $p \in [1, +\infty)$.

Définition 4 On définit *l'homologie et la cohomologie ℓ_p (resp. l'homologie et la cohomologie ℓ_p réduites) de* Γ, comme étant égales à celles de X. D'après l'invariance par quasi-isométrie (Th.3) elles sont bien définies. On les désigne par $\ell_p H_\bullet(\Gamma)$, $\ell_p H^\bullet(\Gamma)$ (resp. $\ell_p \overline{H_\bullet}(\Gamma)$, $\ell_p \overline{H^\bullet}(\Gamma)$).

On se propose à présent de relier la cohomologie ℓ_p de Γ à la théorie cohomologique classique des groupes (voir [10] pour cette dernière).

Soit $N \lhd \Gamma$ un sous-groupe distingué de Γ. Désignons par $\rho : \Gamma \to$ Isom $\ell_p(\Gamma/N)$ la représentation régulière droite de Γ sur $\ell_p(\Gamma/N)$, c'est-à-dire l'application définie pour toute fonction p-sommable $f : \Gamma/N \to \mathbb{R}$ et tout $g \in \Gamma$ par:

$$\rho(g)f = f \circ R_g,$$

où R_g désigne la multiplication à droite par g dans Γ/N. La proposition suivante identifie la cohomologie de Γ à valeurs dans cette représentation, notée $H^\bullet(\Gamma, \ell_p(\Gamma/N))$, à la cohomologie ℓ_p du complexe $N \backslash X$:

Proposition 5 *Les espaces vectoriels $\ell_p H^\bullet(N \backslash X)$ et $H^\bullet(\Gamma, \ell_p(\Gamma/N))$ sont canoniquement isomorphes. En particulier les espaces $\ell_p H^\bullet(\Gamma)$ et $H^\bullet(\Gamma, \ell_p(\Gamma))$ sont canoniquement isomorphes.*

Preuve de la proposition: Notons Y le complexe $N \backslash X$. Quitte à remplacer X par sa première subdivision barycentrique, on peut supposer que Y est simplicial et géométrique. Puisque Γ agit par automorphismes simpliciaux de manière proprement discontinue sur le complexe simplicial contractile X (^{2}), la cohomologie $H^\bullet(\Gamma, \ell_p(\Gamma/N))$ est isomorphe à celle du complexe de cochaînes $\{U^k, d_k\}_{k \in \mathbb{N}}$ où:

$$U^k := \{\phi : X_k \to \ell_p(\Gamma/N) \; ; \; \forall g \in \Gamma : \phi \circ g = \rho(g) \circ \phi\}.$$

Comme N est distingué, pour tout $\phi \in U^k$, $n \in N$ et $\sigma \in X_k$, on a $\phi(n\sigma) = \phi(\sigma)$. Donc U^k s'identifie canoniquement à

$$V^k := \{\psi : Y_k \to \ell_p(\Gamma/N) \; ; \; \forall g \in \Gamma/N : \psi \circ g = \rho(g) \circ \psi\}.$$

Remarquons que pour $\psi \in V^k$ l'application $\sigma \in Y_k \mapsto \psi(\sigma)(N) \in \mathbb{R}$ est un élément de $\ell_p C^k(Y)$. En effet si $g \in \Gamma/N$ on a :

$$\psi(g\sigma)(N) = [\rho(g)\psi(\sigma)](N) = [\psi(\sigma) \circ R_g](N) = \psi(\sigma)(Ng).$$

Donc en utilisant le fait que l'action de Γ/N sur Y est proprement discontinue et cocompacte, et en partitionnant Y_k en orbites, on obtient la propriété cherchée.

Inversement pour $f \in \ell_p C^k(Y)$ on définit $\psi : Y_k \to \ell_p(\Gamma/N)$ par $\psi(\sigma)(Ng) = f(g\sigma)$. Elle satisfait pour $h \in \Gamma/N$:

$$\psi(h\sigma)(Ng) = f(gh\sigma) = \psi(\sigma)(Ngh) = \psi(\sigma) \circ R_h(Ng).$$

Autrement dit $\psi \circ h = \rho(h) \circ \psi$.

Ainsi on obtient des isomorphismes $V^k \to \ell_p C^k(Y)$, qui clairement commutent aux différentielles. $\qquad\qquad\qquad\qquad\qquad\qquad\qquad\qquad\Box$

2.4 Cas du degré 1

Au paragraphe précédent a été définie la cohomologie ℓ_p d'un groupe, en supposant qu'il agisse par isométries de manière proprement discontinue et cocompacte sur un complexe simplicial uniformément contractile. En degré 1, il y a moyen d'étendre canoniquement la définition à tout groupe de type fini. Cette généralisation apparaît dans [42]. Afin de la présenter considérons d'abord un complexe simplicial géométrique simplement connexe X. Ses 1-cycles sont des bords, donc sa cohomologie en degré 1 s'ecrit:

$$\ell_p H^1(X) = \{\omega \in \ell_p C^1(X) \; ; \; \omega(c) = 0 \text{ pour tout 1-cycle c}\}/d\ell_p(X).$$

En intégrant un tel ω le long de chemins contenus dans le 1-squelette de X et issus d'une origine fixée, on obtient une fonction $f : X_0 \to \mathbb{R}$ telle que

---

^{2} Il n'est pas nécessaire de supposer l'action libre, car l'anneau sous-jacent – ici $\mathbb{R}$ – est un corps de caractéristique nulle.

$df = \omega$. Par suite on obtient un isomorphisme canonique

$$\ell_p H^1(X) \simeq \{f : X_0 \to \mathbb{R} \; ; \; df \in \ell_p(X_1)\}/\ell_p(X_0) + \mathbb{R}, \qquad (1)$$

où $\mathbb{R}$ désigne l'ensemble des fonctions constantes sur X_0, et où la topologie est celle induite par la p-norme de df. De même on a

$$\ell_p \overline{H^1}(X) \simeq \{f : X_0 \to \mathbb{R} \; ; \; df \in \ell_p(X_1)\}/\overline{\ell_p(X_0) + \mathbb{R}}. \qquad (2)$$

Observons que seul le 1-squelette de X intervient dans les espaces de droite des relations (1) et (2). Ils ont donc encore un sens si, au lieu d'être un complexe simplicial simplement connexe, X est un graphe connexe quelconque de valence bornée (l'orientation des arêtes ne joue aucun rôle dans (1) et (2)).

De plus on montre facilement que lorsque $\phi : X \to X'$ est une quasi-isométrie entre graphes connexes de valences bornées, l'application $f \mapsto f \circ \phi$ induit canoniquement un isomorphisme entre les membres de droite de (1) (resp. (2)) associés à X' et X.

Ainsi pour un groupe de type fini Γ et un graphe de Cayley G de Γ, il est naturel de définir la cohomologie ℓ_p de Γ en degré 1 (ordinaire et réduite) par

$$\ell_p H^1(\Gamma) = \{f : \Gamma \to \mathbb{R} \; ; \; df \in \ell_p(G_1)\}/\ell_p(\Gamma) + \mathbb{R},$$
$$\ell_p \overline{H^1}(\Gamma) = \{f : \Gamma \to \mathbb{R} \; ; \; df \in \ell_p(G_1)\}/\overline{\ell_p(\Gamma) + \mathbb{R}}.$$

Ce sont des invariants de quasi-isométrie de Γ. Le théorème suivant est issu de [32] p.247-248. Il figure également en partie dans [45] Prop.10.

Théorème 6 *Soit Γ un groupe infini de type fini et soit $p \in (1, +\infty)$. Il y a équivalence entre:*

(a) *Γ est non moyennable,*
(b) *$\ell_p H^1(\Gamma) = \ell_p \overline{H^1}(\Gamma)$,*
(c) *$\ell_p H_0(\Gamma) = 0$,*
(d) *$\Delta : \ell_p(\Gamma) \to \ell_p(\Gamma)$ est inversible.*

De plus l'équivalence entre (a) et (b) est vraie pour $p = 1$ également.

Dans l'énoncé (d) Δ désigne le Laplacien standard sur le graphe de Cayley de Γ associé à un système de générateurs S, c'est-à-dire l'opérateur linéaire de $\ell_p(\Gamma)$ défini par

$$\forall f \in \ell_p(\Gamma) \, , \; \forall x \in \Gamma \; : \; (\Delta f)(x) = f(x) - \frac{1}{|S|} \sum_{y \sim x} f(y).$$

Preuve du théorème: On munit le graphe de Cayley G d'une structure de 1-complexe simplicial en choisissant une orientation pour chacune de ses arêtes.

Montrons d'abord l'équivalence entre (a) et (b) pour $p \geq 1$. Les arguments présentés sont issus de [32] p.247-248. La condition (b) équivaut à la fermeture du sous-espace

$$\text{Im}(d_0 : \ell_p(\Gamma) \to \ell_p(G_1))$$

dans $\ell_p(G_1)$. Puisque Γ est un groupe infini l'application d_0 est injective, et donc grâce au théorème de Banach la condition (b) équivaut à l'inégalité de Sobolev suivante:

$$\forall f \in \ell_p(\Gamma) \ : \ \|f\|_p \leq C_p \|df\|_p, \tag{3}$$

où C_p est une constante indépendante de f. Appliquée aux fonctions caractéristiques des sous-ensembles finis de Γ, elle implique l'inégalité isopérimétrique suivante:

$$\forall A \subset \Gamma \text{ fini } : \ |A| \leq C|\partial A|, \tag{4}$$

où ∂A désigne le bord de A c'est-à-dire l'ensemble des arêtes de G dont l'une des extrémités se trouve dans A et l'autre en dehors. Cette dernière inégalité caractérise les groupes non moyennables (elle équivaut clairement à l'absence de suites de Følner). Ainsi on obtient que (b) implique (a).

Réciproquement si (4) est satisfaite, alors un résultat de Cheeger et Mazzia (voir [14] Th.II.2.1) montre que (3) l'est aussi pour $p = 1$. On passe alors facilement du cas $p = 1$ au cas $p > 1$ avec l'inégalité de Hölder et l'analogue discret de la formule $d(f^p) = pf^{p-1}df$.

Montrons à présent les implications (d) $\Rightarrow$ (c) $\Rightarrow$ (a) $\Rightarrow$ (d) pour $p > 1$. On remarque que (c) équivaut à la surjectivité de $\partial_1 : \ell_p(G_1) \to \ell_p(\Gamma)$. Observons aussi (par un calcul immédiat) que $\Delta = \frac{1}{|S|}\partial_1 \circ d_0$. L'implication (d) $\Rightarrow$ (c) en découle.

Soit q le conjugué de p. D'après (c) l'image de $\partial_1 : \ell_p(G_1) \to \ell_p(\Gamma)$ est fermée dans $\ell_p(\Gamma)$. Donc l'image de $\partial_1^* : \ell_p(\Gamma)^* \to \ell_p(G_1)^*$ est fermée dans $\ell_p(G_1)^*$ (voir [53] Th.4.14). Or $\partial_1^* = d_0$ et $\ell_p^* = \ell_q$, donc (b) est satisfaite pour l'exposant q. Puisque (b) équivaut à (a) on obtient l'implication (c) $\Rightarrow$ (a).

Soit M l'opérateur de $\ell_p(\Gamma)$ défini par $(Mf)(x) = \frac{1}{|S|}\sum_{y \sim x} f(y)$. Puisqu'il est un opérateur de moyenne on a $\|M\|_{\ell_\infty \to \ell_\infty} \leq 1$. De plus lorsque Γ est non moyennable le théorème de Kesten [36] montre que $\|M\|_{\ell_2 \to \ell_2} < 1$. Appliquons le théorème d'interpolation de Riesz entre 2 et $+\infty$ (voir [12] Th.1.4.3), pour tout $p \in [2, +\infty)$ on obtient: $\|M\|_{\ell_p \to \ell_p} < 1$. Par suite $\Delta = I - M$ est inversible sur $\ell_p(\Gamma)$ pour $p \in [2, +\infty)$. Puisqu'il est autoadjoint il est également inversible sur $\ell_q(\Gamma)$ pour $q \in (1, 2]$. Ainsi (a) $\Rightarrow$ (d) en découle. $\qquad\square$

Le résultat suivant, de représentation harmonique, m'a été signalé par P. Pansu à la fin du siècle dernier.

Corollaire 7 *Si Γ est non moyennable alors pour $p > 1$:*

$$\ell_p H^1(\Gamma) = \{f : \Gamma \to \mathbb{R} \; ; \; \Delta f = 0, \; df \in \ell_p(G_1)\}/\mathbb{R}.$$

Preuve: Soit $f : \Gamma \to \mathbb{R}$ avec $[f] \in \ell_p H^1(\Gamma)$. On cherche une fonction h harmonique (c-à-d qui satisfait $\Delta f = 0$) telle que $f - h \in \ell_p(\Gamma) + \mathbb{R}$. Puisque $d_0 f \in \ell_p(\Gamma)$ on a $\Delta f = \frac{1}{|S|}(\partial_1 \circ d_0)f \in \ell_p(\Gamma)$. D'après le théorème précédent Δ est inversible sur $\ell_p(\Gamma)$, donc il existe une fonction $u \in \ell_p(\Gamma)$ telle que $\Delta u = \Delta f$. Alors la fonction $h = f - u$ convient.

Reste à montrer l'unicité du représentant harmonique. Si $f : \Gamma \to \mathbb{R}$ est une fonction harmonique telle que $[h] = 0$ dans $\ell_p H^1(\Gamma)$, alors elle appartient à $\ell_p(\Gamma) + \mathbb{R}$. En particulier elle est asymptote à une fonction constante lorsque $|g| \to +\infty$, $g \in \Gamma$. Alors le principe du maximum entraîne qu'elle est constante. $\square$

Corollaire 8 *Supposons Γ non moyennable, alors $\ell_p H^1(\Gamma) \subset \ell_r H^1(\Gamma)$ pour $1 < p \leq r$.*

Preuve: Puisque $\ell_p(G_1) \subset \ell_r(G_1)$ le résultat découle du corollaire précédent. $\square$

Corollaire 9 *Supposons Γ non moyennable et $p > 1$. Soit $f : \Gamma \to \mathbb{R}$ avec $df \in \ell_p(G_1)$. Alors $[f] = 0$ dans $\ell_p H^1(\Gamma)$ si et seulement si f est asymptote à une fonction constante (lorsque $|g| \to +\infty$, $g \in \Gamma$).*

Preuve: Si $[f] = 0$ dans $\ell_p H^1(\Gamma)$ alors f appartient à $\ell_p(\Gamma) + \mathbb{R}$, donc elle est asymptote à une fonction constante.

Réciproquement si f est asymptote à une fonction constante alors son représentant harmonique (voir Cor.7) également. Par le principe du maximum ce dernier est constant, donc f appartient à $\ell_p(\Gamma) + \mathbb{R}$. $\square$

2.5 Quelques exemples

On présente des exemples et résultats élémentaires d'annulation et de non annulation de cohomologie.

L'énoncé suivant découle du corollaire de [32] p.227. Pour les groupes de Lie, un énoncé similaire se trouve dans [46] Prop.15.

Proposition 10 *([32] p.227). Supposons que Γ agisse par isométries sur un complexe géométrique contactile de manière proprement discontinue et* cocompacte. *Si le centre de Γ est infini alors $\ell_p\overline{H^k}(\Gamma) = 0$ pour tout $p \in$ $(1, +\infty)$ et tout $k \in \mathbb{N}$.*

Preuve: Soit X le complexe géométrique de l'énoncé. Soit z un élément du centre de Γ, alors l'application $L_z : \Gamma \to \Gamma$ définie par $L_z(g) = zg$ est une isométrie à distance bornée de l'identité. Elle induit une quasi-isométrie de X à distance bornée de l'identité. Donc d'après le théorème 3, L_z^* est égale à l'identité en cohomologie. Soit $[\omega] \in \ell_p\overline{H^k}(X)$. Supposons par l'absurde que $[\omega]$ soit non nulle, soit q est le conjugué de p. La proposition 2 et sa preuve montrent qu'il existe $[\tau] \in \ell_q\overline{H_k}(X)$ avec $\langle \omega, \tau \rangle = 1$. Puisque l'accouplement $\langle \omega, \tau \rangle$ ne dépend que des classes de ω et de τ, on a encore $\langle L_z^*(\omega), \tau \rangle = 1$. Par ailleurs, lorsque $|z| \to +\infty$ les supports de $L_z^*(\omega)$ et de τ s'éloignent l'un de l'autre, donc

$$\lim_{|z| \to +\infty} \langle L_z^*(\omega), \tau \rangle = 0.$$

D'où une contradiction. $\qquad\qquad\qquad\qquad\qquad\qquad\qquad\qquad\qquad\qquad\quad\square$

Puisque tout groupe nilpotent de type fini possède un sous-groupe d'indice fini de centre infini ([3]) on obtient:

Corollaire 11 *Soit Γ un groupe nilpotent de type fini, alors $\ell_p\overline{H^k}(\Gamma) = 0$ pour tout $p \in (1, +\infty)$ et tout $k \in \mathbb{N}$.* $\qquad\qquad\qquad\qquad\qquad\qquad\quad\square$

A l'opposé la 1-cohomologie des groupes libres non abéliens est non nulle, et plus généralement:

Proposition 12 *Soit A et B deux groupes de type fini avec $|A| \geq 2$ et $|B| \geq 3$. Alors $\Gamma = A \star B$ satisfait $\ell_p H^1(\Gamma) \neq 0$ pour tout $p \geq 1$.*

Preuve: Observons que Γ est quasi-isométrique au graphe G défini comme suit. On considère les 2-complexes de présentation de A et de B, on joint le sommet du premier au sommet du second par une arête notée a, le graphe G est le 1-squelette du revêtement universel du 2-complexe ainsi obtenu.

Si $\tilde{a}$ relève l'arête (ouverte) a, le graphe $G \setminus \tilde{a}$ possède deux composantes connexes non bornées. Soit $f : G_0 \to \{0, 1\}$ la fonction qui vaut 0 sur les sommets de l'une, et 1 sur ceux de l'autre. Sa différentielle est supportée par $\tilde{a}$, donc $df \in \ell_p(G_1)$. De plus $[f] \neq 0$ dans $\ell_p H^1(\Gamma)$ car f n'est pas asymptote à une fonction constante. $\qquad\qquad\qquad\qquad\qquad\qquad\qquad\qquad\quad\square$

---

[3] Tout groupe nilpotent de type fini possède un sous-groupe d'indice fini sans torsion (voir [51] Lemma 4.6), de plus le centre d'un groupe nilpotent sans torsion est infini.

Le résultat suivant est une sorte d'extension à la cohomologie ℓ_p d'un théorème de W. Lueck [38] dont l'énoncé est rappelé plus bas. Il est tiré de [7].

Proposition 13 *([7] Prop.2). Soit $N < H < \Gamma$ une chaîne de groupes, avec H et Γ de type fini, et avec N infini et distingué dans Γ. On suppose que pour un certain $p \in [1, +\infty)$ on a $\ell_p H^1(H) = 0$, alors $\ell_p H^1(\Gamma) = 0$.*

Preuve: On reprend les idées de la preuve de [7] Prop.2, en les adaptant à l'approche développée dans cette note. Soit G un graphe de Cayley de Γ et soit $f : \Gamma \to \mathbb{R}$ avec $df \in \ell_p(G_1)$. Considérons la restriction de f à une classe gH ($g \in \Gamma$), et notons-la $f_{|gH}$. Par hypothèse il existe une fonction $u_{gH} \in \ell_p(gH)$ telle que $f_{|gH} - u_{gH}$ soit constante sur gH. Puisque les classes gH forment une partition de Γ on obtient une fonction u de Γ telle que $f - u$ soit constante sur chaque classe gH.

Montrons que $u \in \ell_p(\Gamma)$. L'hypothèse d'annulation de cohomologie et le Th.6 impliquent que H est non moyennable. La première partie de la preuve de ce même théorème montre l'existence d'une constante $C \geq 1$ telle que pour toute fonction $\varphi \in \ell_p(H)$ on ait:

$$\|\varphi\|_p \leq C\|d\varphi\|_p.$$

En ayant pris soin de choisir le système de générateurs S de Γ contenant un système de générateurs de H, on a alors:

$$\|u\|_p^p \leq C^p \sum_{gH \in \Gamma/H} \|d(u_{|gH})\|_p^p = C^p \sum_{gH \in \Gamma/H} \|d(f_{|gH})\|_p^p \leq C^p \|df\|_p^p.$$

Ainsi u appartient à $\ell_p(\Gamma)$ et donc $[f] = [f - u]$.

On montre à présent que $f - u$ est une fonction constante. Pour $s \in S$ et $g \in \Gamma$, considérons les classes gH et gsH. Puisque N est distingué dans Γ, pour tout $n \in N$ l'élément gns appartient à gsN. Donc $gn \in gH$ est lié par une arête de G à un élément de gsH. Comme N est un groupe infini, le graphe G contient une infinité d'arêtes dont une extrémité appartient à gH et l'autre à gsH. Puisque $f - u$ est constante sur gH et sur gsH, sa différentielle en ces arêtes est constante égale à la différence des valeurs de $f - u$ sur gH et gsH. Comme $d(f - u) \in \ell_p(G_1)$ les valeurs de $f - u$ sur gH et sur gsH sont les mêmes. Les éléments $g \in \Gamma$ et $s \in S$ étant arbitraires, on obtient que $f - u$ est constante sur Γ. $\qquad\square$

La proposition précédente permet d'obtenir une preuve simple du théorème suivant de W. Lueck dans le cas particulier où N est non moyennable. Le théorème résout une conjecture de M. Gromov énoncée dans [32] p.235.

Théorème 14 ([38] Th.4.1). *Soit Γ un groupe de type fini. On suppose qu'il possède un sous-groupe distingué N infini de type fini, et que le quotient Γ/N contienne un élément d'ordre infini. Alors $\ell_2 \overline{H^1}(\Gamma) = 0$.*

Réduction du théorème 14 à la proposition 13: Supposons N non moyennable (dans le cas moyennable le théorème découle d'un théorème antérieur de Cheeger-Gromov, voir [15] Th.0.3). Soit $\mathbb{Z}$ un sous-groupe cyclique infini de Γ/N. Désignons par π l'application quotient de Γ sur Γ/H, et posons $H = \pi^{-1}(\mathbb{Z})$. En utilisant les propriétés des nombres de Betti ℓ_2 et le fait que H est un produit semi-direct de N avec $\mathbb{Z}$, W. Lueck obtient par un argument élégant que $\ell_2 \overline{H^1}(H) = 0$ (voir [38] preuve du Th.2.1).

Puisque N est non moyennable, H l'est également, et donc avec le théorème 6 on obtient que $\ell_2 H^1(H) = 0$. La proposition 13 conclut.   □

2.6 Compléments et questions

1) Pour simplifier l'exposé, nous nous sommes restreint à la cohomologie ℓ_p des complexes simpliciaux géométriques et à celle des groupes qui leurs sont attachés (à l'exception toutefois du degré 1, voir la section 2.4). Il y a plusieurs façons de définir la cohomologie ℓ_p d'un groupe de type fini quelconque en tout degré. Cheeger-Gromov [15] définissent la cohomologie ℓ_2 singulière de n'importe quel groupe dénombrable. Dans [43], P. Pansu introduit la cohomologie L_p asymtotique d'un espace métrique mesuré quelconque. G. Elek [24] définit la cohomologie ℓ_p grossière de tout groupe de type fini. Toutes ces cohomologies coïncident pour les groupes qui possèdent un espace $K(\pi, 1)$ fini (voir [24]).

Une autre façon naturelle de procéder est d'utiliser la proposition 5, et de définir la cohomologie ℓ_p d'un groupe Γ comme la cohomologie de Γ à valeurs dans la représentation régulière droite de Γ sur $\ell_p(\Gamma)$. Notons que cette définition ne nécessite pas que Γ soit finiment engendré. De plus elle suggère d'étudier la cohomologie de Γ pour d'autres représentations L^p (par exemple celles qui proviennent d'actions de Γ préservant une mesure).

2) Rappelons que deux groupes dénombrables Γ et Φ sont dits mesurablement équivalents, s'il existe un espace mesuré standard (X, μ), et des actions mesurables libres de Γ et Φ sur X, qui commutent, qui présevent la mesure μ, et qui possèdent chacunes un domaine fondamental de mesure finie non nulle.

D. Gaboriau [25] a démontré que les nombres de Betti ℓ_2 de deux groupes mesurablement équivalents sont proportionnels, avec pour facteur de proportionnalité le rapport des mesures des domaines fondamentaux. En particulier on a en tout degré:

$$\ell_2\overline{H^k}(\Gamma) = 0 \iff \ell_2\overline{H^k}(\Phi) = 0.$$

Une question naturelle est de déterminer si ce dernier phénomème subsiste en cohomologie ℓ_p pour $p \neq 2$.

Avec Damien Gaboriau nous avons exhibé un contre-exemple pour $p = 1$: soit Γ le groupe fondamental d'une surface fermée de genre au moins 2, et soit Φ le groupe libre à 2 générateurs. Ils sont mesurablement équivalents car ce sont des réseaux de $PSL_2(\mathbb{R})$. Mais $\ell_1\overline{H^1}(\Gamma) = 0$ (voir [8] Th.0.1 et Th.4.2), alors que $\ell_1\overline{H^1}(\Phi) \neq 0$ d'après la proposition 12.

3) Pour tout groupe Γ moyennable infini, Cheeger-Gromov [15] ont montré que $\ell_2\overline{H^\bullet}(\Gamma) = 0$. Dans [32] p.226, M. Gromov remarque que ce résultat devrait subsister en cohomologie ℓ_p pour $p \in (1, +\infty)$.

A part les groupes nilpotents, dont traite le corollaire 11, il y a une autre classe de groupes moyennables pour laquelle un résultat d'annulation de cohomologie ℓ_p est connu. Dans [55] R. Tessera démontre l'annulation de la 1-cohomologie ℓ_p réduite des groupes moyennables qui possèdent une suite de Fölner "controlée", c'est-à-dire une suite $\{F_n\}_{n\in\mathbb{N}}$ de parties de Γ pour laquelle il existe un système fini de générateurs S de Γ avec $F_n \subset S^n$ et

$$\max_{s\in S} \frac{|sF_n \triangle F_n|}{|F_n|} = O(1/n).$$

La classe des groupes possédant cette propriété comprend les groupes polycycliques, les Baumslag-Solitar, les groupes d'allumeurs de réverbères, et plus généralement les réseaux des groupes résolubles algébriques sur un corps local [56]. A. Gournay établit dans [30] l'annulation de la 1-cohomologie ℓ_p réduite des groupes moyennables qui possèdent la propriété de Liouville (une groupe de type fini possède la propriété de Liouville s'il admet un système de générateurs fini dont le Laplacien est sans fonction harmonique bornée non constante).

4) D. Gaboriau [25] a amélioré le théorème 14 en affaiblissant l'hypothèse Γ/N possède un sous-groupe cyclique infini, par l'hypothèse Γ/N est infini.

Le théorème 14 n'a pas d'analogue pour $p > 2$. En effet, soit Γ le groupe fondamental d'une 3-variété fermée M, à courbures sectionnelles constantes égales à -1, et qui fibre sur le cercle. Sa fibre est une surface fermée de genre au moins 2 dont le groupe fondamemental N est normal de type fini dans Γ. Le quotient Γ/N est isomorphe à $\mathbb{Z}$. Par invariance quasi-isométrique la cohomologie ℓ_p de Γ est la même que celle du revêtement universel de M, c'est-à-dire $\mathbb{H}^3$ l'espace hyperbolique standard. D'après les calculs de cohomologie ℓ_p effectués par P. Pansu [42] on a:

$$\ell_p H^1(\Gamma) = 0 \iff p \in [1, 2].$$

5) La cohomologie ℓ_p des espaces homogènes est étudiée dans [44, 45, 46, 47, 55]. Toutefois le cas des espaces symétriques de rang supérieur a été peu considéré jusqu'à présent. Il est conjecturé dans [32] que la cohomologie ℓ_p des espaces symétriques, sans facteur compact ni euclidien, de rang $k \geq 2$, est nulle en degrés strictement inférieurs à k. Des éléments de preuve se trouvent dans [32] et [29]. Signalons aussi l'article [22] qui traite de la cohomologie ℓ_p des immeubles.

6) La cohomologie ℓ_p en degré 1 des groupes (Gromov) hyperboliques est étudiée dans [42, 47, 32, 23, 52, 8, 4, 5, 6]. Quelques éléments sont repris aux paragraphes 3.2, 3.4.1, 3.4.2 et 3.4.4 du prochain chapitre.

3 Actions isométriques propres sur les espaces L_p

Dans ce chapitre nous démontrons le théorème 1 de l'introduction. Sa démonstration sera l'occasion de discuter des liens entre la 1-cohomologie ℓ_p d'un groupe hyperbolique et son bord à l'infini. Au paragraphe 3.1 nous rappelons brièvement les quelques propriétés du bord d'un groupe hyperboliques qui nous seront utiles. Au paragraphe 3.2 est présenté un procédé de construction de 1-classes de cohomologie ℓ_p en utilisant le bord. Cette construction est dûe à Elek [23]. Le théorème 1 est démontré au paragraphe 3.3. On termine le chapitre par un survol de quelques résultats complémentaires et par des questions.

Afin de donner l'idée de la preuve du théorème, rappelons que toute action affine de Γ sur un espace vectoriel V s'écrit sous la forme

$$\forall g \in \Gamma, \forall v \in V : \ g \cdot v = \pi(g)v + c(g),$$

où $\pi : \Gamma \to \mathrm{GL}(V)$ est une représentation linéaire, et où $c : \Gamma \to V$ est un cocycle c'est-à-dire qu'il satisfait

$$\forall g, g' \in \Gamma : \ c(gg') = c(g) + \pi(g)c(g').$$

L'action possède un point fixe global si et seulement si le cocycle c est un cobord c'est-à-dire s'il s'écrit $c(g) = \pi(g)v - v$ pour un certain $v \in V$. L'espace des cocycles modulo les cobords est isomorphe à la 1-cohomologie de Γ à valeurs dans la représentation π (voir [10] p.19). Lorsque V est un espace vectoriel normé et l'action isométrique, elle est propre si et seulement si $\|c(g)\| \to \infty$ quand $|g| \to \infty$.

L'idée de la démonstration du théorème consiste à se souvenir que la 1-cohomologie ℓ_p de Γ est isomorphe à la 1-cohomologie à valeurs dans la représentation régulière droite de Γ sur $\ell_p(\Gamma)$ notée ρ (voir Prop.5). Elle décrit donc toutes les actions isométriques de Γ sur $\ell_p(\Gamma)$ dont la partie linéaire est ρ.

3.1 Bord d'un groupe hyperbolique, dimension conforme

Soit Γ un groupe hyperbolique non élémentaire et soit $\partial\Gamma$ son bord à l'infini (voir par exemple [31, 28, 21, 9, 2] pour ces notions standards). Soit G le graphe de Cayley de Γ associé à un système de générateurs S. Une métrique d sur $\partial\Gamma$ est dite *visuelle (relativement à G)* s'il existe des constantes $a > 1$ et $C \geq 1$ telles que pour tout $\xi, \eta \in \partial\Gamma$ on ait:

$$C^{-1}a^{-L} \leq d(\xi, \eta) \leq Ca^{-L},$$

où L est la distance dans G entre 1 (l'élément neutre de Γ) et la géodésique $(\xi\eta) \subset G$. Pour a suffisamment proche de 1 une telle métrique existe d'après un théorème de Gromov. De plus en vertu d'un théorème de M. Coornaert [20] elle est *Ahlfors régulière*, en d'autres termes il existe des constantes $C \geq 1$ et $Q > 0$ et une mesure μ sur $\partial\Gamma$, telles que pour toute boule $B(r) \subset (\partial\Gamma, d)$ de rayon $r < \mathrm{diam}(\partial\Gamma)$ on ait

$$C^{-1}r^Q \leq \mu(B(r)) \leq Cr^Q.$$

Cette relation entraine que Q est égale à la dimension de Hausdorff de $(\partial\Gamma, d)$ et que μ est équivalente à la mesure de Haudorff de $(\partial\Gamma, d)$ (voir par exemple [34]).

Les métriques visuelles font partie d'une famille de métriques appelée la *jauge conforme Ahlfors régulière* de $\partial\Gamma$, et notée $\mathcal{J}(\partial\Gamma)$. Afin de la définir rappelons que l'*ombre* d'une boule $B(x, R) \subset G$ est le sous-ensemble suivant de $\partial\Gamma$

$$O(x, R) = \{\xi \in \partial\Gamma \; ; \; [1, \xi) \cap B(x, R) \neq \emptyset\}.$$

La jauge $\mathcal{J}(\partial\Gamma)$ est constituée des métriques Ahlfors régulières sur $\partial\Gamma$ dont les boules ressemblent aux ombres. Formellement:

Définition 15 Une métrique d appartient à $\mathcal{J}(\partial\Gamma)$ si elle est Ahlfors régulière et si elle satisfait aux deux conditions suivantes pour tout $R > 0$ fixé assez grand:

(i) Il existe une fonction croissante $\varphi : [1, +\infty) \to [0, +\infty)$ telle que pour toute paire de boules $B_1 \subset B_2$ de rayons r_1, r_2, on puisse trouver deux ombres $O(x_1, R)$ et $O(x_2, R)$ avec $O(x_1, R) \subset B_1 \subset B_2 \subset O(x_2, R)$ et $|x_1 - x_2| \leq \varphi(\frac{r_2}{r_1})$;

(ii) Il existe une fonction croissante $\psi : [0, +\infty) \to [1, +\infty)$ telle que pour toute paire d'ombres $O(x_1, R) \subset O(x_2, R)$ on puisse trouver des boules B_1, B_2 de rayons r_1, r_2, avec $B_1 \subset O(x_1, R) \subset O(x_2, R) \subset B_2$ et $\frac{r_2}{r_1} \leq \psi(|x_1 - x_2|)$.

La jauge est un invariant complet de quasi-isométrie de Γ. Plus précisément, toute quasi-isométrie $F : \Gamma \to \Phi$ entre deux groupes hyperboliques se prolonge canoniquement en un homéomorphisme $\partial F : \partial \Gamma \to \partial \Phi$, et l'application qui à une métrique d sur $\partial \Phi$ associe la métrique $d(\partial F(\cdot), \partial F(\cdot))$ sur $\partial \Gamma$ est une bijection de $\mathcal{J}(\partial \Phi)$ sur $\mathcal{J}(\partial \Gamma)$. Réciproquement, tout homéomorphisme de $\partial \Gamma$ sur $\partial \Phi$ qui possède cette dernière propriété est l'extension aux bords d'une quasi-isométrie de Γ sur Φ. (Voir [31, 49, 2] pour ces derniers résultats.)

Définition 16 La *dimension conforme (Ahlfors régulière)* de $\partial \Gamma$ est

$$\text{Confdim}(\partial \Gamma) = \inf\{\text{Hdim}(\partial \Gamma, d) \,;\, d \in \mathcal{J}(\partial \Gamma)\},$$

où Hdim désigne la dimension de Hausdorff.

C'est un invariant numérique de quasi-isométrie de Γ qui a été introduit par P. Pansu (voir [39] pour plus de détails sur la dimension conforme, voir aussi la remarque 2 à la fin du chapitre).

3.2 Une construction de G. Elek

On imite une construction de G. Elek [23] qui permet d'exhiber de nombreux éléments du $\ell_p H^1(\Gamma)$. Soit $d \in \mathcal{J}(\partial \Gamma)$ et soit $u : (\partial \Gamma, d) \to \mathbb{R}$ une fonction Lipschitz. On lui associe une fonction $f_u : \Gamma \to \mathbb{R}$ comme suit. Soit $R > 0$ fixé assez grand. Pour tout $g \in \Gamma$ on choisit $\xi_g \in O(g, R)$ et on pose

$$f_u(g) = u(\xi_g).$$

La proposition suivante est implicite dans [23]. La définition de la 1-cohomologie ℓ_p utilisée dans l'énoncé est celle du paragraphe 2.4.

Proposition 17 *Pour $p > \text{Hdim}(\partial \Gamma, d)$ on a $df_u \in \ell_p(G_1)$. De plus si u est non constante alors $[f_u] \neq 0$ dans $\ell_p H^1(\Gamma)$.*

Preuve: Puisque les boules de d sont semblables aux ombres, notons $r(g)$ le rayon minimal d'une boule de $(\partial \Gamma, d)$ qui contient $O(g, 2R)$. Soit $[g_- g_+]$ une arête de G. Les ombres $O(g_-, 2R)$ et $O(g_+, 2R)$ s'intersectent, donc ξ_{g_-} et ξ_{g_+} sont à distance inférieure ou égale à $2(r(g_-) + r(g_+))$. Si λ désigne la constante de Lipschitz de u on obtient:

$$|df_u([g_- g_+])| \leq 2\lambda(r(g_-) + r(g_+)).$$

Par suite il existe une constante $C_1 \geq 1$ telle que

$$\|df_u\|_p^p \leq C_1 \sum_{g \in \Gamma} r(g)^p \leq C_1 \sum_{n \in \mathbb{N}} \Big(\sup_{|g|=n} r(g)^{p-Q} \Big) \sum_{|g|=n} r(g)^Q. \qquad (5)$$

La famille d'ombres $\{O(g, 2R) \,;\, |g| = n\}$ est un recouvrement uniformément localement fini de $\partial\Gamma$. De plus d est Ahlfors régulière par hypothèse. En notant Q sa dimension de Hausdorff et μ sa mesure de Haudorff, il vient pour tout $n \in \mathbb{N}$:

$$\sum_{|g|=n} r(g)^Q \leq C_2 \sum_{|g|=n} \mu(O(g, 2R)) \leq C_3 \mu(\partial\Gamma), \qquad (6)$$

où C_2, C_3 sont des constantes indépendantes de n. Deux métriques quelconques de $\mathcal{J}(\partial\Gamma)$ sont Hölder équivalentes (voir [34] Cor. 11.5). Donc d est Hölder équivalente à une métrique visuelle. Ainsi il existe $b > 1$ tel que l'on ait à une constante multiplicative près : $r(g) \leq b^{-|g|}$. En combinaison avec les inégalités (5), (6) on obtient que $\|df_u\|_p$ est finie. La fonction f_u étant asymptote à u, on a $[f_u] \neq 0$ lorsque u est non constante. $\qquad \square$

3.3 Preuve du théorème 1

Soit $p > \mathrm{Confdim}(\partial\Gamma)$ et soit $d \in \mathcal{J}(\partial\Gamma)$ avec $\mathrm{Hdim}(\partial\Gamma, d) < p$. Considérons une fonction Lipschitz $u : \partial\Gamma \to \mathbb{R}$ et la fonction associée $f_u : \Gamma \to \mathbb{R}$ définie au paragraphe précédent. Notons ρ la représentation régulière droite de Γ sur $\ell_p(\Gamma)$, et définissons un cocycle $c : \Gamma \to \ell_p(\Gamma)$ par "$c(g) = \rho(g)f_u - f_u$" (cette écriture est abusive puisque $f_u \notin \ell_p(\Gamma)$). Autrement dit pour $g, \gamma \in \Gamma$: $c(g)(\gamma) = f_u(\gamma g) - f_u(\gamma)$.

Vérifions que $c(g) \in \ell_p(\Gamma)$. Soit $1 = g_0, g_1, ..., g_n = g$ les sommets successifs d'un rayon géodésique de G reliant 1 à g. On a

$$\|c(g)\|_p^p = \sum_{\gamma \in \Gamma} |f_u(\gamma g) - f_u(\gamma)|^p,$$

avec

$$|f_u(\gamma g) - f_u(\gamma)| \leq \sum_{i=1}^{|g|} |f_u(\gamma g_i) - f_u(\gamma g_{i-1})| \leq \sum_{i=1}^{|g|} |df_u([\gamma g_i, \gamma g_{i-1}])|.$$

Puisque Γ agit librement sur l'ensemble des arêtes orientées de G on obtient avec Hölder que

$$\|c(g)\|_p \leq |g| \cdot \|df_u\|_p.$$

Donc d'après la proposition précédente $c(g) \in \ell_p(\Gamma)$.

Etudions à présent la limite de $\|c(g)\|_p$ lorsque $|g| \to +\infty$. En notant à nouveau $1 = g_0, g_1, \ldots, g_n = g$ les sommets successifs d'un segment géodésique reliant 1 à g, on a

$$\|c(g)\|_p^p \geq \sum_{k=1}^{|g|} |c(g)(g_k^{-1})|^p = \sum_{k=1}^{|g|} |f_u(g_k^{-1}g) - f_u(g_k^{-1})|^p$$

$$= \sum_{k=1}^{|g|} |u(\xi_{g_k^{-1}g}) - u(\xi_{g_k^{-1}})|^p.$$

Le segment géodésique $[g_k^{-1}, g_k^{-1}g]$ contient 1. Donc l'hyperbolicité de Γ entraîne l'existence de constantes $C \geq 0$, $k_0 \in \mathbb{N}$ indépendantes de g, telles que pour $k \in \{k_0, \ldots, |g| - k_0\}$ la géodésique $(\xi_{g_k^{-1}}\xi_{g_k^{-1}g})$ passe dans la boule $B(1, C)$. Convenons d'appeler deux points $\xi, \eta \in \partial\Gamma$ C-*diamétralement opposés* si $(\xi\eta)$ passe dans $B(1, C)$.

Si la fonction u est telle que pour toute paire de points C-diamétralement opposés on ait $|u(\xi) - u(\eta)| \geq 1$, alors on obtient

$$\|c(g)\|_p \geq (|g| - 2k_0)^{1/p},$$

et donc $\|c(g)\|_p \to +\infty$ lorsque $|g| \to +\infty$.

Si une telle fonction n'existe pas ([4]), on peut tout de même trouver une collection finie $u_1, \ldots, u_n$ de fonctions Lipschitz de $(\partial\Gamma, d)$, telle que pour toute paire de points C-diamétralement opposés $\xi, \eta \in \partial\Gamma$ il existe $i \in \{1, \ldots, n\}$ avec $|u_i(\xi) - u_i(\eta)| \geq 1$. On considère alors la représentation diagonale de Γ sur $\ell_p(\Gamma)^n = \ell_p(\sqcup_1^n \Gamma)$ et le cocycle $c = (c_1, \ldots, c_n)$, où c_i est associé à la fonction u_i comme précédamment. A nouveau $\|c(g)\|_p \geq (|g| - 2k_0)^{1/p}$, donc l'action induite de Γ sur $\ell_p(\sqcup_1^n \Gamma)$ est propre. □

3.4 Compléments et questions

1) L'idée de relier la cohomologie ℓ_p au bord à l'infini (idée sur laquelle repose la construction d'Elek décrite en 3.2) apparaît dans Pansu [42] et dans Gromov [32].

Dans [42] P. Pansu montre que la 1-cohomologie ℓ_p des espaces symétriques de rang 1 s'identifie à des espaces de Besov au bord. Ce phénomème se généralise aux espaces hyperboliques de Gromov, voir [8]. Pour les espaces homogènes à courbure négative, il se généralise en partie à la cohomologie de degré supérieur [44, 46, 47].

---

[4] Sur la sphère il n'existe pas de fonction u continue telle que $|u(x) - u(-x)| \geq 1$, car sinon la fonction $h(x) = u(x) - u(-x)$ serait impaire et partout non nulle.

Dans l'esprit de [32] p.258, M. Puls [50] démontre que si Γ est un groupe de type fini, tel que pour un certain $a > 1$ le bord de Floyd de Γ associé à la fonction $\varphi : g \in \Gamma \mapsto a^{-|g|}$ possède au moins deux points distincts, alors $\ell_p \overline{H^1}(\Gamma) \neq 0$ pour les $p \geq 1$ tels que $\varphi \in \ell_p(\Gamma)$. Ce résultat, combiné aux travaux de V. Gerasimov [26] sur le bord de Floyd, implique la non-annulation de la 1-cohomologie ℓ_p réduite des groupes relativement hyperboliques (pour p assez grand). A ma connaissance c'est une question ouverte (due à Cornélia Drutu) de déterminer quels sont les groupes relativement hyperboliques qui possèdent une action isométrique propre sur un espace L_p.

Afin d'élargir la classe des groupes connus dont la 1-cohomologie ℓ_p réduite est non nulle, il serait intéressant de disposer de nouvelles constructions de classes de cohomologie ℓ_p. Dans [6] on décrit un procédé de construction basé sur la cohomologie relative et l'excision. Elle s'applique en particulier à certains groupes de Coxeter.

2) P. Pansu montre que, contrairement aux groupes discrets, la non-annulation de la 1-cohomologie ℓ_p réduite (p assez grand) caractérise l'hyperbolicité parmi les groupes de Lie [45]. C'est encore le cas pour les groupes algébriques sur les corps locaux de caractéristique nulle, voir [18].

3) Il y a une autre classe de groupes, introduite par Haglund-Paulin [33] et généralisée par Cherix-Martin-Valette [17], dont les actions isométriques propres sur les espaces L_p ont été abondamment étudiées, il s'agit des groupes agissant proprement sur les espaces à murs mesurés. Cette classe contient en particulier les groupes de Coxeter, les groupes moyennables, les réseaux de $SO(n, 1)$ et de $SU(n, 1)$. Il découle de [17, 13] que les propriétés suivantes sont équivalentes:

- Γ est un groupe agissant proprement sur un espace à murs mesurés,
- Γ est a-T-menable (voir l'introduction pour la définition),
- Il existe $p \in [1, 2]$ telle que Γ admette une action isométrique propre sur un espace L_p,
- Pour tout $p \in [1, +\infty)$, Γ admet une action isométrique propre sur un espace L_p.

Les premiers exemples de groupes hyperboliques qui n'appartiennent pas à la classe ci-dessus sont les réseaux cocompacts de $Sp(n, 1)$. Pour ceux-ci il est communément conjecturé que la borne du théorème 1 est optimale, c'est-à-dire qu'ils ne possèdent pas d'actions isométriques propres (ni même sans point fixe) sur L_p pour p inférieur ou égal à la dimension conforme.

4) Les groupes hyperboliques dont on connait précisément la dimension conforme de bord sont les réseaux cocompacts des groupes de Lie simples de rang un [48] et les réseaux des immeubles fuchsiens [3]. Pour les premiers la dimension conforme est égale à celle du bord de l'espace symétrique sous-jacent (grâce à l'invariance par quasi-isométrie), de plus on a:

$$\text{Confdim}(\partial \mathbb{H}_K^n) = nk + k - 2, \text{ où } K = \mathbb{R}, \mathbb{C}, \mathbb{H}, \mathbb{C}a, \text{ et où } k = \dim_{\mathbb{R}} K.$$

Il est intéressant de noter que le théorème 1 combiné au Th.3.8 de [6] permet d'envisager que la dimension conforme soit caractérisée par la propriété suivante: $p > \text{Confdim}(\partial \Gamma)$ *si et seulement si Γ possède une action isométrique propre sur ℓ_p, dont la partie linéaire est un multiple de la représentation régulière droite.*

5) Rappelons que tous les groupes dénombrables infinis Γ possèdent une action isométrique propre sur ℓ_∞. En effet étant donné $\gamma_0 \in \Gamma$ l'application $c : \Gamma \to \ell_\infty(\Gamma)$ définie par

$$\forall g, \gamma \in \Gamma \ : \ c(g)(\gamma) = |\gamma g - \gamma_0| - |\gamma - \gamma_0|,$$

est un cocycle propre pour la représentation régulière droite de Γ sur $\ell_\infty(\Gamma)$.

Brown et Guentner [11] ont montré que tous les groupes infinis dénombrables possèdent une action isométrique propre sur un Banach strictement convexe. A l'opposé V. Lafforgue [37] a montré que les actions isométriques des réseaux de SL$(3, \mathbb{Q}_p)$ sur les Banach uniformément convexes possèdent toutes un point fixe. (Un Banach V est *strictement convexe* si pour tout $x, y \in V, x \neq y, \|x\| = \|y\| = 1$ on a $\|\frac{x+y}{2}\| < 1$. Il est *uniformément convexe* si pour tout $\epsilon > 0$ il existe un $\delta > 0$ tel que $\|x\| = \|y\| = 1$ et $\|x - y\| \geq \epsilon$ entraînent que $\|\frac{x+y}{2}\| < 1 - \delta$.)

Dans le même esprit, Bader-Furman-Gelander-Monod [1] ont établi que les actions isométriques des réseaux de rang supérieur sur les espaces L_p ($p \in (1, +\infty)$) ont toutes un point fixe. Ils conjecturent que cette propriété persiste pour les Banach uniformément convexes. Voir [41] pour un aperçu des actions de groupes sur les espaces de Banach.

6) Soit π une représentation linéaire isométrique d'un groupe Γ sur un espace de Banach V. On dit qu'elle est c_0 si pour tout $f \in V^*$ et tout $v \in V$ on a $f(\pi(g)v) \to 0$ lorsque $|g| \to \infty$. Par exemple si Γ est infini alors la représentation régulière droite de Γ sur $\ell_p(\Gamma)$ est c_0.

De Cornulier, Tessera et Valette [19] ont introduit la notion suivante: un groupe Γ a la propriété (BP_0) si toute Γ-action isométrique sur un Banach, de partie linéaire c_0, est ou bien propre ou bien possède une orbite bornée. Ils démontrent dans [19] que les groupes résolubles, les groupes de Lie connexes, les groupes à centre infini ont cette propriété (Y. Shalom [54] l'avait auparavant établie pour les actions de $SO(n, 1)$ et $SU(n, 1)$ sur les espaces de Hilbert).

Marc Bourdon
Laboratoire Painlevé,
UMR CNRS 8524, Université de Lille 1,
59655 Villeneuve d'Ascq, FRANCE
E-mail: bourdon@math.univ-lille1.fr

References

[1] U. BADER, A. FURMAN, T. GELANDER, N. MONOD, *Property (T) and rigidity for actions on Banach spaces,* Acta Math. **198** (2007), no. 1, 57-105.

[2] M. BONK, J. HEINONEN, P. KOSKELA, *Uniformizing Gromov hyperbolic spaces,* Astérisque **270**, 2001.

[3] M. BOURDON, *Sur les immeubles fuchsiens et leur type de quasi-isométrie,* Ergodic theory and Dynamical Systems **20** (2000), 343-364.

[4] M. BOURDON, *Cohomologie ℓ_p et produits amalgamés,* Geometriae Dedicata **107** (2004), 85-98.

[5] M. BOURDON, B. KLEINER, *Combinatorial modulus, the Combinatorial Loewner Property, and Coxeter groups,* Groups Geom. Dyn. **7** (2013), no. 1, 39-107.

[6] M. BOURDON, B. KLEINER, *Some applications of ℓ_p-cohomology to boundaries of Gromov hyperbolic spaces,* Groups Geom. Dyn. **9** (2015), no. 2, 435-478.

[7] M. BOURDON, F. MARTIN, A. VALETTE, *Vanishing and non-vanishing for the first L^p-cohomology of groups,* Comment. Math. Helv. **80** (2005), 377-389.

[8] M. BOURDON, H. PAJOT, *Cohomologie ℓ_p et espaces de Besov,* J. reine angew. Math. **558** (2003), 85-108.

[9] M. BRIDSON, A. HAEFLIGER, *Metric spaces of non-positive curvature,* Grundlheren der Mathematischen Wissenschaften 319, Springer, 1999.

[10] K.S. BROWN, *Cohomology of groups,* Graduate texts in Mathematics, Springer-Verlag, 1982.

[11] N. BROWN, E. GUENTNER, *Uniform embedding of bounded geometry spaces into reflexive Banach spaces,* Proc. Amer. Math. Soc. **133** (2005), no.7, 2045-2050.

[12] YU. A. BRUDNYI, N. YA. KRUGLJAK, *Interpolation functors and interpolation spaces, Vol. 1,* North-Holland Mathematical Library, **47** (1991).

[13] I. CHATTERJI, C. DRUTU, F. HAGLUND, *Kazhdan and Haagerup properties from the median viewpoint,* Adv. Math. **225** (2010), no.2, 882-921.

[14] I. CHAVEL, *Isoperimetric inequalities. Differential geometric and analytic perspectives,* Cambridge Tracts in Maths 145, 2001.

[15] J. CHEEGER, M. GROMOV, *L_2-cohomology and group cohomology,* Topology **25** (1986), p 189-215.

[16] P-A. CHERIX, M. COWLING, P. JOLISSAINT, P. JULG, A. VALETTE, *Groups with the Haagerup property. Gromov's a-T-menability,* Progress in Mathematics **197**, Birkhäuser, 2001.

108   *References*

[17] P-A. CHERIX, F. MARTIN, A. VALETTE, *Spaces with measured walls, the Haagerup property and property (T)*, Ergod. Th. Dynam. Sys. (2004), **24**, 1895-1908.

[18] Y. DE CORNULIER, R. TESSERA, *Contracting automorphisms and L_p-cohomology in degree one*, Arkiv for Matematiks **49**, no. 2, (2011), 295-324.

[19] Y. DE CORNULIER, R. TESSERA, A. VALETTE, *Isometric group actions on Banach spaces and representations vanishing at infinity*, Transform. Groups **13** (2008), no. 1, 125-147.

[20] M. COORNAERT, *Mesures de Patterson-Sullivan sur le bord d'un espace hyperbolique au sens de M. Gromov*, Pacific Journal of Mathematics **159** (1993), 241-270.

[21] M. COORNAERT, T. DELZANT, A. PAPADOPOULOS, *Géométrie et théorie des groupes, les groupes hyperboliques de M. Gromov*, Lecture Notes in Mathematics **1441**, Springer-Verlag, 1991.

[22] J. DYMARA, T. JANUSZKIEWICZ, *Cohomology of buildings and their automorphism groups*, Invent. Math. **150** (2002), 579-627.

[23] G. ELEK, *The ℓ_p-cohomology and the conformal dimension of hyperbolic cones*, Geometriae Dedicata **68** (1997), 263-279.

[24] G. ELEK, *Coarse cohomology and ℓ_p-cohomology*, K-Theory **13** (1998), no. 1, 1-22.

[25] D. GABORIAU, *Invariants ℓ^2 de relations d'équivalence et de groupes*, Inst. Hautes Etudes Sci. Publ. Math. **95** (2002), 93-150.

[26] V. GERASIMOV, *Floyd maps for relatively hyperbolic groups*, Geom. Funct. Anal. **22** (2012), no. 5, 1361-1399.

[27] S. M. GERSTEN, *Isoperimetric functions of groups and exotic cohomology*, In: Combinatorial and Geometric Group Theory, Edinburgh 1993, London Math. Soc. Lecture Notes Ser. 204, Cambridge Univ. Press, 1995, 87-104.

[28] E. GHYS, P. DE LA HARPE, (Eds), *Sur les groupes hyperboliques, d'après Gromov*, Progress in Mathematics **83**, Birkhäuser, 1990.

[29] V. GOLDSTEIN, V. KUZMINOV, I. SHVEDOV, *The Kuenneth formula for L_p cohomology of warped products*, Sib. Math. J. **32** (1991), 749-760.

[30] A. GOURNAY, *Boundary values, random walks and ℓ_p-cohomology in degree one*, Groups, Geom. Dyn., to appear.

[31] M. GROMOV, *Hyperbolic groups, Essays in Group theory*, Ed. S.M. Gersten, Springer 1987, 72-263.

[32] M. GROMOV, *Asymptotic invariants for infinite groups*, London Mathematical Society Lecture Note Series 182, Eds G.A. Niblo and M.A. Roller, 1993.

[33] F. HAGLUND, F. PAULIN, *Simplicité de groupes d'automorphismes d'espaces à courbure négative*, Geom. Topol. Monograph **1** (1998), 181-248.

[34] J. HEINONEN, *Lectures on analysis on metric spaces*, Universitext, Springer, 2001.

[35] G. KASPAROV, G. YU, *The coarse geometric Novikov conjecture and uniform convexity*, Adv. Math. 206 (2006), no. 1, 1-56.

[36] H. KESTEN, *Full Banach mean values on countable groups*, Math. Scand. **7**, (1959), 146-156.

[37] V. LAFFORGUE, *Propriété (T) renforcée banachique et transformation de Fourier rapide*, J. Topol. Anal. **1** (2009), no. 3, 191-206.

[38] W. LUECK, *L^2-Betti numbers of mapping tori and groups*, Topology, **33** (1994), 203-214.

[39] J. MACKAY, J. TYSON, *Conformal dimension. Theory and application,* University Lecture Series, **54**, American Mathematical Society, Providence, RI, 2010.

[40] B. NICA, *Proper isometric actions of hyperbolic groups on L^p-spaces,* Compos. Math. **149** (2013), no. 5, 773-792.

[41] P. NOWAK, *Group actions on Banach spaces,* In *Handbook of Group Actions*; vol. II, 121-149. L. Ji, A. Papadopoulos, S.-T. Yau eds., ALM **32**, International Press, Somerville; Higher Education Press, Beijing, 2015.

[42] P. PANSU, *Cohomologie L^p des variétés à courbure négative, cas du degré un,* PDE and Geometry 1988, Rend. Sem. Mat. Torino, Fasc. Spez. (1989), 95-120.

[43] P. PANSU, *Cohomologie L^p : invariance sous quasiisométries,* Preprint Université Paris-Sud (1995).

[44] P. PANSU, *Cohomologie L^p, espaces homogènes et pincement,* Preprint Université Paris-Sud (1999).

[45] P. PANSU, *Cohomologie L^p en degré 1 des espaces homogènes,* Potential Anal. **27**, (2007), 151-165.

[46] P. PANSU, *Cohomologie L^p et pincement,* Comment. Math. Helv. **83** (2008), no.2, 327-357.

[47] P. PANSU, *L^p-cohomology of symmetric spaces,* Geometry, analysis and discrete groups, 305-326, Adv. Lect. Math. (ALM), 6, (2008).

[48] P. PANSU, *Dimension conforme et sphère à l'infini des variétés à courbure négative,* Annales Academiae Scientiarum Fennicae, Series A.I. Mathematica **14** (1990), 177-212.

[49] F. PAULIN, *Un groupe hyperbolique est déterminé par son bord,* Journal of the London Math. Soc. **54** (1996), 50-74.

[50] M. PULS, *The first L^p-cohomology of some groups with one end,* Arch. Math. (Basel) **88** (2007), no. 6, 500-506.

[51] M. S. RAGHUNATHAN , *Discrete subgroups of Lie groups,* Ergebnisse der mathematik und ihrer Grenzgebiete **68**, Springer-Verlag, 1972.

[52] A. REZNIKOV, *Analytic Topology of Groups, Actions, Strings and Varieties,* Geometry and dynamics of groups and spaces, 3-93, Prog. Math., **265**, Birkauser, Basel, 2008.

[53] W. RUDIN, *Functional Analysis,* McGraw-Hill, 1973.

[54] Y. SHALOM, *Rigidity, unitary representations of semisimple groups, and fondamental groups of manifolds with rank one transformation group,* Ann. of Math. (2) **152** (2000), 113-182.

[55] R. TESSERA, *Vanishing of the first reduced cohomology with values in an L^p representation,* Ann. Inst. Fourier (Grenoble) **59**, no.2, 851-876.

[56] R. TESSERA, *Isoperimetric profile and random walks on locally compact solvable groups,* Rev. Mat. Iberoam. **29** (2013), no. 2, 715-737.

[57] G. YU, *Hyperbolic groups admit proper affine isometric actions on l_p-spaces,* GAFA, **15** (2005), 1144-1151.

4

Compact Clifford–Klein Forms – Geometry, Topology and Dynamics

DAVID CONSTANTINE

Abstract

We survey results on compact Clifford–Klein forms of homogeneous spaces, with a focus on recent contributions and organized around approaches via topology, geometry and dynamics. In addition, we survey results on moduli spaces of compact forms.

1 Introduction

1.1 The basic questions, and a conjecture

This survey will be devoted to two questions. The first, the existence question, has been studied fairly extensively. The second, the deformation question, has so far received less attention, though there have been several recent developments.

Question 1 (The existence question for compact Clifford–Klein forms) Let $H \backslash G$ be a homogeneous space of a noncompact Lie group G. Does there exist a discrete subgroup Γ in G such that $H \backslash G / \Gamma$ is a compact manifold locally modeled on $H \backslash G$?

A compact manifold $H \backslash G / \Gamma$ is generally called a compact Clifford–Klein form. The alternate terminology '$H \backslash G$ has a tessellation' has been suggested by the series of papers [40, 41, 17]. This terminology may be preferable in that it avoids the confusion with compact forms of Lie groups, but it is not standard.

The ultimate goal is to understand all pairs (G, H) which have compact Clifford–Klein forms. There is a Goldilocks tension at play in looking for Γ: it must not be too big, so that it can act freely and properly discontinuously on $H \backslash G$; it must not be too small or else $H \backslash G / \Gamma$ will not be compact. In addition, Γ must be 'well-positioned' in G relative to H, as we will see most clearly in the approaches to the problem in Section 3.2 below.

For all of the results we will outline, some sort of restriction must be placed on G and H. It is very common to assume G semisimple, or even simple, and that H is reductive in G, i.e. its adjoint action on the Lie algebra of G is reducible. These are the main cases of geometric interest and will be the focus of this survey. As exceptions to this rule, however, I would note: the flat Minkowski spacetime $\mathbb{R}^{n,1} = O(n, 1)\backslash(O(n, 1) \ltimes \mathbb{R}^{n+1})$ and the affine space $\mathbb{R}^n = GL_n(\mathbb{R})\backslash(GL_n(\mathbb{R}) \ltimes \mathbb{R}^n)$, especially as studied in relation to the Auslander Conjecture (see [9] and [1] for surveys); and the 'tangential symmetric spaces,' studied in [31] by Kobayashi and Yoshino.

The study of such forms of Riemannian homogeneous spaces goes back to the work of Borel on discrete subgroups of Lie groups (see [6]). As we will soon see, however, the main focus of Question 1 is now pseudo-Riemannian homogeneous spaces, i.e. those for which H is noncompact. The first work on such spaces was that of Calabi and Markus [8]. Since the late '80's, the driving force behind considering the problem for reductive, non-Riemannian homogeneous spaces – our specific concern here – has been Toshiyuki Kobayashi (see the many references below), and in subsequent years a wide variety of mathematicians have contributed to the field.

Let us note two related problems. First, one can ask which homogeneous spaces have finite-volume Clifford–Klein forms. A few of the results below are general enough to deal with this situation, but to the author's knowledge nothing specific to the finite-volume case has been done. It would be quite surprising to find a homogeneous space possessing a finite-volume form but no compact one.

Second, one can also ask whether a compact (G, X)-manifold where $X = H\backslash G$ exists. Such a structure consists of the following: a *developing map* $dev : \tilde{M} \to X$ and a *holonomy homomorphism hol* $: \pi_1(M) \to G$ such that the developing map is a local diffeomorphism and equivariant with respect to the holonomy map. (G, X)-structures will be used in Section 6.2 below, and are treated in more detail there. A compact (G, X)-manifold need not be of the form $H\backslash G / \Gamma$. The author knows of two results flexible enough to treat this more general existence question ([5] and [34], Sections 4.1 and 5.1 below), but again, most cases are open.

Naturally, the second question we survey is the uniqueness counterpart of the first:

Question 2 (The deformation question for compact forms) For homogeneous spaces possessing a compact form, can one describe the space of all compact forms? Specifically, can compact forms be deformed locally?

Work on this question is, generally speaking, more recent, and includes some very recent results which I will survey in Section 6.

There is one central conjecture in this field, having to do with the following algebraic construction of compact forms. Let us suppose that L and H are Lie subgroups of G such that:

- $HL = G$
- $H \cap L$ is compact
- Γ is a uniform lattice in L

Then a compact form of $H\backslash G$ may be formed by taking

$$H\backslash G/\Gamma \cong (H \cap L)\backslash L/\Gamma.$$

The requirement above that $HL = G$ can be loosened – we need only that HL is cocompact in G. These algebraically-constructed compact forms will be called *standard forms*. Kobayashi conjectures the following:

Conjecture 3 (Kobayashi, [31] Conj. 3.3.10) *If $H\backslash G$ possesses a compact form, for G and H reductive, it possesses a standard form.*

The reader should note that Kobayashi's conjecture is not simply 'all compact forms are standard.' We will see in Section 6 that the answer to the deformation question is sometimes yes, and that not all the forms are standard. The first such results were due to Goldman, Ghys and Kobayashi ([14], [13], [29]). In addition, there are nonstandard forms which are not deformations of standard ones, first obtained by Salein ([45], see Section 6.2).

Remark 4 I would like to note here a result that bears on the construction mentioned above. In [10] Lemma 9.2.2, we prove that when G is simple and H is contained in a maximal proper parabolic subgroup of G, then the condition $HL = G$ forces L to be semisimple. Assuming H reductive, one can then apply the work of Oniščik in [42], where all decompositions of simple G as products of reductive subgroups are classified.[1] Most examples in this classification do not satisfy $H \cap L$ compact; the remainder of Oniščik's list provides a useful catalogue of spaces potentially having compact forms.

We will survey below some of the results that give evidence for Conjecture 3. However, as the reader will gather from the wide variety of approaches to the problem (each of which has its own particular realm of usefulness), there is no unified approach at the moment.

---

[1] Oniščik's work assumes that H is reductive but not that H is contained in a proper parabolic subgroup.

1.2 Motivation and special examples

We motivate the existence question with a special case. Let H be a compact subgroup of G. Then it is easy to see that finding a suitable Γ reduces to finding a uniform lattice in G. For G semisimple, Borel found such lattices (see [6]). These examples include the Riemannian symmetric spaces, spaces of great geometric interest.

As we allow H noncompact, we can still form symmetric spaces by taking H to be an open subgroup of the points of G fixed by some involution. In general, however, these will be only *pseudo-Riemannian* symmetric spaces (see Prop 7). The main motivation for the existence question is to understand these symmetric spaces as we do the Riemannian ones, although they are, generally speaking, much harder to study. The main difficulty is that $H\backslash G/\Gamma$, with both H and Γ noncompact, usually does not inherit the main structures present on G; a good example is a right-invariant Riemannian metric. Hence the need for the many new techniques developed below.

To motivate the deformation question, consider the special case $G = SL_2(\mathbb{R})$, $H = SO(2)$. Here the deformation spaces of compact forms (up to conjugation of Γ in G) are Teichmüller spaces of marked hyperbolic structures on closed, genus $g \geq 2$ surfaces. Some of the deformation results we will examine below are actually on a similar problem, examining again constant curvature -1 structures, but this time on three-dimensional Lorentzian manifolds.

A few special examples are worth mentioning. The first example one might consider – simply because of the straight-forward structure of the groups involved – is $SL_k(\mathbb{R})\backslash SL_n(\mathbb{R})$, where $SL_k(\mathbb{R})$ is embedded in the upper left-hand corner of $SL_n(\mathbb{R})$. This can be generalized by embedding $SL_k(\mathbb{R})$ via other representations. Amazingly, the existence question for these spaces is not entirely solved, although all work so far gives a negative answer to that question. A variety of results deal with small-dimensional versions of this problem; the main results for general dimension follow from the topological approaches discussed in Section 3 and the dynamical approaches in Section 5.

The second example is geometrically motivated, and predates the example above. Call a pseudo-Riemannian manifold M with signature (p, q) a (p, q)-*space form* if it has constant sectional curvature κ. We are familiar, of course, with the Riemannian $(p, 0)$-space forms; $(p, 1)$-space forms are the Lorentzian case (see, e.g., [51]). The first motivation for the existence problem was Calabi and Markus's study of $(3, 1)$-space forms as possible models for space-time ([8], see Sec. 3.1). If we assume as well that M is geodesically complete, then M must be a quotient of one of the following spaces (see [51]):

- $\kappa < 0$: $\mathbb{H}^{p,q} := O(p,q)\backslash O(p, q+1)$ for $q \geq 2$, or $\widetilde{O(p,1)}\backslash\widetilde{O(p,2)}$
- $\kappa = 0$: $\mathbb{R}^{p,q} := O(p,q)\backslash O(p,q) \ltimes \mathbb{R}^{p+q}$
- $\kappa > 0$: $\mathbb{S}^{p,q} := O(p,q)\backslash O(p+1,q)$ for $p \geq 2$, or $\widetilde{O(1,q)}\backslash\widetilde{O(2,q)}$.

Kobayashi has the following conjecture about pseudo-Riemannian space forms:

Conjecture 5 (Kobayashi, [30], Conj 2.6) *There exists a compact space form with signature* (p,q) $(p, q \neq 0)$ *and sectional curvature* κ *if and only if:*

- $\kappa < 0$ *and* (p,q) *belongs to*

p	1	3	7
q	2N	4N	8

- $\kappa = 0$ *and* (p,q) *is arbitrary*

- $\kappa > 0$ *and* (p,q) *belongs to*

p	2N	4N	8
q	1	3	7

The 'if' part of this conjecture is proven (see [31] and references therein); the 'only if' part is open, though we will see several results below that address portions of it. The $\kappa = 0$ portion is trivial; the $\kappa < 0$ and $\kappa > 0$ portions are equivalent as $\mathbb{S}^{p,q} \simeq \mathbb{H}^{q,p}$.

As preparation for a few of the techniques below, let me record a definition of an important class of homogeneous spaces and a fact about their pseudo-Riemannian structure.

Definition 6 The homogeneous space $H\backslash G$ is of reductive type if G is reductive, H is a closed subgroup with finitely many connected components and H is stable under a Cartan involution[2] of G.

Note that $H\backslash G$ of reductive type implies that H is reductive; the converse holds if G is reductive and H is semisimple or algebraic.

Proposition 7 (See, e.g., [31] Prop 3.2.7) *Let* $H\backslash G$ *be a homogeneous space of reductive type. Then* $H\backslash G$ *has a right-G-invariant pseudo-Riemannian metric of signature*

$$(d(G) - d(H), dim(G) - dim(H) - d(G) + d(H))$$

induced by the Killing form on $\mathfrak{g}$. *Here* $d(-)$ *is the dimension of the Riemannian symmetric space associated to the Lie group.*

---

[2] Recall that a Cartan involution of the Lie algebra $\mathfrak{g}$ is an involution θ such that $-\kappa(X, \theta Y)$ is positive definite as bilinear form in X and Y, where κ is the Killing form. It plays a central role in the theory of semisimple and reductive Lie groups. H is stable under this involution if its Lie algebra is stable.

1.3 Organization and aims of the survey

I would be very remiss not to mention that there are already several excellent surveys of the existence question for compact forms. In particular, let me note the survey "Quelques résultats récents sur les espaces localement homogènes compacts" [35] by Labourie, the survey "Compact Clifford–Klein forms of symmetric spaces – revisited" [31] by Kobayashi and Yoshino, and the introduction to Kassel's thesis [18]. Kobayashi and Yoshino's survey in particular is very extensive and for the results related to Kobayashi's large amount of work on the problem, no better survey could be desired.

It is not my intention to replace any of these excellent papers, but rather to supplement them. In particular I would like to point out the following features that I hope will make this survey a serviceable supplement:

- We survey a few results (eg. [10], [45], [20], [16]) which have been published since the most recent survey [31], or were not included there.
- This survey gives the first complete presentation of the representation-theoretic approaches to the problem (see Section 4.2).
- We will treat the extensive work on the problem by Kobayashi and his collaborators lightly; [31] is still the best resource on this work.
- We provide the first survey of material on the deformation question, much of which has been developed recently (Section 6).

As this survey is being written as part of the proceedings for the conference Geometry, Topology and Dynamics in Negative Curvature (Bangalore, 2010), I have organized the presentation around the themes of the conference. In Section 3 we survey approaches to the existence question that can be broadly classed as topological. In Section 4 we survey geometric approaches and in Section 5 we survey approaches primarily using dynamics. The reader will find, of course, that many of our results straddle these categories. Finally, in Section 6 we survey results on the deformation question.

1.4 Thanks

I would like to thank the organizers of the conference Geometry, Topology and Dynamics in Negative Curvature (Bangalore, 2010) for inviting me to speak at that excellent event, and for inviting me to contribute this survey to the proceedings. I would especially like to thank Fanny Kassel for reading an earlier draft of this survey and providing numerous and very helpful comments and suggestions, and an unnamed referee for a very careful reading of an earlier draft and helpful suggestions too numerous to detail. Remaining mistakes are my own.

2 Notation

The following notation will be fixed throughout the survey for consistency. Notation which only makes an appearance relating to one result will conform, as much as possible, to the authors' original notation, for ease of reference.

- G and H will be (real) Lie groups; Γ will be a discrete subgroup of G.
- $H \backslash G / \Gamma$ will be a compact Clifford–Klein form of the homogeneous space $H \backslash G$.
- $Z_G(H)$ will be the centralizer of H in G; often $J < Z_G(H)$ will be a particular subgroup of this centralizer, usually a semisimple group.
- German letters ($\mathfrak{g}$, $\mathfrak{j}$, etc.) denote the Lie algebras associated to the corresponding Lie groups.
- For a Lie group G, let $d(G)$ denote the dimension of the (Riemannian) symmetric space associated to G.

3 Topology

In basic formulation, the compact forms question is a topological one, and the first approach to it is via the algebra and topology of Lie groups. Some of what is surveyed in this section is not strictly topological, but there is a significant contrast with the approaches found in Section 4, where stronger geometric structures are used.

The topological approach has a long history, with the most extensive contribution being provided by Toshiyuki Kobayashi. In fact, no better survey of the topological approaches to this problem can be found than his survey [31] with Yoshino. Below I will present an abbreviated look at the basic features of the topological approaches; greater detail can be found in Kobayashi and Yoshino's survey.

3.1 Rank restrictions – the Calabi–Markus phenomenon

Several of the main nonexistence results for compact forms utilize restrictions on the real ranks of the Lie groups G and H. In fact, the first nonexistence result is in this vein – the so-called Calabi–Markus phenomenon. In [8], Calabi and Markus investigate what they call relativistic space forms, that is, complete Lorentz manifolds with constant curvature. Of these there are three models, the Minkowski plane $\mathbb{R}^{n,1}$ with curvature zero, the de Sitter space $dS^n = \mathbb{S}^{n-1,1}$ with curvature $+1$, and the anti-de Sitter space $AdS^n = \mathbb{H}^{n-1,1}$ with curvature -1 (take $n \geq 2$ here). The de Sitter space is formed by taking the standard Minkowski form on $\mathbb{R}^{n,1}$, namely $ds^2 = -x_1^2 + x_2^2 + \cdots + x_{n+1}^2$ and restricting this form to the quadric hypersurface $-x_1^2 + x_2^2 + \cdots + x_{n+1}^2 = 1$. The group $G = SO(1, n)$ preserves this hypersurface and the stabilizer of the point $(0, 1, 0, \ldots, 0)$ is $SO(1, n-1)$ so $dS^n = SO(1, n-1)\backslash SO(1, n)$. Topologically, dS^n is the product of the real line with a sphere, hence noncompact.

Calabi and Markus prove the following, of which the nonexistence of a compact (or finite-volume) form is a clear corollary.

Theorem 8 (Calabi–Markus, [8] Thm 1) *Any* Γ *which acts properly discontinuously by isometries on* $dS^n = SO(1, n-1)\backslash SO(1, n)$ *is finite.*

Idea of proof Consider the equatorial sphere in dS^n described by setting $x_1 = 0$. It is an easy exercise using the symmetry about the origin to show that the image of this sphere under any isometry intersects the original sphere. As this sphere is compact, these intersections cannot occur for infinitely many elements of Γ without violating proper discontinuity of the action. □

The general idea here – that $SO(1, n-1)$ is 'large enough' with respect to $SO(1, n)$ to leave no room for an infinite Γ – is codified more generally in the following criterion.

Theorem 9 (Calabi–Markus phenomenon, see [8], [50], [24]) *If the $\mathbb{R}$-rank of H is the same as the $\mathbb{R}$-rank of G, only a finite Γ can act properly discontinuously on $H\backslash G$. Thus $H\backslash G$ admits a compact form only if it is already compact.*

3.2 Generalizations of Calabi–Markus

The most comprehensive approach to this situation has been provided by Kobayashi and Benoist (see [24], [26], [27], [28] and [3], [4]). Benoist and Kobayashi formulate a very general criterion for proper action which suppresses even the group structures of H and Γ and treats them symmetrically.

Definition 10 ([28], Defn 1.11.1; [4], Section 3.1) Let A and B be subsets of G.

- The pair (A, B) is *transversal*, denoted $A \bar{\pitchfork} B$, if $A \cap SBS$ is relatively compact for any compact $S \subset G$.[3]

---

[3] Benoist calls the pair (A, B) 'G-proper.'

- The pair (A, B) is *similar*, denoted $A \sim B$, if there exists a compact subset S in G such that $A \subset SBS$ and $B \subset SAS$.

Note that Γ acts properly on $H \backslash G$ if and only if $\Gamma \bar{\pitchfork} H$, that transversality is preserved by similarity, and that if $H \sim G$, then $H \bar{\pitchfork} \Gamma$ only if Γ is compact. For $G \cong \mathbb{R}^n$, and A and B closed cones in G, the definitions are particularly simple: similarity corresponds to equality and transversality corresponds to $A \cap B = \{0\}$. To exploit these facts, recall the concept of Cartan decomposition for a reductive Lie group: $G = KAK$, where K is a maximal compact subgroup and A is a Cartan subgroup. Let $\pi : G \to \mathfrak{a}/W$ be the associated Cartan projection (W is the Weyl group); it is continuous, surjective and proper. Write $\mathfrak{a}(S) = \pi(S)W \subset \mathfrak{a}$ for any subset $S \subset G$. The considerations above indicate that properness can be checked by examining $\mathfrak{a}(H)$ and $\mathfrak{a}(\Gamma)$. Specifically, $H \bar{\pitchfork} \Gamma$ (resp. $H \sim \Gamma$) in G if and only if $\mathfrak{a}(H) \bar{\pitchfork} \mathfrak{a}(\Gamma)$ (resp. $\mathfrak{a}(H) \sim \mathfrak{a}(\Gamma)$). Another way to state the properness condition is that Γ acts properly on $H \backslash G$ if and only if $\mathfrak{a}(\Gamma)$ goes away from $\mathfrak{a}(H)$ at infinity, i.e. for any compact subset C of $\mathfrak{a}$, one has that $\mathfrak{a}(\Gamma) \cap (\mathfrak{a}(H) + C)$ is compact. The result is the following theorem, which strengthens Theorem 9 by providing a converse:

Theorem 11 (General Calabi–Markus phenomenon) *Let $H \backslash G$ be of reductive type. Then the following are equivalent:*

- $\mathbb{R}$-*rank*$(H) = \mathbb{R}$-*rank*(G)
- *Only a finite Γ acts properly discontinuously on $H \backslash G$.*

In [4], Benoist develops a further use of the Cartan projection to put a restriction on any properly acting Γ. Assume H is stable under the Cartan involution defining K. Define $B_W = \{aW \in \mathfrak{a}/W : aW = -aW\}$.

Theorem 12 (Benoist, [4] Section 7.5) *Let $H \backslash G$ be of reductive type and let H be stable under the Cartan involution defining K as above. Only a virtually abelian Γ may act properly discontinuously on $H \backslash G$ if and only if $B_W \subset \mathfrak{a}(H)$. In particular, if $H \backslash G$ is noncompact and $B_W \subset \mathfrak{a}(H)$, then $H \backslash G$ has no compact form.*

Table 1 lists a few spaces to which this theorem applies.

Note that the first example in Table 1, with $k = 1$, is $\mathrm{SL}_{n-1}(\mathbb{R}) \backslash \mathrm{SL}_n(\mathbb{R})$ when n is odd.[4] When n is even, Benoist constructs non-virtually abelian groups which act properly discontinuously (though not cocompactly) on $\mathrm{SL}_{n-1}(\mathbb{R}) \backslash \mathrm{SL}_n(\mathbb{R})$; specifically, he constructs nonabelian, free, Schottky

---

[4] As reported in [17], the case $\mathrm{SL}_2(\mathbb{R}) \backslash \mathrm{SL}_3(\mathbb{R})$, was also proven by Margulis, though never published.

Table 1 *Some homogeneous spaces $H\backslash G$ without compact Clifford–Klein forms*

G	H	conditions
$\mathrm{SL}_n(\mathbb{R})$	$(\mathrm{SL}_k(\mathbb{R}) \times \mathrm{SL}_{n-k}(\mathbb{R}))$	$1 \leq k \leq n$, and $k(n-k)$ even
$\mathrm{SL}_{2n}(\mathbb{R})$	$\mathrm{Sp}(n, \mathbb{R})$	$n \geq 1$
$\mathrm{SL}_{2n}(\mathbb{R})$	$\mathrm{SO}(n, n)$	$n \geq 1$
$\mathrm{SL}_{2n+1}(\mathbb{R})$	$\mathrm{SO}(n, n+1)$	$n \geq 1$
$\mathrm{SO}(n+1, m)$	$\mathrm{SO}(n, m)$	$n \geq m$, n even
$G_{\mathbb{C}}$	$H_{\mathbb{C}}$	G simple, complex; $H_{\mathbb{C}}$ the fixed points of a complex involution of $G_{\mathbb{C}}$; except $(\mathrm{SO}(4n-k, \mathbb{C}) \times \mathrm{SO}(k, \mathbb{C}))\backslash \mathrm{SO}(4n, \mathbb{C})$ for $n \geq 2$, and k odd.

groups.[5] The case for n even is an interesting open question; one expects that no compact form of this space exists.

We will close this subsection with a mention of some of the work of Kassel on the problem. She works on homogeneous spaces $H\backslash G$ with $\mathbb{R}$-rank$(H) =$ $\mathbb{R}$-rank$(G) - 1$, i.e. examples just outside the reach of the Calabi–Markus phenomenon. In these cases (and under some natural assumptions on G and H) she is able to describe the Cartan projection of any Γ acting properly discontinuously on $H\backslash G$ ([19] Thm 1.1). Two applications of the main result in [19] are a structure theorem for properly discontinuous Γ for $diag(G')\backslash(G' \times G')$ ([19] Theorem 1.3; $diag$ refers throughout to the diagonal embedding), and a simplified proof of Benoist's nonexistence result for $\mathrm{SL}_{n-1}(\mathbb{R})\backslash \mathrm{SL}_n(\mathbb{R})$ for n odd. Kassel also extends this work to Lie groups over local fields. We will return to some of Kassel's further work along this line in Section 6.2.

3.3 Results in small dimensions

There are scattered results on the existence question when we restrict to small dimension. I would like to mention here the case $\mathrm{SL}_2(\mathbb{K})\backslash \mathrm{SL}_3(\mathbb{K})$ where $\mathbb{K}$ is any of the real numbers, the complex numbers or the quaternions. Benoist's work covers these cases; earlier, Kobayashi dealt with the latter two cases (see [25]). Oh and Witte-Morris generalized Benoist's work to homogeneous spaces of $\mathrm{SL}_3(\mathbb{R})$ with H nonreductive in [41], proving that only the obvious examples have compact forms:

---

[5] Benoist's construction works for all reductive $H\backslash G$ admitting proper actions by non-virtually abelian Γ; he characterizes these in [4].

Theorem 13 (Oh–Witte-Morris, [41] Prop 1.10) *Let H be a closed, connected subgroup of* $SL_3(\mathbb{K})$. *If* $H \backslash SL_3(\mathbb{K})$ *has a compact form, either H or* $H \backslash SL_3(\mathbb{K})$ *is compact.*

There is a nice treatment of this problem and a simplification of Oh and Witte-Morris's proof in Section 6 of [17].

3.4 A dimension criterion

In the approaches to the problem just detailed, only the properness of the action of Γ is considered. In [24], Kobayashi uses the ideas of Definition 10 to examine conditions under which a compact form does exist. He produces the following theorem.

Theorem 14 ([24], Thm 4.7) *Let G be a reductive linear group; H and L subgroups having finitely many components and stable under some Cartan involution* θ *of G. If*

- $\mathfrak{a}(H) \cap \mathfrak{a}(L) = \{0\}$ *and*
- $d(H) + d(L) = d(G)$,

then both $H \backslash G$ *and* G/L *admit compact forms.*

The first condition of this theorem arises from the properness issues discussed above. The second is a consideration for compactness. Its necessity follows from the following Proposition.

Proposition 15 *Let G be reductive and linear with connected closed subgroups* $H \pitchfork L$. *Then* $d(H) + d(L) \le d(G)$ *with equality if and only if* $H \backslash G/L$ *is compact.*

Table 2 lists some homogeneous spaces that *do* admit compact forms under the criterion of Theorem 14 together with the relevant subgroup L. We encounter here the compact forms that are evidence for Conjecture 3. As we noted in Remark 4, up to switching H and L, these are the only homogeneous spaces of reductive type with G simple and linear admitting standard forms (up to conjugation of H in G and perhaps taking connected components).

The standard compact forms with $G = SO(2, 2n)$ and $G = SO(4, 4n)$ were first found by Kulkarni in [32]; the rest of the examples in Table 2 are due to Kobayashi in [24], [25] and [28].

Table 2 *Some homogeneous spaces $H\backslash G$ with compact Clifford–Klein forms*

G	H	L
SU(2, 2n)	Sp(1, n)	U(1, 2n)
SO(2, 2n)	U(1, n)	SO(1, 2n)
SO(4, 4n)	SO(3, 4n)	Sp(1, n)
SO(4, 4)	SO(4, 1) × SO(3)	Spin(4, 3)
SO(4, 3)	SO(3, 1) × SO(2)	$G_{2(2)}$
SO(8, 8)	SO(7, 8)	Spin(1, 8)
SO(8, ℂ)	SO(7, ℂ)	Spin(1, 7)
SO(8, ℂ)	SO(7, 1)	Spin(7, ℂ)
SO(8, ℂ)	SO(7, ℂ)	Spin(1, 7)
SO*(8)	U(3, 1)	Spin(1, 6)
SO*(8)	SO*(6) × SO*(2)	Spin(1, 6)

3.5 Virtual cohomological dimension

The central thrust of the Calabi–Markus phenomenon and the related results above is that there must be some coherence among the sizes (measured by real rank or by dimension of the corresponding Riemannian symmetric space) of G, H, and L. This idea was used further by Kulkarni in examining the spaces $O(p, q)\backslash O(p + 1, q)$. When p or q are 1, we recover the relativistic space forms Calabi and Markus studied in [8]. Kulkarni proves a number of results in his article. Toward our purposes here, he uses cohomological dimension of groups and a generalized Gauss-Bonnet formula to prove the following:

Theorem 16 (Kulkarni, [32] Cor 2.10) *If p and q are odd, then there is no compact Clifford–Klein form of $O(p, q)\backslash O(p + 1, q)$.*

This approach has found its most complete statement in the work of Kobayashi:

Theorem 17 (Kobayashi [26], Thm 1.5) *Let $H\backslash G$ be of reductive type. If there exists a reductive $H' < G$ such that $H' \sim H$ and $d(H') > d(H)$, then $H\backslash G$ has no compact form.*

From the definition of $\sim$ (10) it is easy to check that any discrete Γ acting properly on $H\backslash G$ will also act properly on $H'\backslash G$. Thus, we can construct a restriction on the sizes of G, H and Γ for proper actions in the vein of Theorem 14 but with the discrete group Γ replacing L. This is supplied by the following Lemma, in which virtual cohomological dimension (vcd) replaces the

Table 3 *Some homogeneous spaces $H\backslash G$ without compact Clifford–Klein forms. Throughout, $n = p + q$ and $p, q > 1$.*

G	H	$\mathfrak{h}'$
$\mathrm{SL}_{2n}(\mathbb{R})$	$\mathrm{SO}(n, n)$	$\mathfrak{sp}(n, \mathbb{R})$
$\mathrm{SU}^*(2n)$	$\mathrm{SO}^*(2n)$	$\mathfrak{sp}(\lfloor\frac{n}{2}\rfloor, n - \lfloor\frac{n}{2}\rfloor)$
$\mathrm{SU}(2n, 2n)$	$\mathrm{SO}^*(4n)$	$\mathfrak{sp}(n, n)$
$\mathrm{Sp}(2n, \mathbb{R})$	$\mathrm{U}(n, n)$	$\mathfrak{sp}(n, \mathbb{C})$
$\mathrm{SO}(2n, 2n)$	$\mathrm{SO}(2n, \mathbb{C})$	$\mathfrak{u}(n, n)$
$\mathrm{SO}^*(2n)$	$\mathrm{SO}^*(2p) \times \mathrm{SO}^*(2q)$	$\mathfrak{so}(2) + \mathfrak{so}^*(2n - 2)$
$\mathrm{SL}_n(\mathbb{C})$	$\mathrm{SO}(n, \mathbb{C})$	$\mathfrak{u}(\lfloor\frac{n}{2}\rfloor, n - \lfloor\frac{n}{2}\rfloor)$
$\mathrm{SO}(n, \mathbb{C})$	$\mathrm{SO}(p, \mathbb{C}) \times \mathrm{SO}(q, \mathbb{C})$	$\mathfrak{so}(n - 1, \mathbb{C})$
$\mathrm{SO}^*(2n)$	$\mathrm{U}(p, q)$	$\mathfrak{so}^*(2r)$
	$n \geq 3, \quad 3p \leq 2n \leq 6p$	$r = \min(n, 2p + 1, 2q + 1)$
$\mathrm{SL}_{2n}(\mathbb{C})$	$\mathrm{SU}(n, n)$	$\mathfrak{sp}(n, \mathbb{C})$
$\mathrm{Sp}(n, \mathbb{R})$	$\mathrm{Sp}(p, \mathbb{R}) \times \mathrm{Sp}(q, \mathbb{R})$	$\mathfrak{sp}(n, \mathbb{R})$

dimension of the corresponding symmetric space. The proof of Theorem 17 from this lemma is clear. Recall that the cohomological dimension of a group Γ is the maximal n for which the group cohomology $H^n(\Gamma, M)$, where M is an arbitrary $\mathbb{R}\Gamma$-module, does not vanish. The *virtual* cohomological dimension is this value for any finite-index subgroup of Γ (see [7]).

Lemma 18 (Serre [46] when $H\backslash G$ is simply connected, Kobayashi [26] in general) *If discrete $\Gamma < G$ satisfies $\Gamma \pitchfork H$, then*

- $vcd(\Gamma) + d(H) \leq d(G)$ and
- $vcd(\Gamma) + d(H) = d(G)$ if and only if $H\backslash G / \Gamma$ is compact.

This approach supplies numerous other examples of homogeneous spaces without compact forms. The examples in Table 3 are taken from [26], Table 4.4; see that table for a number of examples involving exceptional Lie groups.

3.6 Clifford algebras

As a final entry in this catalogue of 'topological' approaches, let me record the following theorem of Kobayashi and Yoshino:

Theorem 19 (Kobayashi–Yoshino, [31] Thm 4.2.1) *The following triples of Lie groups satisfy the conditions of Thm 14, and hence $H\backslash G$ and G/L have compact forms:*

G	H	L
$GL_2(\mathbb{H})$	$GL_1(\mathbb{H})$	$Spin(1, 5)$
$O^*(8)$	$O^*(6)$	$Spin(1, 6)$
$O(8, \mathbb{C})$	$O(7, \mathbb{C})$	$Spin(1, 7)$
$O(8, 8)$	$O(7, 8)$	$Spin(1, 8)$

Some of these have appeared already in Table 2, and several can be obtained by other approaches. What is particularly interesting in this theorem is the unified treatment of the $Spin(1, q)$ cases and the introduction of a new technique, namely, the use of Clifford algebras to calculate Cartan projections and verify the first condition of Theorem 14. I will not attempt to survey this approach here, but instead refer the reader to [31].

4 Geometry

4.1 A contribution from symplectic geometry

In [5], Benoist and Labourie provide a useful restriction on H via symplectic geometry. The restriction is algebraic – the center of H cannot be 'large' in that it cannot contain any hyperbolic elements.

A simplified version of Benoist and Labourie's proof is available in the following special case. Suppose T is a one-parameter subgroup of semisimple G generated by a hyperbolic element t in $\mathfrak{g}$; let $Z_G(T)$ be the centralizer of T. We will show that there is no compact manifold with a $(G, G/Z_G(T))$-structure. first, note that $Z_G(T)$ is isomorphic to a direct product $M \times T$ for some subgroup M of G. Then one has a fiber bundle $G/M \to G/Z_G(T)$ with $\mathbb{R}$-fibers corresponding to the fibration by T-orbits. The Killing form $\langle \, , \, \rangle$ gives a 1-form on G/M, namely $Y \mapsto \langle t, Y \rangle$ for any $Y \in \mathfrak{g}$, which descends to $G/Z_G(T)$. One can show that this 1-form is a connection whose curvature is a symplectic form, specifically $\omega_t(Y, Z) = \langle t, [Y, Z] \rangle$, for $Y, Z \in \mathfrak{g}$. As ω_t is symplectic, $\omega_t^{\wedge d}$ is the volume form on $G/Z_G(T)$. However, after perhaps passing to a finite cover, there is a non-vanishing section of the line bundle over $G/Z_G(T)$, and ω_t will be exact – the differential of the connection 1-form associated to that section. This in turn implies that the volume form obtained from ω_t is exact, which is impossible for a compact manifold and proves there is no compact form. Note that this entire argument deals with local, G-invariant objects, so it can be transferred over to a manifold with a $(G, G/Z_G(T))$-structure.

Benoist and Labourie adapt this argument for the more general situation $H \subseteq Z_G(T)$, providing restrictions on the center of H. Their argument involves using fibrations by weight spaces for elements in H to build a connection with corresponding curvature tensor which is symplectic and cohomologically trivial. They obtain the following results. Note that as their approach deals only with local, G-invariant objects, Theorem 20 and some of its corollaries apply to compact manifolds locally modeled on $H \backslash G$, not just to Clifford–Klein forms.

Theorem 20 (Benoist–Labourie, [5] Thm 1) *Let G be a connected, semisimple real Lie group, H a connected unimodular subgroup. If there exists a compact manifold with a $(G, H \backslash G)$-structure, then the Lie algebra of the center of H does not contain any hyperbolic elements.*

Corollary 21 (Benoist–Labourie, [5] Cor 1) *Let G be algebraic, semisimple and real. If H is algebraic and reductive and there exists a compact manifold with a $(G, H \backslash G)$-structure, then the center of H is compact.*

Corollary 22 (Benoist–Labourie, [5] Cor 2) *If G is connected, semisimple, with finite center and H is connected and reductive and $H \backslash G$ admits a compact Clifford–Klein form, then the center of H is compact.*

4.2 Two approaches via representation theory

In [38], Margulis presents an approach to the existence question built on representation theory of the groups involved. Specifically, he gives a criterion for nonexistence based on matrix coefficients for unitary representations of H on certain L^2 functions on H.

Let G be unimodular, locally compact, K a compact subgroup and H a closed subgroup. Matrix coefficients enter via his definition of (G, K, H)-tempered actions on some space X; for our purposes the following is the relevant application of this definition.

Definition 23 ((G, K)-tempered subgroups, [38] Defn 2) We call H a (G, K)-tempered subgroup of G if there exists a function $q \in L^1(H)$ such that

$$|\langle \psi(h)w_1, w_2 \rangle| \le q(h) \|w_1\| \|w_2\|$$

for all $h \in H$, all $\psi(K)$-invariant functions $w_1, w_2 \in L^2(H)$ and all unitary representations ψ of G on $L^2(H)$.[6]

---

[6] For $L^1(H)$ and $L^2(H)$ the implied measure is the Haar measure on H.

Margulis's result is the following:

Theorem 24 (Margulis, [38] Thm 1) *Let G, H and K be as above, and let F be any noncompact closed subgroup of H. Suppose that H is (G, K)-tempered. Then there is no compact form of $F \backslash G$. (In particular, there is no compact form of $H \backslash G$.)*

The main idea of Margulis's approach is the following. Suppose there is a compact form of $H \backslash G$. Then one may take a compact set $M \subset G/\Gamma$ such that $HM = G/\Gamma$. Margulis constructs a pair of L^2 functions, one supported around M and one supported on some compact set far from M in the H-direction. Applying the (G, K)-tempered condition with these functions, he shows that the measure of $G/\Gamma - HM$ is positive; in particular that HM cannot be all of G/Γ.

The (G, K)-tempered criterion is not always easy to check, but Margulis provides several nice examples. Further examples and more detailed exposition can be found in [39]. First, if G is connected semisimple with Kazhdan's Property (T) (see, e.g. [52] Chapt. 7), K maximal compact and H abelian and diagonalizable, standard arguments showing exponential decay of matrix coefficients can be applied to show H is (G, K)-tempered. Perhaps of more interest to us here, however, are two results involving the case $G = \mathrm{SL}_n(\mathbb{R})$, $K = \mathrm{SO}(n)$. By direct computations and some knowledge of representation theory, one can show that $H = \alpha_n(\mathrm{SL}_2(\mathbb{R}))$ is (G, K)-tempered, where α_n is the irreducible n-dimensional representation of $\mathrm{SL}_2(\mathbb{R})$ and $n \geq 4$. Likewise, for $n \geq 3$, let $\phi : L \to \mathrm{SL}_n(\mathbb{R})$ be a representation of a connected simple Lie group, with irreducible components ϕ_i. If the sum of the highest weights for the nontrivial ϕ_i is larger than the sum of the positive roots for L, then $\phi(L)$ is (G, K)-tempered. Thus, if there are many nontrivial ϕ_i, or high-dimensional ϕ_i in comparison to the 'size' of L (as measured by the sum of its positive roots) then $\phi(L) \backslash \mathrm{SL}_n(\mathbb{R})$ has no compact form.

It is perhaps instructive to put these results alongside the results for the examples $\mathrm{SL}_{n-k}(\mathbb{R}) \backslash \mathrm{SL}_n(\mathbb{R})$ discussed in connection with Kobayashi and Benoist's results (in particular Thm 12). For Margulis's approach, we need $\mathrm{SL}_{n-k}(\mathbb{R})$ to be quite small, but to be embedded in $\mathrm{SL}_n(\mathbb{R})$ in a 'large' way – irreducibly, perhaps. Benoist's result holds when $\mathrm{SL}_{n-k}(\mathbb{R})$ is large ($k = 1$) but is not embedded irreducibly in $\mathrm{SL}_n(\mathbb{R})$.

To close this section on 'geometric' approaches to the problem, let us note a second approach that uses representation theory. This approach is due to Shalom [47], and again relies on decay of matrix coefficients, although in a different way than Margulis. Shalom studies unitary representations of $\mathrm{SO}(1, n)$

and SU(1, n) and their cohomology. He is able to reproduce some of the rigidity results associated to groups of higher rank in these cases. A full description of his results is far beyond our present scope, but presenting their application to the compact forms question will serve as a bridge between the representation-theoretic approach and the dynamical approach to be described in the next section.

Shalom proves the following theorem.

Theorem 25 (Shalom, [47] Thm 1.7) *Let G be a simple Lie group with finite center, Λ a discrete, infinite subgroup of G admitting a discrete embedding into* SO(1, n) *or* SU(1, n). *Assume there exists a nonamenable, closed subgroup L in G with noncompact center which commutes with Λ. Then Λ\G admits no compact quotient.*

Taking H a closed Lie subgroup of G containing Λ yields results on compact forms of homogeneous spaces. One particular example is the following:

Corollary 26 *Let* $G = SL_n(\mathbb{R})$, $n \geq 4$, $H = SL_2(\mathbb{R})$ *embedded in the upper left-hand corner of G. Then H\G has no compact quotient. The same holds with* $\mathbb{C}$ *replacing* $\mathbb{R}$.

By way of explaining how this theorem serves as a bridge between our exposition of the approach via representation theory and the approach via dynamics, let us give the briefest of overviews of the proof. Following an idea developed by Zimmer, and which will be used repeatedly in the next section, Shalom studies the L-action by left-multiplication on the Γ bundle $G/\Gamma \to \Lambda\backslash G/\Gamma$ over a compact form. Shalom then uses the representation theory he has developed to study cocycles associated to this action to complete the proof. A similar idea will appear again in the next section, where dynamics will fully take over with the appearance of ergodic theory.

5 Dynamics

5.1 The work of Zimmer, Labourie and Mozes

The approach to the compact forms problem via dynamics was pioneered by Zimmer in [53]. In this paper he deals with cases where $Z_G(H)$ is a large group; in particular, he works under the following assumption:

- There is a semisimple group of higher rank $J < Z_G(H)$.

An illustrative example of this case is given by taking $G = \mathrm{SL}_n(\mathbb{R})$ and $H = \mathrm{SL}_{n-k}(\mathbb{R})$ with $k \geq 3$, embedded in the upper left-hand corner of G. Here $J = \mathrm{SL}_k(\mathbb{R})$ and the assumption above requires only that $k \geq 3$.

The key element to Zimmer's approach is the application of his cocycle superrigidity theorem. Specifically, J acts by left-multiplication on the H-bundle $G/\Gamma \to H\backslash G/\Gamma$. If $\sigma : H\backslash G/\Gamma \to G/\Gamma$ is a measurable section of this bundle, $j \in J$ and $x \in H\backslash G/\Gamma$, one calculates that

$$\sigma(j \cdot x) = \alpha(j, x)j \cdot \sigma(x) \tag{1}$$

where $\alpha : J \times H\backslash G/\Gamma \to H$ is a measurable map. It is easy to verify that α satisfies the (dynamical) cocycle equation $\alpha(j_1 j_2, x) = \alpha(j_1, j_2 \cdot x)\alpha(j_2, x)$. Zimmer's cocycle rigidity theorem (see [52]) states that if J is semisimple of higher rank, H is an algebraic group, the algebraic hull of α in G is connected semisimple, and the J-action on $H\backslash G/\Gamma$ is irreducible, then the cocycle α is equivalent to a trivial cocycle. That is, up to a new choice of the measurable section, α is independent of the space variable x, in which case the cocycle equation reduces to the assertion that α is a homomorphism from J to H.

Two remarks are in order here. First, the *algebraic hull* is the unique (up to conjugacy) minimal algebraic group in which a cocycle equivalent to α takes values. Questions about the algebraic hull of α are dealt with in the final section of [34]. The assumptions on the hull detailed above are ensured, perhaps after moving to a finite ergodic cover of $H\backslash G/\Gamma$, provided that the algebraic hull is not compact. In the compact case, simpler arguments are deployed in [34]. Second, the action of J on $H\backslash G/\Gamma$ is *irreducible* for a J-ergodic probability measure μ if the action by any noncentral normal subgroup of J is also ergodic. This is a significant issue for the compact forms question; it has thus far been addressed by assuming that all simple factors of J are higher rank on their own so that superrigidity applies to each individually. See Question 31 below for a comment on a situation in which the irreducibility issue obstructs the solution of a natural compact forms question by these methods.

Zimmer first approaches the compact forms question in [53] and continues with Labourie and Mozes in [34]. Their most general result is the following:

Theorem 27 (Labourie–Mozes–Zimmer, [34] Thm 1.1) *Let H and G be real algebraic and unimodular, and suppose that there is a semisimple $J < Z_G(H)$ all of whose simple factors are higher rank. Suppose that J is not contained in a proper, normal subgroup of G and that*

(i) *the image of every nontrivial homomorphism $\tilde{J} \to H$ has compact centralizer in H, where $\tilde{J}$ is the universal cover of J;*

(ii) there is a nontrivial, $\mathbb{R}$-split, 1-parameter subgroup B in $Z_G(HJ)$ that is not contained in a normal subgroup of G.

If there is a compact form $H\backslash G/\Gamma$, then H is compact.

In [34], Labourie, Mozes and Zimmer actually address a more general situation than Clifford–Klein forms. Theorem 27 applies to any compact manifold locally modeled on $H\backslash G$, the variant of Question 1 mentioned in the introduction.

The proof of Theorem 27 falls into two cases, as dictated by condition *(i)*: either ρ is trivial (i.e. the cocycle for the J-action takes values in a compact subgroup of H), or its image has compact centralizer. In the second case a short argument allows the authors to consider the cocycle for the B-action (utilizing *(ii)*) and conclude that *it* takes values in a compact group (now $Z_H(\rho(J))$). In either case one lifts the volume measure m from $H\backslash G/\Gamma$ to G/Γ using the section σ and averages it over a compact set in the fiber direction to obtain a finite, J-invariant measure which must by construction be the restriction of Haar measure on G/Γ to a positive measure subset. An application of Moore's ergodicity theorem (see, e.g. [52]) shows that this can only occur if H is compact.

The conditions *(i)* and *(ii)* of Theorem 27 are specific to details of the proof but not to the overall philosophy of applying superrigidity to the compact forms question. Note that in the test case of $SL_{n-k}(\mathbb{R})\backslash SL_n(\mathbb{R})$, condition *(i)* restricts application of the theorem to $k \geq n/2$. In [37], Labourie and Zimmer provide a separate argument to prove nonexistence for the case $k \geq 3$:

Theorem 28 (Labourie–Zimmer, [37]) *There is no compact form of $SL_{n-k}(\mathbb{R})\backslash SL_n(\mathbb{R})$ for $k \geq 3$.*

Sketch of proof We prove by contradiction; $G = SL_n(\mathbb{R})$ and $H = SL_{n-k}(\mathbb{R})$. Cocycle superrigidity provides a trivial cocycle for the $J = SL_3(\mathbb{R})$-action – i.e. $\alpha = \rho$, a homomorphism from J to G, up to an error in a compact subgroup K of H. Lift the ergodic measure μ to a finite measure $\sigma_*(\mu)$ on G/Γ and average this over K to provide a finite measure $\hat{\mu}$ which covers μ. Examining equation (1) one sees that $\hat{\mu}$ is invariant under $gr(\rho)(J)$, the graph of ρ applied to J. Likewise, $h_*\hat{\mu}$ is invariant under $h\,gr(\rho)(J)h^{-1}$, for $h \in H$. If ρ is irreducible as a representation of $SL_3(\mathbb{R})$ into $SL_{n-k}(\mathbb{R})$, Labourie and Zimmer fall back on the arguments of [34]; if it is reducible they show that a properly chosen H-conjugate of $gr(\rho)$ intersects H in a noncompact subgroup. This noncompact group acting in the fiber direction of the bundle $G/\Gamma \to H\backslash G/\Gamma$ cannot preserve the finite measure $h_*\hat{\mu}$, providing a contradiction. $\qquad\square$

The final step of this argument involves algebraic structure that is fairly specific to the case of SL's – namely that there is an element of the Weyl group for $SL_n(\mathbb{R})$ that exchanges any two diagonal entries while leaving the rest fixed. To remove the algebraic conditions of Theorem 27 it is necessary to adapt a more general approach as below. It also proceeds by (eventually) reducing the question to a question about subgroups of G.

5.2 A recent improvement

The following theorem presents an improvement to Theorem 27 in that it removes the algebraic conditions (i) and (ii) and the requirement that J is not contained in a proper, normal subgroup of G. To achieve this, we must sacrifice slightly by requiring G simple, but this is not a terrible loss. Indeed, all examples Labourie, Mozes and Zimmer provide are for G simple.

Theorem 29 (Constantine, [10] Main Theorem) *Let G be a connected, simple Lie group with finite center and H a connected, noncompact, reductive Lie subgroup. Suppose that $J < Z_G(H)$ is a simple Lie group with real rank at least two. Then there is no compact form of $H \backslash G$.*

Sketch of proof Proceed as above, applying cocycle superrigidity and producing the $gr(\rho)$-invariant measure $\hat{\mu}$ on G/Γ. An additional result of cocycle superrigidity is that ρ is a rational map, and hence it takes unipotent subgroups to unipotent subgroups. We may assume that J is generated by unipotents, then take an ergodic component of $\hat{\mu}$ for the $gr(\rho)$-action and apply Ratner's measure classification for unipotent flows to this measure ([43]). Ratner provides that this measure is the image of the Haar measure for some subgroup $L \supset gr(\rho)$ of G along an L-orbit in G/Γ.

It is clear from its construction that the measure described by the L-orbit covers μ and its support extends only compactly in the H-fiber direction. The latter fact implies that $L \cap H$ is compact. To utilize the former, one studies the dynamics of the J-action on $H \backslash G / \Gamma$ and the measure μ. The application of superrigidity allows one to calculate Lyapunov exponents for this action, recording exponential expansion and contraction of orbits under various flows in the J-action. Standard tools from the theory of hyperbolic dynamics imply that (roughly speaking) the support of μ extends in directions with nonzero exponents. Separate arguments using dynamics of unipotent elements in J and the pseudo-Riemannian structure of $H \backslash G / \Gamma$ show the support also extends in directions with zero exponents. The result of all this is that since the L-orbit covers a measure whose support extends in all directions in $H \backslash G / \Gamma$, one also has that $HL = G$. The proof is completed by showing that under the conditions

Table 4 *Some homogeneous spaces $H \backslash G$ without compact Clifford–Klein forms*

G	H	conditions
$SL_n(\mathbb{R})$	$SL_{n-k}(\mathbb{R})$	$k \geq 3$
$SL_n(\mathbb{C})$	$SL_{n-k}(\mathbb{C})$	$k \geq 3$
$SO(n, m)$	$SO(n - k, m - l)$	$k \geq 2, l \geq 3$
$PSO(2n, \mathbb{C})$	$PSO(2(n - k), \mathbb{C})$	$k \geq 2$
$SO(2n + 1, \mathbb{C})$	$PSO(2n - k), \mathbb{C})$	$k \geq 2$
$SO(2n + 1, \mathbb{C})$	$SO(2(n - k) + 1, \mathbb{C})$	$k \geq 2$
$SU(p, q)$	$SU(p - k, q - l)$	$k, l \geq 2$
$Sp(2m, \mathbb{R})$	$Sp(2(m - k), \mathbb{R})$	$k \geq 2$
G listed above	H'	H' a noncompact reductive subgroup of the corresponding H listed above

of the theorem, no subgroups L satisfying $L \cap H$ compact and $HL = G$ exist (see Remark 4). □

One interesting feature of this method of proof is that it proceeds through verifying Conjecture 3 for the homogeneous spaces under consideration; this holds with slightly loosened restrictions on G:

Theorem 30 (Constantine, [10] Characterization Theorem) *Let G and H be as above, but with G allowed to be semisimple rather than simple. Assume that there is a semisimple Lie group $J < Z_G(H)$ such that:*

(1) All simple factors of J have real-rank at least two
(2) The vector space sum of $\mathfrak{h}$ and the Lie algebra generated by all nonzero weight spaces for a Cartan subgroup $A < J$ is $\mathfrak{g}$.

Then any compact form of $H \backslash G$ is standard.

Table 4 lists a few examples of homogeneous spaces which these theorems imply do not have compact forms. In their full generality, most arise from Theorem 29, with the exception of $SL_{n-k}(\mathbb{R}) \backslash SL_n(\mathbb{R})$, which is due to Labourie and Zimmer as noted above. Many are also proven by Kobayashi as noted in [28], but with stronger restrictions on k and l.

To close this section, the author would like to pose the following question:

Question 31 Does $SO(n - 2, m - 2) \backslash SO(n, m)$ have a compact Clifford–Klein form?

The reader will notice that this case is excluded from Table 4. This is because the semisimple part of $Z_G(H)$ is $SO(2, 2)$ which is semisimple and of higher rank, but not simple ($SO(2, 2)_o \cong (SL_2(\mathbb{R}) \times SL_2(\mathbb{R}))/\{\pm(1, 1)\}$). To prove nonexistence of a compact form, one needs to address the irreducibility condition for superrigidity – if this is accomplished, the argument which proves Theorem 29 will apply.

6 Deformations and moduli spaces of compact forms

We close this survey by taking up some results, many of them quite recent, on the second question of the introduction. Namely, when a homogeneous space has a compact form, what can we say about the space of all possible compact forms? We will begin by collecting the evidence for Kobayashi's Conjecture 3.

6.1 Evidence for the 'standard forms' conjecture

The reader will note that the only positive results on the existence question which we have seen so far are those provided by Table 2 and due to the algebraic construction of Theorem 14. That is, they are all standard forms. This fact is the main empirical evidence for Conjecture 3.

We can note a small amount of further evidence. It is (trivially) true in the Riemannian case. It is true for homogeneous spaces of $SO(2, n)$ by work of Oh, Witte-Morris and Iozzi which will be reported on below. Theorem 30 states that the conjecture is true in the stronger sense that all compact forms are standard when there is a higher-rank action present (and the rest of the requirements of that theorem are fulfilled). Note, however, that the (now purely algebraic question) of whether any such forms exist is still open[7], and their existence may be unlikely, given the situation when G is simple.

This evidence is slight, of course; the main argument for the conjecture is the empirical one. It seems to the author that this will be an extremely difficult conjecture to prove in general, in large part because, as we have seen, there is no over-arching approach to the problem.

6.2 Moduli spaces of compact forms

The reader will have noticed that Kobayashi's conjecture is not that every compact form is standard. This simpler situation is ruled out by the following result, arrived at by Goldman, Ghys and Kobayashi.

---

[7] Remark 4 does not apply when G is only semisimple.

Theorem 32 ([14], [13], [29]) *There are nonstandard compact forms.*

There are a number of proofs to this theorem stated as such. Let us begin with Goldman's and Kobayashi's original approaches and then proceed to look at some very recent results in this direction. Ghys's work deals with the cases of $G = SL_2(\mathbb{R})$ and $SL_2(\mathbb{C})$. He constructs some interesting examples.

The work of Kobayashi and Goldman

Recall that we have previously defined AdS^3 as $\mathbb{H}^{2,1} = SO(2, 2)/SO(1, 2)$. It can also be defined as $(PSL_2(\mathbb{R}) \times PSL_2(\mathbb{R}))/diag(PSL_2(\mathbb{R}))$ up to a covering of order 2, or as $PSL_2(\mathbb{R})$ endowed with the metric given by its Killing form, at least up to a finite cover. The second description is more common among the authors whose work is discussed in this subsection and fits better with their generalizations of work on AdS^3, so I will adopt it now. The description as $PSL_2(\mathbb{R})$ is favored by Salein, Kassel and Guéritaud below.

In [14], Goldman shows that not all compact forms of AdS^3 are standard, an issue raised by the work of Kulkarni and Raymond in [33]. In this case, the question of whether all forms are standard becomes whether all forms have Γ conjugate into $SL_2(\mathbb{R}) \times SO(2)$ or $SO(2) \times SL_2(\mathbb{R})$. He proves

Theorem 33 (Goldman, [14] Thm 1) *Let M be a standard compact form of AdS^3 with $H^1(M; \mathbb{R}) \neq 0$. Then there is a nontrivial deformation space of nonstandard compact form structures on M.*

Goldman recasts compact forms of $SL_2(\mathbb{R})$ in the language of geometric structures. Let G be a Lie group and X a homogeneous space of G. As noted earlier, a (G, X)-structure on a manifold M consists of a holonomy homomorphism $hol : \pi_1(M) \to G$ and a developing map $dev : \tilde{M} \to X$ which is a local diffeomorphism and which is equivariant with respect to hol. The pair (hol, dev) is well-defined up to the natural G-action: namely, $g \cdot (hol, dev) = (ghol(\cdot)g^{-1}, g \cdot dev)$. The developing map provides a well-defined X-coordinate patch structure on M. If G preserves a pseudo-Riemannian metric on X and we further require that the developing map be a local isomorphism; this enforces a unique pseudo-Riemannian metric on M. The (G, X)-structure is *complete* if the developing map is a covering map, in which case the (G, X)-structure gives M the structure $\tilde{X}/\Gamma$ where $\Gamma = \pi_1(M)$ and $\tilde{X}$ is the universal cover of X.[8]

In this setting, let us take $X = SL_2(\mathbb{R})$ and $G = SL_2(\mathbb{R}) \times SL_2(\mathbb{R})$, acting by left- and right-multiplication, which preserves the metric given by the Killing

---

[8] This completeness is equivalent to a notion of geodesic completeness; see [14].

form. To find nonstandard forms, we search for holonomy representations that take unbounded image in both factors of G. It is a general fact about (G, X)-structures – the Ehresmann–Thurston principle, see [48] and [11] – that given any holonomy representation $h_0 \in \text{Hom}(\Gamma, G)$ for Γ the fundamental group of a fixed compact manifold, there is an open set U containing h_0 in the variety $\text{Hom}(\Gamma, G)$ consisting of holonomy representations. Let us take h_0 a holonomy representation of the form $(1, \pi)$ for some $\pi : \Gamma \to G$. The (G, X)-structure defined by h_0 is complete and identifies M with a quotient of X. What remains for us to show is that we can find a nearby representation h which still provides a *free and properly discontinuous* action of Γ on X.

Goldman proves that h of the following form works. Let B be any hyperbolic or parabolic one-parameter subgroup of $\text{SL}_2(\mathbb{R})$ and let v be a nonconstant representation of Γ into B which is sufficiently close to the constant representation (which exists because of the assumption $H^1(M; \mathbb{R}) \neq 0$). Then $h(\gamma) = (v(\gamma), h_0(\gamma))$ gives a free and properly discontinuous action and hence a (G, X)-structure on a compact manifold M which is nonstandard.

In [29], Kobayashi extends Goldman's work significantly. We record a definition first:

Definition 34 A deformation $\phi_t(\Gamma)$ of $\Gamma < G$ is called trivial if each $\phi_t(\Gamma)$ is conjugate to Γ in G. If all sufficiently small deformations of Γ are trivial, Γ is called locally rigid in G.

Theorem 35 (Kobayashi, [29] Thm A) *Let $H \backslash G = diag(G') \backslash (G' \times G')$ where G' is a simple linear Lie group. Then*

(1) For any uniform lattice $\Gamma' < G'$, the quotient $H \backslash G / \Gamma$ remains a compact form for Γ any sufficiently small deformation of $\Gamma' \times \{1\}$ in G. That is to say, any sufficiently small deformation of $\Gamma' \times \{1\}$ still acts properly discontinuously and cocompactly on $H \backslash G$.

(2) It is possible to find uniform lattices with nontrivial deformations of this type in G if and only if G' is locally isomorphic to $\text{SO}(1, n)$ or $\text{SU}(1, n)$.

This generalizes Goldman's work by taking $G' = \text{PSL}_2(\mathbb{R}) \cong \text{SO}(1, 2)_o$ and Ghys's by taking $G' = \text{PSL}_2(\mathbb{C}) \cong \text{SO}(1, 3)_o$. To prove (2) $\implies$ (1), Kobayashi proves that for G' locally isomorphic to $\text{SO}(1, n)$ or $\text{SU}(1, n)$, there are uniform lattices in $G' \times 1$ with nontrivial deformations, and any sufficiently small deformation preserves the properly discontinuous character of the action on $diag(G') \backslash (G' \times G')$. That (1) $\implies$ (2) follows from Weil's local rigidity theorem [49] and some vanishing theorems for Betti numbers (see citations in [29], p. 406).

Kobayashi also observes the following:

Proposition 36 (Kobayashi, [29] Thm B and Section 1.8) *The following homogeneous spaces admit compact forms that have nontrivial deformations* $(n \geq 1)$:

$$SO(1, 2n)\backslash SO(2, 2n),\ Sp(1, n)\backslash SU(2, 2n),$$
$$G_2(\mathbb{R})\backslash SO(3, 4),\ Spin(3, 4)\backslash SO(4, 4).$$

There are locally rigid standard forms for the following homogeneous spaces $(n \geq 1, m \geq 2)$:

$$U(1, 2n)\backslash SU(2, 2n),\ SO(3, 4m)\backslash SO(4, 4m),\ Sp(1, n)\backslash SO(4, 4n).$$

Note that all these examples are taken from Table 2, which lists homogeneous spaces with compact forms constructed from a triple of Lie groups G, H, L (H and L play symmetric roles in this construction). The key to Theorem 35 and to the first set of examples in Proposition 36 is the existence of a nontrivial centralizer of L in G. When $Z_G(L)$ is nontrivial, an embedding of some lattice Γ in L into G can be deformed by finding a nonconstant homomorphism $\rho : \Gamma \to Z_G(L)$ and mapping $\gamma \mapsto \gamma\rho(\gamma)$. Denote these deformed embeddings of Γ by Γ_ρ; Kobayashi shows that for ρ in a sufficiently small neighborhood of the constant homomorphism, Γ_ρ still acts properly discontinuously, thus yielding a compact form. In Proposition 36 these centralizers are compact and hence the deformations do not affect proper discontinuity. In Theorem 35 on the other hand, the centralizer is noncompact, making the proof that proper discontinuity of the action survives much more difficult. For the second set of examples in Prop. 36 Kobayashi observes that any uniform lattice in the L corresponding to the given pair is locally rigid in G. One feature of this circle of results is that the deformation of Γ does not need to be by the specific one-parameter subgroups Goldman uses.

In addition to the papers of Goldman and Kobayashi, we note that similar deformations have also been given by Ghys [12, 13] and Salein [44]. We will have more to say about Salein's extension of this work below.

The work of Salein, Kassel and Guéritaud

Recently Salein, Kassel and Guéritaud have continued the work done on moduli spaces of compact forms by providing new sufficient conditions for understanding when deformed embeddings of Γ in G still give rise to compact forms. In this work, the new compact forms presented are no longer small, continuous deformations of standard forms. Rather, they are far from the standard examples – even topologically different. Salein's demonstration of this fact was surprising and the work of these authors has greatly increased our understanding of moduli spaces of compact forms.

Salein's work on the problem can be found in [45]. He again studies AdS^3 which is a model space for all Lorentzian 3-manifolds of constant curvature -1 and is identified with $PSL_2(\mathbb{R})$ with the Lorentzian metric given by its Killing form. Recall that $G := PSL_2(\mathbb{R}) \times PSL_2(\mathbb{R})$ acts on this space isometrically. We will write $(\rho, \rho_0)(\Gamma)$ for the embedding of a discrete group in G, and view a compact form of this homogeneous space as $PSL_2(\mathbb{R})/(\rho, \rho_0)\Gamma$ where Γ acts on $PSL_2(\mathbb{R})$ by $(\rho(\gamma), \rho_0(\gamma))x = \rho(\gamma)x\rho_0(\gamma)^{-1}$.

Now let $\rho_0(\Gamma)$ be a Fuchsian group in $PSL_2(\mathbb{R})$; let g be the genus of the usual quotient $PSL_2(\mathbb{R})/\rho_0(\Gamma)$. We will call a representation ρ from $\text{Hom}(\Gamma, PSL_2(\mathbb{R}))$ ρ_0-*admissible* if $(\rho, \rho_0)(\Gamma)$ acts properly discontinuously on $PSL_2(\mathbb{R})$.[9] Salein proves the following:

Theorem 37 (Salein, [45] Thm 2.1.1) *The set $Adm(\rho_0)$ of ρ_0-admissible homomorphisms is an open subset of $\text{Hom}(\Gamma, PSL_2(\mathbb{R}))$. For certain choices of ρ_0, $Adm(\rho_0)$ is disconnected, and has components in every connected component of $\text{Hom}(\Gamma, PSL_2(\mathbb{R}))$, except the two extremal ones (which are copies of Teichmüller space).*

The openness part of this theorem is a consequence of the completeness of constant curvature compact Lorentz manifolds, which is due to Klingler [23], and of the Ehresmann–Thurston principle. The second statement is Salein's contribution, and was quite surprising at the time in that it shows that the moduli space of compact forms is not connected. In particular, there are nonstandard forms which are not deformations of standard ones.

The key step in Salein's approach is the following Lemma, which provides a criterion for admissibility of the pair (ρ_0, ρ):

Lemma 38 (Salein, [45], Lemma 2.1.3) *If there exists a function $f : \mathbb{H}^2 \to \mathbb{H}^2$ which is uniformly strictly contracting and $(\rho_0, \rho)(\Gamma)$ equivariant in that*

$$f(\rho_0(\gamma)z) = \rho(\gamma)f(z) \text{ for all } \gamma \in \Gamma, z \in \mathbb{H}^2,$$

then ρ is ρ_0-admissible.

Techniques in Fuchsian groups and hyperbolic geometry, such as studying isometries via translation length, underpin Salein's approach to the problem.

---

[9] Kulkarni and Raymond prove for any pair of representations that if $(\rho, \rho_0)(\Gamma)$ acts properly on $SL_2(\mathbb{R})$ then either ρ or ρ_0 is injective with discrete and cocompact image [33].

Salein's criterion allows him to construct his examples in the following simple way. Position a regular $4g$-gon which is a fundamental domain for $\rho_0(\Gamma)$ about the center of the Poincaré disk model for $\mathbb{H}^2$. Let f be a contraction of the fundamental domain towards this center, chosen so that the $4g$-gon maps to a $4g$-gon with geodesic sides, but now with angle sum $2\pi m$ with $1 < m < 2g$. One can extend f to be Γ-equivariant so that it satisfies Lemma 38; the new angle sum implies the Euler number of an associated surface, which specifies which component of $\mathrm{Hom}(\Gamma, \mathrm{PSL}_2(\mathbb{R}))$ the representation belongs to.

Guéritaud and Kassel have also studied criteria for admissible pairs in [15], building on Kassel's earlier work in [18]. They study (ρ_0, ρ)-equivariant maps f such as those Salein's criterion calls for, but in much greater generality. They generalize to the isometry group of n-dimensional hyperbolic space and define for each pair (ρ_0, ρ) of representations into $\mathrm{Isom}(\mathbb{H}^n) = \mathrm{PSO}(1, n)$ the following constant:

Definition 39

$$C(\rho_0, \rho) := \inf\{\mathrm{Lip}(f) : f \text{ Lipschitz and } (\rho_0, \rho)\text{-equivariant}\}$$

where $\mathrm{Lip}(f)$ is the Lipschitz constant of f.

They then prove:

Theorem 40 (Guéritaud–Kassel; see Chapter 5 of [18] for the case $n = 2$ and [15] Theorem 1.8 for the full result) *The pair (ρ_0, ρ), with ρ geometrically finite, is admissible if and only if, up to switching ρ_0 and ρ, the representation ρ_0 is injective and discrete and $C(\rho_0, \rho) < 1$.*

This result comes out of a larger project of understanding the import of the constant $C(\rho_0, \rho)$. In particular, they study f achieving $C(\rho_0, \rho)$ as their Lipschitz constant, and the geometry of the *stretch locus* for such f, i.e. those points in $\mathbb{H}^n$ which have no neighborhood over which the Lipschitz constant for f is smaller than the global constant $\mathrm{Lip}(f)$. The 'if' part of Theorem 40 is a consequence of Lemma 38 with proof adapted to the case of $\mathbb{H}^n$; the 'only if' follows from careful understanding of the structure of the stretch locus when $C(\rho_0, \rho) \geq 1$.

In [20], Kassel improves on Kobayashi's Proposition 36:

Theorem 41 (Kassel, [20] Thm 1.1) *Let G be a real, reductive, linear Lie group; let H and L be closed, reductive subgroups with $\mathbb{R}$-rank$(L) = 1$ with L acting properly discontinuously and cocompactly on $H \backslash G$. For any* *uniform lattice* Γ *of L, there exists a neighborhood* $\mathcal{U}$ *of the natural inclusion in* Hom(Γ, G) *such that any* $\phi \in \mathcal{U}$ *satisfies:*

- $\phi(\Gamma)$ *is discrete in G,*
- $\phi(\Gamma)$ *acts properly discontinuously and cocompactly on* $H \backslash G$.

As a corollary, Kassel obtains

Corollary 42 (Kassel, [20] Corollary 1.2) *There are Zariski-dense* Γ *in* $SO(2, 2n)$ *providing compact forms of* $U(1, n) \backslash SO(2, 2n)$.

Prior to Kassel's work, the only Zariski-dense Γ known (for H noncompact) were for homogeneous spaces of the form $diag(G') \backslash (G' \times G')$. The $\mathbb{R}$-rank = 1 condition is natural as Margulis superrigidity implies that uniform lattices $\Gamma < L$ of higher rank are locally rigid in G.

The idea of Kassel's proof is to study the Cartan projection for elements of Γ. She then uses some interesting dynamics of the action of G on $\mathbb{P}(V)$ for representations of G on V. The study of these dynamics allows one to decompose any $\gamma \in \Gamma$ as a product of elements from a finite set whose Cartan projections can be carefully controlled.

The work of Guichard and Wienhard

Recent work by Guichard and Wienhard on Anosov representations has yielded some corollaries about deformations of compact forms. In [16], they prove the following:

Theorem 43 (Guichard–Wienhard, [16] Section 13) *Let* $L = SO(1, 2n)$, *embedded in* $G = SO(2, 2n)$ *in the standard way. Let* Γ *be a uniform lattice in* $SO(1, 2n)$ *and* ρ *the induced embedding of* Γ *into* $SO(2, 2n)$. *Then any representation in the connected component of* ρ *in* Hom(Γ, $SO(2, 2n)$)) *yields a compact form of* $U(1, n) \backslash SO(2, 2n)$.

That non-trivial deformations of compact forms for these spaces exist is not new (see Prop. 36 and Thm. 41). However, Guichard and Wienhard produce them in a new way. Their paper deals with Anosov representations of discrete groups into semisimple Lie groups G' – a definition due to Labourie ([36]) that they prove unifies many previous examples of special representations. To each such representation they can associate a flag variety of G' (formed by quotienting out by a parabolic subgroup) and on this flag variety there is a domain of discontinuity on which Γ acts properly discontinuously and cocompactly. In some special cases, this domain has the structure of a homogeneous space for some $G < G'$, giving rise to their examples. For the theorem above, they use

Barbot's result ([2]) that the full connected component of ρ consists of Anosov
representations (relative to a certain parabolic subgroup) to obtain a stronger
conclusion than Theorem 41 and Corollary 42 have obtained: *all* of these
representations give compact forms.

These examples are a small part of an extensive paper. Guichard and Wienhard do not undertake an exhaustive search for applications to compact forms;
it is likely that there are more to be found.

6.3 Deformations of the homogeneous space

To close this section, I would like to briefly touch on the work of Oh, Witte-
Morris and Iozzi on deformations of compact forms in another sense. Rather
than deforming the embedding of Γ in G to produce new compact forms
of $H\backslash G$, they deform the embedding of H in G, thereby producing new
homogeneous spaces that admit nonstandard compact forms. In this approach
the analogy with Teichmüller theory breaks down. We are no longer studying
the deformation space of compact forms of a given homogeneous space; rather,
the underlying homogeneous space changes. These results, rather, speak to the
wide (and likely wild) world of compact forms that exist.

Oh and Witte-Morris's work is announced in the paper [40] and treated in
full detail in [41]. Iozzi and Witte-Morris continue this work in [17]. Oh and
Witte-Morris deal with homogeneous spaces of $SO(2, n)$ and their main result
divides into two pieces depending on the parity of n:

Let $G = SO(2, 2m)$, presented as the group preserving the form $2x_1x_{n+2} +
2x_2x_{n+1} + \sum_{i=3}^{n} x_i^2$. Let H_{SU} be the intersection of $SU(1, m)$ (embedded in G
in the standard way) with AN where A consists of the diagonal elements in G
with positive entries and N consists of the upper-triangular matrices with ones
along the diagonal. Note that $SU(1, m)/H_{SU}$ is compact and that, in contrast to
the reductive situation discussed through most of this survey, H_{SU} is solvable.
Let Γ be a uniform lattice in $SO(1, 2m)$; it will act properly discontinuously and
cocompactly on $H_{SU}\backslash SO(2, 2m)$ because it does so on $SU(1, m)\backslash SO(2, 2m)$.

Theorem 44 (Oh–Witte-Morris [41] Thms 1.5 & 1.7)

*(1) For a specific family of deformations H_B of H_{SU}, $H_B\backslash G/\Gamma$ yields a compact
form. Moreover, for almost all choices of the deforming parameter B, the
group H_B is not conjugate into $SU(1, m)$, so these are truly new examples.
(See their paper for details on H_B.)*

*(2) Among closed, connected, upper triangular, noncompact subgroups H of
the Borel subgroup of G such that $H\backslash G$ is noncompact, the only $H\backslash G$*

possessing compact forms are those which are conjugate to a cocompact subgroup of SO(1, 2m) *or to one of the* H_B. *(See section 3.9 of their paper for an argument that reduces any closed, connected H to the upper triangular case.)*

A streamlined proof of this result is provided in [17]; the key advance is an *a priori* lower bound on the dimension of H, which makes the subsequent case-by-case analysis quicker. For specifics on H_B we refer the reader to [41], but remark that the constructions are entirely explicit. The upshot of this theorem is that the homogeneous spaces of SO(2, 2m) are entirely understood. Those H conjugate into SO(1, 2m) return us to the examples studied by Kobayashi in [29] with their nontrivial deformation space.

Now let $G = $ SO(2, 2m + 1). As the algebraic construction of Conjecture 3 does not hold for this G and $H = $ SU(1, m) it is conjectured that there are no compact forms of SU(1, m)\$G$. This problem is still open, but Oh and Witte-Morris show that its solution will settle all questions for homogeneous spaces of SO(2, 2m + 1):

Theorem 45 (Oh–Witte-Morris [41] Thm 1.9) *Let* $G = $ SO(2, 2m + 1) *and let H be closed, connected, noncompact with* $H \backslash G$ *noncompact. If* SU(1, m)\$G$ *has no compact form, then neither does* $H \backslash G$.

Finally, Iozzi and Witte-Morris obtain the following analogous result for homogeneous spaces of $G = SU(2, 2m)$:

Theorem 46 (Iozzi–Witte-Morris, [17] Thm 11.5″) *Let* $G = $ SU(2, 2m) *and let H be a closed, connected, noncompact subgroup of G with* $H \backslash G$ *noncompact. Then* $H \backslash G$ *admits a compact Clifford–Klein form if and only if* $d(H) = 4m$ *and H belongs to a specific family of deformations of* SU(1, 2m) *or* Sp(1, m) *in G. (Recall that $d(H)$ was defined in Prop 7.)*

Again, the reader is referred to the paper for specifics of the deformations, but they are explicitly given.

7 The road ahead

Despite the extensive work on the existence question for compact forms, there are still many open cases. The author's favorite is $SL_{n-2}(\mathbb{R}) \backslash SL_n(\mathbb{R})$.[10] Two

---

[10] Recall that the case $SL_2(\mathbb{R}) \backslash SL_4(\mathbb{R})$ is known to have no compact form by Corollary 26.

very different approaches to the problem – the 'topological' approach via Cartan projections of Benoist and Kobayashi, and the dynamical approach initiated by Zimmer – come right up to this problem, but both fail critically. We certainly expect there are no compact forms, but a new idea seems necessary, even for such an algebraically simple example. There is no shortage of other homogeneous spaces for which the existence question is open.

The deformation question is somewhat less developed, and there is certainly plenty more to be done. The work of Salein indicates how interesting a moduli space for even the simplest examples will be, and one expects that there is far more to the full moduli space than he and Guéritaud–Kassel have discovered. The work of Oh, Witte-Morris and Iozzi indicates that there are plenty of new compact forms to be discovered as we loosen the restrictions on the homogeneous space.

I would like to close this survey, however, by briefly mentioning a very recent result of Kassel and Kobayashi in which they have gone beyond the existence and deformation questions, and begun to study the spectral theory of these spaces.

Let σ be an involutive automorphism of G and let $H = (G^\sigma)_o$ be the identity component of the set of fixed points, so that $H\backslash G$ is a pseudo-Riemannian symmetric space. Let θ be a Cartan involution of G commuting with σ and K the corresponding maximal compact subgroup. Let $\mathrm{Spec}_d(H\backslash G/\Gamma)$ be the set of eigenvalues associated to L^2 eigenfunctions of the pseudo-Riemannian Laplacian, i.e. the discrete spectrum of this operator. Kobayashi and Kassel introduce the notion of 'sharpness' in their paper [22]. Roughly speaking, we say that the pair (H, Γ) satisfies the *sharpness condition* if the Cartan projection of Γ diverges *linearly* from the Cartan projection of H as one heads to infinity in the Cartan subgroup. A precise formulation can be found in [22] Section 1.6.

Theorem 47 (Kassel–Kobayashi, announced in [21], detailed proofs in [22]) *Suppose that* $\mathrm{rank}(H\backslash G) = \mathrm{rank}((K \cap H)\backslash H)$ *where by rank we mean the dimension of a maximal, semisimple, abelian subspace in the set of fixed points of* $-d\sigma$. *Suppose that the pair* (H, Γ) *satisfies the sharpness condition. Then:*

(1) *For any compact form* $H\backslash G/\Gamma$, *the spectrum* $\mathrm{Spec}_d(H\backslash G/\Gamma)$ *is infinite.*

(2) *For standard compact forms with* $\Gamma < L$ *and* $\mathbb{R}\text{-}rank(L) = 1$ *and for all compact forms of* $\mathrm{diag}(\mathrm{SO}(1, n))\backslash(\mathrm{SO}(1, n) \times \mathrm{SO}(1, n))$, *there is an infinite subset of* $\mathrm{Spec}_d(H\backslash G/\Gamma)$ *which is stable under any small deformation of* $H\backslash G/\Gamma$.

For standard forms, the sharpness condition is always satisfied. In certain other cases – for example AdS^3-manifolds – it is known that all compact forms are sharp [18]. Kassel and Kobayashi conjecture that it is always satisfied.

This nice result gives us an indication of one road ahead for the study of Clifford–Klein forms. There are many basic, unanswered geometric questions about these spaces; they are likely to be much more difficult in the pseudo-Riemannan case than in the Riemannian case for the reasons we have noted above. Very little work has been done in this direction, but the results surveyed here provide a very wide variety of tools to address such problems, as well as a library of examples on which to test them.

David Constantine
Department of Mathematics and Computer Science,
Wesleyan University, Middletown, CT 06459 U.S.A.
E-mail: dconstantine@wesleyan.edu

References

[1] Abels, H. 2001. Properly discontinuous groups of affine transformations, a survey. *Geometriae Dedicata*, **87**, 309–333.

[2] Barbot, Thierry. 2013. *Deformations of Fuchsian AdS representations are quasi-Fuchsian*. arXiv:1301.4309.

[3] Benoist, Yves. 1994. Actions propres de groupes libres sur les espaces homogènes réductifs. *Comptes Rendus de l'Académie des Sciences. Série I. Mathématique Comptes Rendus de l'Académie des Sciences - Series I - Mathematics*, **319**(9), 937–940.

[4] Benoist, Yves. 1996. Actions propres sur les espaces homogènes réductifs. *Annals of Mathematics*, **144**, 315–347.

[5] Benoist, Yves, and Labourie, François. 1992. Sur les espaces homogènes modèles de variétés compactes. *Publications Mathématiques de l'I.H.É.S.*, **76**, 99–109.

[6] Borel, Armand. 1963. Compact Clifford-Klein forms of symmetric spaces. *Topology*, **2**, 111–122.

[7] Brown, Kenneth S. 1994. *Cohomology of groups*. Graduate Texts in Mathematics, vol. 87. New York: Springer-Verlag.

[8] Calabi, E., and Markus, L. 1962. Relativistic space forms. *Annals of Mathematics*, **75**, 63–76.

[9] Charette, Virginie, Drumm, Todd, Goldman, William, and Morrill, Maria. 2003. Complete flat affine and Lorentzian manifolds. *Geometriae Dedicata*, **97**, 187–198.

[10] Constantine, David. 2011. *Compact forms of homogeneous spaces and higher-rank semisimple group actions*. arXiv:1209.3940.

[11] Ehresmann, C. 1950. Les connexions infinitésimales dans un espace fibré différentiable. Pages 29–55 of: *Colloque de topologie (espaces fibrés)*. Bruxelles.

[12] Ghys, É. 1987. Flots d'Anosov dont les feuilletages stables sont différentiables. *Annales Scientifiques de l'École Normale Supérieure*, **20**(2), 251–270.

[13] Ghys, É. 1995. Déformation des structures complexes sur les espaces homogènes de $SL(2, \mathbb{C})$. *Journal für die reine und angewandte Mathematik*, **468**, 113–138.

[14] Goldman, William M. 1985. Nonstandard Lorentz space forms. *Journal of Differential Geometry*, **21**(2), 301–308.

142

[15] Guéritaud, François, and Kassel, Fanny. 2013. *Maximally stretched laminations on geometrically finite hyperbolic manifolds.* arXiv:1307.0250.

[16] Guichard, Olivier, and Wienhard, Anna. 2012. Anosov Representations: domains of discontinuity and applications. *Inventiones Mathematicae*, **190**(2), 357–438.

[17] Iozzi, Alessandra, and Witte-Morris, Dave. 2004. Tessellations of homogeneous spaces of classical groups of real rank two. *Geometriae Dedicata*, **103**, 115–191.

[18] Kassel, Fanny. *Quotients compacts d'espaces homogènes réels ou p-adiques.* Ph.D. Thesis, Université Paris-Sud 11, November, 2009. (available at http://math.univ-lille1.fr/~kassel/These.pdf).

[19] Kassel, Fanny. 2008. Proper actions on corank-one reductive homogeneous spaces. *Journal of Lie Theory*, **18**, 961–978.

[20] Kassel, Fanny. 2012. Deformation of proper actions on reductive homogeneous spaces. *Mathematische Annalen*, **353**(2), 599–632.

[21] Kassel, Fanny, and Kobayashi, Toshiyuki. 2011. Stable spectrum for pseudo-Riemannian locally symmetric spaces. *Comptes Rendus Mathématique. Académie des Sciences. Paris*, **349**, 29–33.

[22] Kassel, Fanny, and Kobayashi, Toshiyuki. 2012. *Discrete spectrum for non-Riemannian locally symmetric spaces. I. Construction and stability.* arXiv:1209.4075.

[23] Klingler, Bruno. 1996. Complétude des variétés Lorentziennes à courbure constante. *Mathematische Annalen*, **306**(2), 353–370.

[24] Kobayashi, Toshiyuki. 1989. Proper action on a homogeneous space of reductive type. *Mathematische Annalen*, **285**, 249–263.

[25] Kobayashi, Toshiyuki. 1992a. Discontinuous groups acting on homogeneous spaces of reductive type. Pages 59–75 of: Kawazoe, T., Oshima, T., and Sano, S. (eds), *Representation Theory of Lie Groups and Lie Algebras at Fuji-Kawaguchiko, 1990 August September.* World Scientific.

[26] Kobayashi, Toshiyuki. 1992b. A necessary condition for the existence of compact Clifford-Klein forms of homogeneous spaces of reductive type. *Duke Mathematical Journal*, **67**, 653–664.

[27] Kobayashi, Toshiyuki. 1996a. Criterion for proper action on homogeneous spaces of reductive groups. *Journal of Lie Theory*, **6**(2), 147–163.

[28] Kobayashi, Toshiyuki. 1996b. Discontinuous groups and Clifford-Klein forms of pseudo-Riemannian homogeneous manifolds. Pages 99–165 of: Schlichtkrull, H., and Ørsted, B. (eds), *Algebraic and Analytic Methods in Representation Theory.* Perspectives in Mathematics, vol. 17. Academic Press.

[29] Kobayashi, Toshiyuki. 1998. Deformation of compact Clifford-Klein forms of indefinite-Riemannian homogeneous manifolds. *Mathematische Annalen*, **310**, 395–409.

[30] Kobayashi, Toshiyuki. 2001. Discontinuous groups for non-Riemannian homogeneous spaces. Pages 723–747 of: Engquist, B., and Schmid, W. (eds), *Mathematics Unlimited – 2001 and Beyond.* Springer.

[31] Kobayashi, Toshiyuki, and Yoshino, Taro. 2005. Compact Clifford-Klein forms of symmetric spaces – revisited. *Pure and Applied Mathematics Quarterly*, **1**(3), 591–663.

[32] Kulkarni, R. 1981. Proper actions and pseudo-Riemannian space forms. *Advances in Mathematics*, **40**, 10–51.

[33] Kulkarni, R., and Raymond, F. 1985. 3-dimensional Lorentz space-forms and Seifert fiber spaces. *Journal of Differential Geometry*, **21**, 231–268.

[34] Labourie, F., Mozes, S., and Zimmer, R.J. 1995. On Manifolds Locally Modelled on Non-Riemannian Homogeneous Spaces. *Geometric and Functional Analysis*, **5**(6), 955–965.

[35] Labourie, François. 1996. Quelques résultats récents sur les espaces localement homogènes compacts. Pages 267–283 of: *Manifolds and Geometry (Pisa 1993)*. Sympos. Math., no. XXXVI. Cambridge: Cambridge University Press.

[36] Labourie, François. 2006. Anosov flows, surface groups and curves in projective space. *Inventiones Mathematicae*, **165**(1), 51–114.

[37] Labourie, François, and Zimmer, Robert J. 1995. On the Existence of Cocompact Lattices for $SL(n)/SL(m)$. *Mathematical Research Letters*, **2**, 75–77.

[38] Margulis, Gregory. 1997. Existence of compact quotients of homogeneous spaces, measurably proper actions, and decay of matrix coefficients. *Bulletin de la Société Mathématique de France*, **125**, 447–456.

[39] Oh, Hee. 1998. Tempered subgroups and representations with minimal decay of matrix coefficients. *Bulletin de la Société Mathématique de France*, **126**, 355–380.

[40] Oh, Hee, and Witte, Dave. 2000. New examples of compact Clifford-Klein forms of homogeneous spaces of $SO(2, n)$. *International Mathematics Research Notices*, 235–251.

[41] Oh, Hee, and Witte, Dave. 2002. Compact Clifford-Klein forms of homogeneous spaces of $SO(2, n)$. *Geometriae Dedicata*, **89**, 25–57.

[42] Oniščik, A. L. 1969. Decompositions of reductive Lie groups. *Mathematics of the USSR-Sbornik*, **9**(4), 515–554.

[43] Ratner, Marina. 1991. On Ragunathan's measure conjecture. *Annals of Mathematics*, **134**(3), 545–607.

[44] Salein, François. 1997. Variétés anti-de Sitter de dimension 3 possédant un champ de Killing non trivial. *Comptes Rendus Mathématique. Académie des Sciences. Paris*, **324**, 525–530.

[45] Salein, François. 2000. Variétés anti-de Sitter de dimension 3 exotiques. *Annales de l'institut Fourier*, **50**(1), 257–284.

[46] Serre, J.-P. 1971. Cohomologie des groupes discrets. Pages 337–350 of: *Séminaire Bourbaki, 23ème année (1970/1971), Exp. No. 399*. Lecture Notes in Math, vol. 244. Springer.

[47] Shalom, Yehuda. 2000. Rigidity, unitary representations of semisimple groups, and fundamental groups of manifolds with rank one transformation group. *Annals of Mathematics*, **152**, 113–182.

[48] Thurston, W. P. 1980. *Geometry and topology of three-manifolds.* (unpublished; available from library.msri.org/books/gt3m/).

[49] Weil, André. 1964. Remarks on the cohomology of groups. *Annals of Mathematics*, **80**, 149–157.

[50] Wolf, Joseph A. 1962. The Clifford-Klein space forms of indefinite metric. *Annals of Mathematics*, **75**, 77–80.

[51] Wolf, Joseph A. 2011. *Spaces of constant curvature*. 6th edn. Providence, RI: AMS Chelsea Publishing.

[52] Zimmer, Robert J. 1984. *Ergodic theory and semisimple groups*. Monographs in Mathematics, vol. 81. Basel: Birkhäuser.

[53] Zimmer, Robert J. 1994. Discrete Groups and Non-Riemannian Homogeneous Spaces. *Journal of the American Mathematical Society*, **7**(1), 159–168.

5

A survey on noncompact harmonic and asymptotically harmonic manifolds

GERHARD KNIEPER

Abstract

Harmonic manifolds provide an interesting class of Riemannian manifolds which can be characterized by the mean value property of harmonic functions. In 1990 S.I Szabo finished the classification of compact harmonic manifolds confirming the Lichnerowicz conjecture which states that harmonic manifolds are locally symmetric spaces of rank 1. In 1992 Damek and Ricci found new examples of noncompact harmonic manifolds which are homogeneous but not symmetric. Very recently new interesting global results where derived for noncompact harmonic manifolds and more general asymptotically harmonic manifolds. The purpose of this survey is to provide an update of the latest developments in this field.[1]

1 Introduction

It is a classical result in Analysis that in the Euclidean space $\mathbb{R}^n$ a solution of the Laplace equation $\Delta\varphi = 0$ depending only on the distance to the origin is given by

$$\varphi : \mathbb{R}^n\backslash\{0\} \to \mathbb{R},$$

$$\varphi(x) = \begin{cases} \log\|x\|, & \text{if } n = 2, \\ \|x\|^{2-n}, & \text{if } n \neq 2. \end{cases}$$

Around 1930, Ruse [Ru] investigated *radial* solutions of $\Delta\varphi = 0$ in open balls $B_r(p)$ of arbitrary Riemannian manifolds (X, g), i.e. solution depending only on the Riemannian distance to the center p. The Laplace operator on a

[1] **Keywords:** harmonic manifolds, geodesic flows, Lichnerowicz conjecture.
AMS codes: Primary 37C40, Secondary 53C12, 37C10.

Riemannian manifold (X, g) is defined as $\Delta = \text{div} \circ \text{grad}$ where the divergence of a vector field Y its given by $\text{div} Y = \text{tr}(v \mapsto \nabla_v Y)$. It turns out that nontrivial radial solutions can only appear if for any $p \in X$ the volume density $\theta_p(q) = \sqrt{\det g_{ij}(q)}$ in normal coordinates, centered at any point p, is a radial function. This observation led to the definition of a *harmonic space*. In 1950, Willmore [Wi1] gave an analytic characterization of harmonic spaces as those Riemannian manifolds where all harmonic functions satisfy the *mean value property (MVP)*. A continuous function $\varphi : X \to \mathbb{R}$ satisfies MVP if the average of φ over arbitrary geodesic spheres $S_r(p) \subset X$ (with the induced Riemannian metric) agrees with the value $\varphi(p)$.

An important problem consists in the classification of all harmonic spaces. Since harmonic spaces are Einstein manifolds, i.e. the Ricci curvature is constant, all harmonic spaces of dimension 2 and 3 have constant sectional curvature. Examples of harmonic manifolds are given by the Euclidean and rank one symmetric spaces, sometimes called model spaces. In 1944 Lichnerowicz [Li] conjectured – in dimension 4 – that these are the only examples of nonflat harmonic spaces. The proof of this conjecture was given in 1949 by Walker [Wal]. This together with Ledger's result [Led] stating that symmetric spaces are harmonic if and only if they are of rank one, led to the following conjecture in arbitrary dimensions, famously known as *Lichnerowicz's Conjecture*: "Every simply connected harmonic space is flat or a rank one symmetric space."

In a celebrated paper Szabó proved in 1990 the Lichnerowicz's Conjecture for *compact* (simply connected) harmonic spaces. In 1992 Damek and Ricci [DR] constructed, beginning at dimension 7, simply connected one-dimensional extensions of Heisenberg-type groups which are non-symmetric harmonic spaces. Like rank one symmetric spaces, these so-called *Damek-Ricci spaces* have nonpositive sectional curvature, but unlike in the case of symmetric spaces, nonsymmetric Damek-Ricci spaces do not admit compact quotients. The existence of Damek-Ricci spaces disproved Lichnerowicz's Conjecture in the case of *noncompact* harmonic spaces. The Einstein condition implies that harmonic manifolds fall in one of the three categories.

(a) $\lambda = \text{Ric}_X > 0$: Then X is a compact rank one symmetric space.
(b) $\lambda = \text{Ric}_X = 0$: Then X is the Euclidean space $\mathbb{R}^n$ which is an easy consequence of the Riccati equation.
(c) $\lambda = \text{Ric}_X < 0$: Then X is noncompact and the Damek-Ricci spaces provide homogeneous but nonsymmetric examples of harmonic spaces in this class.

Since for harmonic spaces the volume density is a radial function geodesic spheres have constant mean curvature. Furthermore, noncompact harmonic spaces have no conjugate points and therefore *horospheres* obtained as geometric limits of families of geodesic spheres with radii increasing to infinity have constant mean curvature (denoted by h) as well. Motivated by results from rigidity theory Ledrappier [Le] introduced the notion of asymptotically harmonic manifolds. These are complete Riemannian manifolds without conjugate points and horospheres of constant mean curvature $h \geq 0$. However, under this condition properties like the Einstein condition or the mean value property of harmonic functions are not known to hold. On the other hand there are no examples of asymptotically harmonic manifolds which are not harmonic.

Let us state what is known for those spaces and what we like to discuss in this survey. In 2002, Ranjan and Shah showed that noncompact harmonic spaces with polynomial volume growth are flat [RSh1] (see Section 5). Nikolayevsky [Ni] proved in 2005 that the density functions $f(r)$ of all noncompact harmonic manifolds are exponential polynomials. In particular, subexponential volume growth implies polynomial volume growth. Since subexponential volume growth is guaranteed by the minimality of horospheres, $h = 0$ implies flatness as well (see Section 4).

In 2006, Heber [He] proved that among the homogeneous harmonic spaces (and asymptotically harmonic Einstein manifolds) only the model spaces and the Damek-Ricci spaces occur. Therefore, it remains to study nonhomogeneous harmonic manifolds of exponential volume growth. In particular, these are spaces without conjugate points and horospheres of constant mean curvature $h > 0$.

In 2009, the author [Kn4] began to study, using a notion of rank, the asymptotic geometry and the geodesic flow on such manifolds. In particular, we showed that for noncompact harmonic spaces X properties like purely exponential volume growth, the rank of X is one, the geodesic flow is Anosov (with respect to the Sasaki metric) and X is Gromov hyperbolic are equivalent. We conjectured that all harmonic manifolds with $h > 0$ should be of rank one and we confirmed this conjecture under the additional assumption of no focal points or bounded asymptote and odd dimensions (see Sections 8 and 9).

In 2012 Zimmer [Zi1] extended those results to asymptotically harmonic manifolds but he had to assume the existence of a compact quotient.

In a very recent paper N. Peyerimhoff and myself [KnPe] could even drop the assumption of a compact quotient provided the curvature tensor and its covariant derivative is uniformly bounded (see Sections 6 and 7).

Under the assumption of a compact quotient methods from dynamical systems and rigidity theory become available. In particular, using the celebrated results of Besson, Courtois and Gallot, [BCG] and Benoist, Foulon and Labourie [BFL] in combination with Foulon and Labourie [FL] one obtains: Is X an asymptotically harmonic manifold with compact quotient whose geodesic flow is Anosov then X is a symmetric space of negative curvature. This in turn implies with the above mentioned results that harmonic spaces [Kn4] (see Section 9) and more generally asymptotically harmonic spaces [Zi1] with compact quotient are symmetric if they have no focal points. Replacing no focal points by the more general assumption of bounded asymptote (see Section 8) this remains true provided X is a harmonic space of odd dimensions ([Kn4]).

Finally, we like to mention that there are other recent results on noncompact harmonic and asymptotically harmonic manifolds not covered in this survey. In dimension 3 there is a nearly complete classification of asymptotically harmonic manifolds. It has been shown in [SS] that 3-dimensional asymptotically harmonic manifolds with positive mean curvature of the horospheres are of constant negative sectional curvature. Under the additional assumption of negative sectional curvature this has been previously obtained in [HKS]. However, the natural question whether 3-dimensional asymptotically harmonic manifolds with zero mean curvature of the horospheres are flat is to the knowledge of the author still open. On the other hand, Nikolayaveski [Ni] has shown that nonsymmetric harmonic manifolds need to have at least dimension 6.

In [CS] the authors study asymptotically harmonic spaces under the special condition of negative curvature. In [Zi2] Zimmer investigates the boundary structure of noncompact harmonic spaces which yields under the assumption of a quotient of finite volume the existence of a dense orbit for the geodesic flow on this quotient. In [PeSa] interesting results are obtained on the integral geometry of noncompact harmonic manifolds.

2 Basics on Jacobi tensors on manifolds without conjugate points

Since noncompact harmonic manifolds and asymptotically harmonic manifolds have no conjugate points, it is important to recall their basic properties. We note however, that beyond special subclasses of manifolds without conjugate points, like manifolds of nonpositive curvature or no focal points, general results are rare. The most basic result is the Theorem of Cartan-Hadamard, which states that for complete simply connected Riemannian manifold (X, g) without conjugate points the exponential map $\exp_p : T_p X \to X$ is a diffeormorphism and therefore X is diffeomorphic to $\mathbb{R}^{\dim X}$.

Variational properties of manifolds without conjugate points which can be expressed with the help of Jacobi tensors are of central importance in the study of harmonic manifolds.

Let us first briefly recall some basic facts on the calculus of Jacobi tensors (see e.g. [Es], [Gre], [Kn1] and [Kn3] for more details). Let $c : I \to X$ be a unit speed geodesic and let $N(c)$ denote the normal bundle of c given by a disjoint union

$$N_t(c) := \{w \in T_{c(t)}X \mid \langle w, \dot{c}(t)\rangle = 0\}.$$

A $(1, 1)$-tensor along c is a differentiable section

$$Y : I \to \text{End } N(c) = \bigcup_{t \in I} \text{End}(N_t(c)),$$

i.e., for all orthogonal parallel vector fields x_t along c the covariant derivative of $t \to Y(t)x_t$ exists. The derivative $Y'(t) \in \text{End}(N_t(c))$ is defined by

$$Y'(t)(x_t) = \frac{D}{dt}(Y(t)x_t).$$

Y is called parallel if $Y'(t) = 0$ for all t. If Y is parallel we have $Y(t)x_t = (Y(0)x)_t$ and, therefore, $\langle Y(t)x_t, y_t\rangle$ is constant for all parallel vector fields x_t, y_t along c. In particular, Y is parallel if and only if Y is a constant matrix with respect to parallel frame field in the normal bundle of c. Therefore, parallel $(1, 1)$-tensors are also called constant.

The curvature tensor R induces a symmetric $(1, 1)$-tensor along c given by

$$R(t)w := R(w, \dot{c}(t))\dot{c}(t).$$

A $(1, 1)$-tensor Y along c is called a Jacobi tensor if it solves the Jacobi equation

$$Y''(t) + R(t)Y(t) = 0.$$

If Y, Z are two Jacobi tensors along c the derivative of the Wronskian

$$W(Y, Z)(t) := Y'^*(t)Z(t) - Y^*(t)Z'(t)$$

is zero and thus, $W(Y, Z)$ defines a parallel $(1, 1)$-tensor. A Jacobi tensor Y along a geodesic $c : I \to X$ is called Lagrange tensor if $W(Y, Y) = 0$. The importance of Lagrange tensors comes from the following Proposition.

Proposition 1 *Let $Y : I \to \text{End } N(c)$ be a Lagrange tensor along a geodesic $c : I \to X$ which is nonsingular for all $t \in I$. Then for $t_0 \in I$ and any other Jacobi tensor Z along c, there exist constant tensors C_1 and C_2 such that*

$$Z(t) = Y(t)\left(\int_{t_0}^{t}(Y^*Y)^{-1}(s)ds\, C_1 + C_2\right) \tag{1}$$

for all $t \in I$. Conversely, every tensor of the form (1) with Y, C_1, C_2 as above is a Jacobi tensor.

Proof One checks that both sides of the equation define Jacobi tensors which agree for suitable choice of C_1, C_2. For a proof see, e.g, [Kn4, Prop. 7.2]. □

Let SX denote the unit tangent bundle of X with fibres $S_p X$, $p \in X$ and $\pi : SX \to X$ be the canonical footpoint projection. For every $v \in SX$, let $c_v : \mathbb{R} \to X$ denote the unique geodesic satisfying $\dot{c}_v(0) = v$. Define A_v to be the Jacobi tensor along c_v with $A_v(0) = 0$ and $A'_v(0) = $ id. Then the volume of a geodesic sphere $S(p, r)$ of radius r about p is given by

$$\text{vol } S(p, r) = \int\limits_{S_p X} \det A_v(r) d\theta_p(v),$$

where $d\theta_p(v)$ is the volume element of $S_p X$ induced by the Riemannian metric.

Definition 2 Let (X, g) be a complete, simply connected manifold without conjugate points. X is a *harmonic manifold* if the volume density $\det A_v(t)$ does not depend on $v \in SX$. We call the function

$$f(t) = \det A_v(t)$$

the *density function* of the harmonic space X.

Remark If (X, g) is a harmonic space, we have

$$\text{vol } S(p, r) = \omega_n f(r),$$

where ω_n is the volume of the sphere in the Euclidean space $\mathbb{R}^n$. Since

$$\frac{f'(r)}{f(r)} = \frac{(\det A_v(r))'}{\det A_v(r)} = \text{tr}(A'_v(r) A_v(r)^{-1})$$

is the mean curvature of the geodesic sphere of radius $r > 0$ about $\pi(v)$ in $c_v(r)$, X is harmonic if and only if the mean curvature of all spheres is a function depending only on the radius.

Proposition 3 *Let (X, g) be a simply connected manifold without conjugate points and $v \in SX$. Let B_v be the Jacobi tensor along c_v with $B_v(0) = $ id and $B'_v(0) = 0$. Since (X, g) has no conjugate points the tensor*

$$Q_v(t) = A_v^{-1}(t) B_v(t)$$

is well defined for $t \neq 0$. Furthermore, Q_v satisfies

$$Q_v(t) - Q_v(s) = - \int\limits_s^t (A_v^* A_v)^{-1}(u) du$$

for $0 < s \leq t$.

Proof Since A_v is a Lagrange tensor along c_v and nonsigular for $t \neq 0$ $((X, g)$ has no conjugate points), Proposition 1 yields

$$B_v(t) = A_v(t) \left(\int\limits_s^t (A_v^* A_v)^{-1}(u)du \, C_1 + C_2 \right). \tag{2}$$

For $t = s$, we obtain $B_v(s) = A_v(s)C_2$, and therefore $C_2 = Q_v(s)$. Differentiation at $t = s$ yields

$$\begin{aligned} B_v'(s) &= A_v'(s)C_2 + A_v(s)(A_v^* A_v)^{-1}(s)C_1 \\ &= A_v'(s)Q_v(s) + (A_v^*)^{-1}(s)C_1 \\ &= (A_v'(s)A_v^{-1}(s))B_v(s) + (A_v^*)^{-1}(s)C_1 \\ &= (A_v^*)^{-1}(s)(A_v')^*(s)B_v(s) + (A_v^*)^{-1}(s)C_1. \end{aligned}$$

Here we used that $A_v'(s)A_v^{-1}(s)$ is symmetric since it is the second fundamental form of a geodesic sphere with radius s around p. This implies that

$$A_v^*(s)B_v'(s) = (A_v')^*(s)B_v(s) + C_1,$$

i.e., the constant tensor C_1 satisfies $C_1 = A_v^*(s)B_v'(s) - (A_v')^*(s)B_v(s) = -W(A_v, B_v)(s)$. At $s = 0$, we conclude $C_1(0) = -\text{id} \cdot \text{id} = -\text{id}$. Plugging this into (2), we obtain

$$Q_v(t) = A_v^{-1}(t)B_v(t) = -\int\limits_s^t (A_v^* A_v)^{-1}(u)du + Q_v(s).$$

□

Proposition 4 *Let $A_{v,s}$ be Jacobi tensor along c_v with $A_{v,s}(s) = 0$ and $A_{v,s}'(s) = \text{id}$. Then*

$$A_{v,s}(t) = A_v(t) \int\limits_s^t (A_v^* A_v)^{-1}(u)du \, A_v^*(s). \tag{3}$$

Note that $A_v^*(s)$ in the above formula is a constant tensor.

Proof $A_{v,s}$ and the right hand side of (3) are both Jacobi tensors, because of Proposition 1. Since

$$A_{v,s}(s) = 0 = A_v(s) \int\limits_s^s (A_v^* A_v)^{-1}(u)du \, A_v^*(s) = 0$$

and

$$A'_{v,s}(s) = \text{id} = A'_v(s) \cdot 0 + A_v(s)(A_v^* A_v)^{-1}(s)A_v^*(s),$$

both Jacobi tensors agree. □

Corollary 5 *We have*

$$Q_v(s) - Q_v(t) = A_v^{-1}(t)A_{v,s}(t)(A_v^*)^{-1}(s)$$

and therefore

$$\det(Q_v(s) - Q_v(t)) = \frac{\det A_{v,s}(t)}{\det A_v(t) \cdot \det A_v(s)}$$

fo all $0 < s \le t$

Proof Using Proposition 1 and 4, we conclude

$$Q_v(s) - Q_v(t) = \int_s^t (A_v^* A_v)^{-1}(u)du = A_v^{-1}(t)A_{v,s}(t)(A_v^*)^{-1}(s).$$

□

Of fundamental importance are the stable and unstable Jacobi tensors (see e.g. [Es], [Kn1] or [Kn3] for details). For a general complete simply connected manifold without conjugate points X they are defined as follows. For $v \in SX$ and $r > 0$ denote by $S_{v,r}$ and $U_{v,r}$ the Jacobi tensors along c_v such that

$$S_{v,r}(0) = U_{v,r}(0) = \text{id} \quad \text{and} \quad S_{v,r}(r) = 0, \; U_{v,r}(-r) = 0.$$

Let

$$S_v = \lim_{r \to \infty} S_{v,r} \quad \text{and} \quad U_v = \lim_{r \to \infty} U_{v,r}$$

be the stable and unstable Jacobi tensors. For each $r > 0$ we have

$$W(S_{v,r}, S_{v,r})(0) = S_{v,r}^{\prime*}(0) - S_{v,r}'(0) = W(S_{v,r}, S_{v,r})(r) = 0$$

and

$$W(U_{v,r}, U_{v,r})(0) = U_{v,r}^{\prime*}(0) - U_{v,r}'(0) = W(U_{v,r}, U_{v,r})(-r) = 0,$$

which implies that the Jacobi tensors $S_{v,r}$ and $U_{v,r}$ are Lagrangian and the endomorphisms $S_{v,r}'(0)$ and $U_{v,r}'(0)$ are symmetric. By passing to the limit, the stable and unstable Jacobi tensors S_v and U_v are Lagrangian and the endomorphisms $S_v'(0)$ and $U_v'(0)$ are symmetric as well.

Lemma 6 *(see [Kn4, Lemma 2.3]) Let X be a complete simply connected manifold without conjugate points. Let $c_v : \mathbb{R} \to X$ be a geodesic with $\dot{c}_v(0) = v \in SX$ and $s, r > 0$. Then we have*

$$(U'_{v,r}(0) - S'_{v,s}(0))^{-1} = \int_0^s (U^*_{v,r} U_{v,r})^{-1}(u)\,du$$

and

$$(U'_v(0) - S'_{v,s}(0))^{-1} = \int_0^s (U^*_v U_v)^{-1}(u)\,du.$$

Similarly, for $0 < s < r$ we have

$$(S'_{v,r}(0) - S'_{v,s}(0))^{-1} = \int_0^s (S^*_{v,r} S_{v,r})^{-1}(u)\,du$$

and

$$(S'_v(0) - S'_{v,s}(0))^{-1} = \int_0^s (S^*_v S_v)^{-1}(u)\,du.$$

Furthermore, the function

$$\det\left(\int_0^s (U^*_v U_v)^{-1}(u)\,du\right) = \frac{1}{\det(U'_v(0) - S'_{v,s}(0))}$$

is strictly monotonically increasing.

Lemma 7 *(see [Kn4, Lemma 7.3]) Let M be a manifold without conjugate points. Then for all $v \in SM$ we have*

$$S_{\phi^u(v)}(t) = S_v(t+u)S_v^{-1}(u) \text{ and } U_{\phi^u(v)}(t) = U_v(t+u)U_v^{-1}(u) \quad (4)$$

$$S'_{\phi^t(v)}(0) = S'_v(t)S_v^{-1}(t) \text{ and } U'_{\phi^t(v)}(0) = U'_v(t)U_v^{-1}(t) \quad (5)$$

$$U'_{\phi^t(v)}(0) - S'_{\phi^t(v)}(0) = U_v^{*-1}(t)(U'_v(0) - S'_v(0))S_v^{-1}(t) \quad (6)$$

$$= S_v^{*-1}(t)(U'_v(0) - S'_v(0))U_v^{-1}(t).$$

Furthermore,

$$U'_{\phi^t(v)}(0) - S'_{\phi^t(v)}(0) = S_v^{*-1}(t)\left(\int_{-\infty}^t (S^*_v S_v)^{-1}(u)\,du\right)^{-1} S_v^{-1}(t). \quad (7)$$

Note, that $\operatorname{tr} U'_{v,r}(0) = \operatorname{tr}(A'_{\phi^{-r}v}(r)A^{-1}_{\phi^{-r}v}(r))$. In particular, if X is a complete noncompact harmonic manifold the remark after definition 2 implies $\operatorname{tr} U'_{v,r}(0) = \frac{f'(r)}{f(r)}$ is converging to the mean curvature $\operatorname{tr} U'_v(0) =: h \geq 0$ of the horospheres. Hence,

$$\lim_{r \to \infty} \frac{\log \operatorname{vol} S(p, r)}{r} = \lim_{r \to \infty} \frac{\log f(r)}{r} \lim_{r \to \infty} \frac{f'(r)}{f(r)} = h. \tag{8}$$

Of importance in the study of manifolds without conjugate points are the Busemann functions. For $v \in SX$ and $r > 0$ one defines $b_{v,r} : X \to \mathbb{R}$ by $b_{v,r}(q) = d(q, c_v(r)) - r$. The levels of $b_{v,r}$ are geodesic spheres about $c_v(r)$. Furthermore, the triangle inequality implies that the family of functions $b_{v,r}$ is monotonically decreasing in r. The limit

$$b_v(q) = \lim_{r \to \infty} b_{v,r}$$

is called Busemann function. The levels are called horospheres. In general Busemann function are of class $C^{1,1}$, i.e. the gradient $\operatorname{grad} b_v$ is a Lipschitz continuous vector field (see [Kn1]). Busemann function are of class class C^2 if X is of continuous asymptote (see [Es]), i.e. the map $v \mapsto U(v) := U'_v(0)$ is continuous. If $S(v) := S'_v(0) = -U(-v)$ then the map $v \mapsto S(v)$ is continuous as well. $Z_r(q) := \operatorname{grad} b_{v,r}(q)$ defines on $X \setminus c_v(r)$ the vector field outward normal to the geodesic spheres about $c_v(r)$. The hessian $\nabla \operatorname{grad} b_{v,r}(q) : T_q X \to T_q X$ of $b_{v,r}$ is given by

$$\nabla_w \operatorname{grad} b_{v,r}(q) = \begin{cases} U'_{Z_r(q),d(q,c_v(r))}(0)w & \text{if } w \perp Z_r(q) \\ 0 & \text{if } w \in\, < Z_r(q) > \end{cases}$$

In case of continuous asymptote the hessian $\nabla \operatorname{grad} b_v(q) : T_q X \to T_q X$ of b_v is given by

$$\nabla_w \operatorname{grad} b_v(q) = \begin{cases} U'_{Z(q)}(0)w & \text{if } w \perp Z(q) \\ 0 & \text{if } w \in\, < Z(q) > \end{cases}$$

where $Z(q) = \lim_{r \to \infty} Z_r(q)$. The sets

$$W^s(v) = \{\operatorname{grad} b_{-v}(q) \mid b_{-v}(q) = 0\} \text{ resp.} \tag{9}$$

$$W^u(v) = \{\operatorname{grad} b_v(q) \mid b_v(q) = 0\}$$

are called stable resp. unstable manifolds through v and are submanifolds of SX which are of class C^1 in the case of continuous asymptote. We need the following estimates whose proofs can be found in [Kn3, Chapter 1, Cor. 2.12 and Lemma 2.17]:

Lemma 8 *Assume that there exists a constant $R_0 > 0$ such that $-R_0$ id $\leq$ $R_v(t)$ for all $v \in SX$ and $t \in \mathbb{R}$. Then we have*

$$-\sqrt{R_0} \leq A'_{\phi^{-r}v}(r)A^{-1}_{\phi^{-r}v}(r) = U'_{v,r}(0) \leq \sqrt{R_0}\coth(r\sqrt{R_0})$$

for all $t > 0$. By taking limits one obtains

$$-\sqrt{R_0} \leq S'_v(0) \leq U'_v(0) \leq \sqrt{R_0}$$

for all $v \in SX$.

It turns out that the notion of rank which has been introduced for manifolds of nonpositive curvature by Ballmann, Brin and Eberlein [BBE] and is also of the central importance in the study of noncompact harmonic manifolds. This notion can be generalized to manifolds without conjugate points as follows.

Definition 9 Let (X, g) be a Riemannian manifold without conjugate points. The rank of $v \in SM$ is defined by

$$\text{rank}(v) = \dim \mathcal{L}(v) + 1,$$

where $\mathcal{L}(v) = \ker(U'_v(0) - S'_v(0))$. The *rank* of X is defined to be

$$\text{rank}(X) = \min\{\text{rank}(v) \mid v \in SX\}.$$

Remark In the case of nonpositive curvature (or more generally no focal points) the rank(v) equals the dimension of parallel Jacobi fields along the geodesic c_v. However, in general such a relation does not hold. See [Bu] for an explicit example and [Kn1] for general results on manifolds without conjugate points. The rank(v) is a measure for the transversality of stable and unstable manifolds $W^s(v)$ and $W^u(v)$ defined by the equations 9. In particular, rank$(v) = 1$ iff the submanifolds $W^s(v)$ and $W^u(v)$ are transversal at v.

3 Volume growth on asymptotically harmonic manifolds

Definition 10 An *asymptotically harmonic manifold* (X, g) is a complete, simply connected Riemannian manifold without conjugate points such that for all $v \in SX$ we have $\text{tr}\, U(v) = h$ for a constant $h \geq 0$, where as above $U(v) := U'_v(0)$.

Lemma 11 *Let (X, g) be an asymptotically harmonic manifold. Then $U'_{v,t}(0)$ converges uniformly on compact subsets of SX to $U'_v(0)$. In particular X is of continuous asymptote, i.e. the $v \mapsto U(v)$ is continuous on SX.*

Proof Since the symmetric endomorphism $U'_{v,t}(0) - U'_v(0)$ is positive, we have for $t > 0$

$$\|U'_{v,t}(0) - U'_v(0)\| \leq \operatorname{tr}(U'_{v,t}(0) - U'_v(0)) = \operatorname{tr}(U'_{v,t}(0)) - h.$$

Note that $\operatorname{tr}(U'_{v,t}(0))$ converges pointwise monotonically to h as $t \to \infty$, which implies by Dini's Theorem that the convergence is uniformly on all compact subsets of SX. Since the maps $v \mapsto U'_{v,t}(0)$ are continuous for all $t > 0$ and the convergence $U'_{v,t}(0) \to U'_v(0) = U(v)$ is uniformly on compact sets, we conclude continuity of $v \mapsto U(v)$. $\qquad\square$

Proposition 12 *Let X be an asymptotically harmonic manifold. Then for all $v \in SX$ the ratio*

$$\frac{\det A_v(t)}{e^{ht}} = \frac{1}{\det(U'_v(0) - S'_{v,t}(0))}$$

is monotonically increasing. In particular,

$$\lim_{t \to \infty} \frac{\det A_v(t)}{e^{ht}} = \frac{1}{\det(U'_v(0) - S'_v(0))},$$

where the right hand side is ∞ if $\det(U'_v(0) - S'_v(0)) = 0$. Moreover,

$$\frac{\operatorname{vol} S_r(p)}{e^{hr}} = \int_{S_pX} \frac{1}{\det(U'_v(0) - S'_{v,r}(0))} d\theta_p(v)$$

$$\geq \int_{S_pX} \frac{1}{\det(U'_v(0) - S'_{v,1}(0))} d\theta_p(v)$$

for all $r \geq 1$. If X is harmonic, i.e. $\det A_v(t) = f(t)$ is independent of v, $\det(U'_v(0) - S'_v(0))$ and $\det(U'_v(0) - S'_{v,t}(0))$ are independent of $v \in SX$ as well.

Proof Let A_v be the Jacobi tensor along the geodesic $c_v : \mathbb{R} \to X$ with $\dot{c}_v(0) = v \in SX$ such that $A_v(0) = 0$ and $A'_v(0) = \operatorname{id}$. Then Proposition 1 implies

$$A_v(t) = U_v(t) \int_0^t (U_v^* U_v)^{-1}(u) du.$$

Since by lemma 7 we have

$$(\log \det U_v)'(t) = \operatorname{tr} U'_v(t) U_v^{-1}(t) = \operatorname{tr} U'_{\phi^t(v)}(0) = h,$$

we obtain

$$\frac{\det A_v(t)}{e^{ht}} = \det\left(\int_0^t (U_v^* U_v)^{-1}(u) du\right) = \frac{1}{\det(U'_v(0) - S'_{v,t}(0))}.$$

$\qquad\square$

Using the result above we obtain that manifolds of constant negative curvature have minimal volume growth among those asymptotically harmonic manifolds with fixed positive mean curvature of the horospheres. This result was obtained in [Kn4] for harmonic manifolds but it carries over without any difficulties to asymptotically harmonic manifolds.

Corollary 13 *Let X be a n-dimensional asymptotically harmonic manifold with mean curvature of the horospheres equal to $h > 0$. Let ω_{n-1} denote the volume of the unit sphere $S^{n-1} \subset \mathbb{R}^n$. Then $r \mapsto \frac{\text{vol } S_r(p)}{e^{hr}}$ is strictly monotonically increasing and*

$$\lim_{r \to \infty} \frac{\text{vol } S_r(p)}{e^{hr}} \geq \omega_{n-1} \left(\frac{n-1}{2h} \right)^{n-1}$$

where the limit might be infinite. Furthermore, equality holds for all $p \in X$ if and only if X has constant negative sectional curvature.

Proof Monotonicity follows immediately from Proposition 12. Furthermore this Proposition together with the Theorem of Beppo Levi yields:

$$\lim_{r \to \infty} \frac{\text{vol } S_r(p)}{e^{hr}} = \int_{S_p X} \frac{1}{\det(U_v'(0) - S_v'(0))} d\theta_p(v).$$

Note, that for a given symmetric matrix B on $\mathbb{R}^k$ with non negative eigenvalues, we have $(\det B)^{1/k} \leq \frac{\text{tr } B}{k}$, where equality holds if and only if $B = \lambda$ id. Applying this to $B = (U_v'(0) - S_v'(0))$, that

$$\lim_{r \to \infty} \frac{\text{vol } S_r(p)}{e^{hr}} \geq \omega_{n-1} \left(\frac{n-1}{2h} \right)^{n-1},$$

where equality holds if and only if $(U_v'(0) - S_v'(0)) = \frac{2h}{n-1}$ id for almost all $v \in S_p X$ and by continuity of $(U_v'(0) - S_v'(0))$ equality holds for all $v \in S_p X$.

Let us now assume that equality holds for all $p \in X$. Hence, $(U_v'(0) - S_v'(0)) = \frac{2h}{n-1}$ id for all $v \in SX$.

Consider $U(v) = U_v'(0)$ and $S(v) = S_v'(0)$ then they are both solutions of the Riccati equation (see e.g. [Gre]), i.e.

$$\frac{d}{dt} U(\phi^t(v)) + U^2(\phi^t(v)) + R_v(t) = 0$$

and

$$\frac{d}{dt} S(\phi^t(v)) + S^2(\phi^t(v)) + R_v(t) = 0$$

where $\phi^t : SX \to SX$ denotes the geodesic flow on the unit tangent bundle SX and $R_v(t)$ is the Jacobi operator induced by the curvature tensor R, i.e.,

$R_v(t)(x) = R(x, \phi^t(v))\phi^t(v)$ for $x \in \phi^t(v)^\perp$. Subtracting the two Riccati equations, we obtain

$$0 = \frac{d}{dt}\Big|_{t=0} U(\phi^t(v)) - \frac{d}{dt}\Big|_{t=0} S(\phi^t(v)) + U^2(v) - S^2(v) = U^2(v) - S^2(v)$$

and hence,

$$U^2(v) = \left(\frac{2h}{n-1} \text{ id} + S(v)\right)^2 = \left(\frac{2h}{n-1}\right)^2 \text{id} + \frac{4h}{n-1}S(v) + S(v)^2.$$

Since $S^2(v) = U^2(v)$, this implies

$$S(v) = -\frac{h}{n-1} \text{ id}$$

and, therefore, using the Riccati equation again we obtain

$$R_v(0) = -S(v)^2 = -\left(\frac{h}{n-1}\right)^2 \text{id},$$

for all $v \in SX$. Hence, the sectional curvature is constant. $\qquad\square$

Now we introduce the asymptotic volume growth or *volume entropy* which is related to the topological entropy of the geodesic flow on a compact if such a quotient exists (see e.g. [Kn3]).

Definition 14 The *volume entropy* $h_{vol}(X)$ of a connected Riemannian manifold X is defined as

$$h_{vol}(X) = \limsup_{r\to\infty} \frac{\log \text{vol } B_r(p)}{r}, \tag{10}$$

where $B_r(p) \subset X$ is the ball of radius r around $p \in X$.

Remark (a) Note that (10) does not depend on the choice of reference point p and $h_{vol}(X)$ is therefore well defined. In case that a compact quotient exists one can replace lim sup by lim.

(b) Since the volume of a geodesic ball $B_r(p)$ of radius $r > 0$ is given by

$$\text{vol } B_r(p) = \int_0^r \text{vol } S_t(p) dt,$$

Corollary 13 implies $h_{vol}(X) \geq h$ for asymptotically harmonic manifolds. For harmonic manifolds $h_{vol}(X) = h$ follows from 8. However, we believe that $h_{vol}(X) = h$ holds for general asymptotically harmonic manifolds. We can prove it under the following condition.

Proposition 15 *If* $\mathrm{tr}(U'_{v,t}(0))$ *converges uniformly on* SX *to* h *we have*

$$h_{vol}(X) = \lim_{r \to \infty} \frac{\log \mathrm{vol}\, B_r(p)}{r} = h.$$

In particular, this holds if X *has a compact quotient or has no focal points or more generally if* X *has bounded asymptote (see section 8).*

Proof By the rules of L'Hospital we have

$$\lim_{r \to \infty} \frac{\log \mathrm{vol}\, B_r(p)}{r} = \lim_{r \to \infty} \frac{\mathrm{vol}\, S_r(p)}{\mathrm{vol}\, B_r(p)}$$

if the second limit exists. Using L'Hospital again we obtain

$$\lim_{r \to \infty} \frac{\mathrm{vol}\, S_r(p)}{\mathrm{vol}\, B_r(p)} = \lim_{r \to \infty} \frac{\int\limits_{S_p X} \det A_v(r)\, d\theta_p(v)}{\int\limits_0^r \int\limits_{S_p X} \det A_v(t)\, d\theta_p(v)\, dt}$$

$$= \lim_{r \to \infty} \frac{\int\limits_{S_p X} \mathrm{tr}\, A'_v(r) A_v^{-1}(r) \det A_v(r)\, d\theta_p(v)}{\int\limits_{S_p X} \det A_v(r)\, d\theta_p(v)}$$

$$= \lim_{r \to \infty} \frac{\int\limits_{S_p X} \mathrm{tr}\, U'_{\phi^r(v),r}(0) \det A_v(r)\, d\theta_p(v)}{\int\limits_{S_p X} \det A_v(r)\, d\theta_p(v)} = h,$$

where the last limit exists since $\mathrm{tr}(U'_{v,t}(0))$ converges uniformly on SX to h. If X has a compact quotient uniform convergence is a consequence of Lemma 11. In case of bounded asymptote this follows from the estimate 24. $\qquad\square$

4 Density functions of harmonic manifolds

In this section we will discuss a result of Nikolayevsky's [Ni], namely, that the density function $f(t) = \det A_v(t)$ of a harmonic manifold is an exponential polynomial. Since $\det A_{v,s}(t) = f(t - s)$ one obtains from Corollary 5

$$\frac{f(t - s)}{f(t) f(s)} = \det(Q_v(s) - Q_v(t)) \quad \text{where } Q_v(s) = A_v^{-1}(s) B_v(s).$$

Using this, Nikolayevsky's result is the consequence of the following general fact.

Proposition 16 *For a function* $\varphi \in C^\infty(\mathbb{R})$, *we denote its translation to* $s \in \mathbb{R}$ *by* φ_s *and define this function as* $\varphi_s(t) = \varphi(t - s)$. *Assume that the vector space,* *spanned by all translates of φ, is finite dimensional. Then φ is an exponential polynomial, i.e., we have*

$$\varphi(t) = \sum_{i=1}^{k} (p_i(t)\sin(\beta_i t) + q_i(t)\cos(\beta_i t))e^{\alpha_i t}$$

where p_i, q_i polynomials and $\beta_i, \alpha_i \in \mathbb{R}$.

Proof Let V be the finite dimensional vector space defined by

$$V := \text{span } \{\varphi_s \mid s \in \mathbb{R}\}.$$

The derivative of φ can be expressed as the following limit of functions:

$$\begin{aligned}
\varphi'(t) &= \lim_{s \to 0} \frac{\varphi(t) - \varphi(t - s)}{s} \\
&= \lim_{s \to 0} \frac{\varphi(t) - \varphi_s(t)}{s} \\
&= \left(\lim_{s \to 0} \frac{\varphi - \varphi_s}{s}\right)(t).
\end{aligned}$$

We have $\frac{1}{s}(\varphi - \varphi_s) \in V$ for all $s > 0$ and $\varphi' = \lim_{s \to 0} \frac{1}{s}(\varphi - \varphi_s)$. Since V is finite dimensional, it is closed and we have $\varphi' \in V$.

Now, the $(\dim V) + 1$ functions $\varphi, \varphi', \ldots, \varphi^{(\dim V)} \in V$ must be linear dependent over $\mathbb{R}$. This shows that φ satisfies a linear ordinary differential equation with constant coefficients and is, therefore, an exponential polynomial. □

Now one obtains:

Theorem 17 *(see [Ni, Theorem 2]) Let (X, g) be a harmonic manifold. Then the density function $f(t)$ is an exponential polynomial.*

Proof As introduced earlier, we define the translations $f_s : \mathbb{R} \to \mathbb{R}$ by $f_s(t) = f(t - s)$, where f is the density function. Since $Q_v(s) - Q_v(t)$ is a matrix with entries of the form $q(s) - q(t)$, we have

$$\begin{aligned}
f_s(t) &= \det(Q_v(s) - Q_v(t)) f(t) f(s) \\
&= \sum_{\alpha=1}^{N} b_\alpha(s)c_\alpha(t),
\end{aligned}$$

for some integer N and with suitable smooth functions b_α, c_α.

In particular, the vector space

$$V := \text{span } \{f_s \mid s \in \mathbb{R}\} \subset \text{span } \{c_\alpha \mid 1 \le \alpha \le N\}$$

has dimension $\leq N$. Applying Proposition 16, we see that $f(t)$ is an exponential polynomial. □

Remark Let (X, g) be a harmonic manifold.

(a) The volume of a geodesic ball $B_t(p)$ of radius $t > 0$ in (X, g) can be expressed by

$$\text{vol } B_t(p) = \int_0^t \text{vol } S_u(p) du$$

$$= \int_0^t \int_{S_pX} f(u) du = \omega_n \int_0^t f(u) du,$$

where ω_n is the volume of the $(n - 1)$-dimensional Euclidean unit sphere.
(b) We have $f(t) = (-1)^{\dim X - 1} f(-t)$, since

$$A_v(t) = -A_{-v}(-t) =: C(t)$$

which follows from the fact that both sides are Jacobi tensors along c_v with $A_v(0) = C(0) = 0$ and $A'_v(0) = C'(0) = \text{id}$. This implies

$$f(t) = \det A_v(t) = \det[-A_{-v}(-t)] = (-1)^{\dim X - 1} \det A_{-v}(-t)$$
$$= (-1)^{\dim X - 1} f(-t).$$

(c) The quotient $\frac{f'}{f}(r) \geq 0$ is the mean curvature of sphere $S_r(p)$ (with respect to the outward normal vector) in (X, g), and $\frac{f'}{f}(r)$ is monotone decreasing to the (constant) mean curvature $h \geq 0$ of the horospheres of (X, g).

Corollary 18 *Let (X, g) be a harmonic manifold. Then the following properties are equivalent:*

- *$h = 0$,*
- *X has polynomial volume growth,*
- *X has subexponential volume growth.*

Proof Assume $h = 0$. Then we have $\lim_{r \to \infty} \frac{f'(r)}{f(r)} = 0$. Using l'Hospital, we obtain

$$\lim_{r \to \infty} \frac{\log f(r)}{r} = \lim_{r \to \infty} \frac{f'(r)}{f(r)} = 0.$$

This shows that $f(r)$ has subexponential volume growth.

Assume that $f(r)$ has subexponential volume growth. Since $f(r)$ is an exponential polynomial, this implies

$$|f(r)| \leq C(1+r)^k \qquad \forall \, r \geq 0,$$

with a suitable $C > 0$ and for some $k \geq 0$. Therefore, (X, g) has polynomial volume growth.

Finally, assume that (X, g) has polynomial volume growth. This implies that

$$\lim_{r \to \infty} \frac{\log f(r)}{r} = 0.$$

On the other hand, we have

$$\lim_{r \to \infty} \frac{f'(r)}{f(r)} = h.$$

Using l'Hospital, again, we conclude that

$$0 = \lim_{r \to \infty} \frac{\log f(r)}{r} = \lim_{r \to \infty} \frac{f'(r)}{f(r)} = h.$$

This shows that $h = 0$. $\qquad\qquad\qquad\qquad\qquad\qquad\qquad\qquad\qquad\qquad$ $\square$

5 Harmonic manifolds with subexponential volume growth are flat

In this section, we discuss a result of Ranjan and Shah ([RSh1]), namely, that noncompact harmonic manifolds with polynomial volume growth are flat. Together with the result of Nikolayevsky one obtains.

Theorem 19 *Let X be a noncompact harmonic manifold with vanishing mean curvature of the horospheres. Then X is flat*

The function μ, which we introduce in the following Proposition, plays a crucial role in the proof of Ranjan and Shah. We begin by collecting some general properties of this function.

Proposition 20 *Let (X, g) be a general harmonic manifold with density function $f(r)$. Then the function $\mu(r) = \frac{\int_0^r f(s)ds}{f(r)}$ satisfies the following properties:*

(a) *We have $\mu(0) = 0$, $\mu'(0) = \frac{1}{n}$, $\mu''(0) = 0$, and $\mu'''(0) = \frac{2\mathrm{Ric}^X}{n(n+2)}$.*
(b) *We have $\mu(r) \geq 0$, for all $r \geq 0$.*
(c) *We have $0 \leq \mu'(r) \leq 1$ for all $r \geq 0$.*
(d) *We have $-\mu''(r)\mu(r) \leq \frac{1}{4}$, for all $r \geq 0$.*

Proof **(a)** Recall from [Wi2, Chapter 3.6] that

$$f(r) = r^{n-1}\left(1 - \frac{\mathrm{Ric}^X}{6n}r^2 + O(r^4)\right).$$

By integration, we obtain

$$\int_0^r f(t)dt = \frac{r^n}{n}\left(1 - \frac{\mathrm{Ric}^X}{6(n+2)}r^2 + O(r^4)\right).$$

This implies that

$$\mu(r) = \frac{\int_0^r f(s)ds}{f(r)} = \frac{r^r}{nr^{n-1}}\frac{1 - \frac{\mathrm{Ric}^X}{6(n+2)}r^2 + O(r^4)}{1 - \frac{\mathrm{Ric}^X}{6n}r^2 + O(r^4)}$$

$$= \frac{r}{n}\left(1 - \frac{\mathrm{Ric}^X}{6(n+2)}r^2 + O(r^4)\right)\left(1 + \frac{\mathrm{Ric}^X}{6n}r^2 + O(r^4)\right)$$

$$= \frac{r}{n}\left(1 + \frac{\mathrm{Ric}^X}{6}r^2\left(\frac{1}{n} - \frac{1}{n+2}\right) + O(r^4)\right)$$

$$= \frac{r}{n} + \frac{\mathrm{Ric}^X}{3n(n+2)}r^3 + O(r^5),$$

which allows us to read off the results of (a).

(b) This follows immediately from $f(r) > 0$ for all $r \geq 0$.

(c) Choose a point $p_0 \in X$. It is straightforward to see that μ satisfies the following differential equation

$$\mu'(r) + \frac{f'}{f}(r)\mu(r) = 1, \tag{11}$$

i.e., $\Delta(\mu \circ d_{p_0}) = 1$, which shows that $\mu \circ d_{p_0}$ is subharmonic. Applying the maximum principle to $\mu \circ d_{p_0}$, we see that the restriction of this function to any closed ball around p_0 assumes its maximum at the boundary of this ball. But $\mu \circ d_{p_0}$ is constant along the boundary of any of these balls, and we conclude that

$$\mu'(r) \geq 0 \qquad \forall r > 0.$$

On the other hand, using the above differential equation for μ again, as well as $(f'/f)(r) \geq 0$ and $\mu(r) \geq 0$, we obtain

$$\mu'(r) = 1 - \frac{f'}{f}(r)\mu(r) \leq 1.$$

(d) Rewriting (11), we obtain

$$\frac{f'}{f}(r) = \frac{1 - \mu'(r)}{\mu(r)},$$

and, consequently, using the fact that $(f'/f)(r)$ converges monotonely decreasing to h,

$$0 \geq \left(\frac{f'}{f}\right)'(r) = \frac{-\mu''(r)\mu(r) - (1 - \mu'(r))\mu'(r)}{\mu^2(r)}.$$

This implies that

$$-\mu''(r)\mu(r) \leq (1 - \mu'(r))\mu'(r).$$

Since $0 \leq \mu'(r) \leq 1$, we conclude that

$$-\mu''(r)\mu(r) \leq \frac{1}{4}.$$

$\square$

We will also make use of the following general fact.

Proposition 21 *Let (X, g) be a general harmonic manifold and $p_0 \in X$. Let $\varphi \in C^2(X)$ be a function satisfying*

$$\Delta\varphi = c, \qquad \varphi(p_0) = 0,$$

for some constant $c \in \mathbb{R}$. Let $g \in C^2(\mathbb{R})$ such that $g \circ d_{p_0} = \pi_{p_0}(\varphi) = \frac{1}{\omega_n} \int_{S_{p_0} X} \varphi(c_w(r)) d\theta_{p_0}(w)$. Then we have

$$g'(r) = c\mu(r).$$

Proof Note that $g(0) = \varphi(p_0) = 0$ and that g is even, since $\pi_{p_0}(\varphi)$ is radial around p_0. Therefore, we have $g'(0) = 0$. The function g satisfies the differential equation

$$g''(r) + \frac{f'(r)}{f(r)} g'(r) = c,$$

i.e,

$$f(r)g''(r) + f'(r)g'(r) = cf(r),$$

which, in turn, transforms into

$$(fg')'(r) = cf(r).$$

Integrating over $[0, r]$, and using $f(0)g'(0) = 0$, leads to

$$g'(r) = c\frac{\int_0^r f(t)dt}{f(r)} = c\mu(r).$$

$\square$

From now on, we assume that (X, g) is a harmonic manifold with polynomial volume growth. From Corollary 18, we know that this is equivalent to $h = 0$, i.e., all its horospheres are minimal. The main step in the proof of flatness is to prove that (X, g) is Ricci flat. We will show that $\mathrm{Ric}^X < 0$ would imply unboundedness of $-\mu''\mu$ from above, in contradiction to Proposition 20(d). On the other hand, we cannot have $\mathrm{Ric}^X > 0$, because this would imply compactness of (X, g), by the Bonnet-Myers Theorem. But we only consider noncompact harmonic manifolds.

Let $v \in S_{p_0}X$, and b_v be the associated Busemann function. Then

$$\Delta b_v = h = 0,$$

i.e., b_v is harmonic. This implies that

$$\Delta b_v^2 = 2\|\operatorname{grad} b_v\|^2 = 2, \tag{12}$$

and we can apply Proposition 21 with $\varphi = b_v^2$. Then the function $g \circ d_{p_0} := \pi_{p_0}(b^2)$ satisfies $g'(r) = 2\mu(r)$. Our next goal is to prove that the function g is a very special exponential polynomial.

Proposition 22 *g has the following properties:*

(a) g is an even function.
(b) We have $0 \le g(r) \le r^2$, for all $r \ge 0$.
(c) We have $0 \le g''(r) \le 2$, for all $r \ge 0$.

Proof **(a)** Since $\pi_{p_0}(b^2)$ is radial around p_0, g must be even.
 (b) $0 \le g(r)$ follows from $g(0) = b_v(p_0)^2 = 0$, $g(r) = \frac{1}{2}\int_0^r \mu(t)dt$ and $\mu(r) \ge 0$, for $r \ge 0$. Moreover,

$$g(r) = \frac{1}{\omega_n}\int_{S_{p_0}X} \underbrace{b_v^2(c_w(r))}_{\le r^2} \le r^2.$$

(c) This follows from Proposition 20(c). $\square$

The next Proposition is a key ingredient for the proof that g is an exponential polynomial.

Proposition 23 *Let (X, g) be a harmonic manifold with minimal horospheres. Consider the vector space*

$$\mathcal{F} = \{\phi : X \to \mathbb{R} \mid \exists\, c, c_1, c_2 > 0 \text{ with } \Delta\phi = c, |\phi(x)| \leq c_1 + c_2 d_{p_0}(x)^2\}.$$

Then $\mathcal{F}$ is finite dimensional.

Proof We know from Theorem 4.2 Li and Wang [LiW] that

$$H_2(p_0) = \{\phi : X \to \mathbb{R} \mid \Delta\phi = 0, \exists\, c_1, c_2 > 0 \text{ with } |\phi(x)| \leq c_1 + c_2 d_{p_0}(x)^2\}$$

is finite dimensional provided X has polynomial volume growth. Since the map $\Phi : \mathcal{F} \to \mathbb{R}$, $\Phi(\phi) = \Delta\phi$ is linear and $\ker \Phi = H_2(p_0)$, we obtain

$$\dim \mathcal{F} \leq \dim H_2(p_0) + 1 < \infty. \qquad \square$$

Next, we introduce the concept of translation of a radial function.

Definition 24 Let $\phi \in C(X)$ be a radial function around p_0 with $\phi(q) = F(d_{p_0}(q))$. The translation of ϕ to another point $p \in X$ is denoted by ϕ_p and defined by

$$\phi_p(q) = F(d_p(q)).$$

We now consider the radial function $\phi = \pi_{p_0}(b_v^2)$. Then $\phi = g \circ d_{p_0}$ and $\phi_p = g \circ d_p$. From (12) we conclude that

$$\Delta\phi = \Delta\pi_{p_0}(b_v^2) = \pi_{p_0}(\Delta b_v^2) = 2.$$

The translation of ϕ to any point $p \in X$ satisfies also

$$\Delta\phi_p = 2,$$

since Δ, applied to a radial function, is in radial coordinates independent of the center. Using 22(b), we have, for all $p \in X$,

$$|\phi_p(x)| \leq d_p(x)^2 \leq (d_p(p_0) + d_{p_0}(x))^2 \leq 2d_p(p_0)^2 + 2d_{p_0}(x)^2,$$

which shows that $\phi_p \in \mathcal{F}$, for all $p \in X$.

Let $\gamma : \mathbb{R} \to X$ be a geodesic with $\gamma(0) = p_0$. For a function $\psi \in C^\infty(X)$, we define $\gamma^*\psi \in C^\infty(\mathbb{R})$ via $(\gamma^*\psi)(t) = \psi(\gamma(t))$. Let $\mathcal{F}$ be the finite dimensional vector space introduced in Proposition 23. Then the vector space $\tilde{\mathcal{F}} = \gamma^*\mathcal{F} \subset C^\infty(\mathbb{R})$ has also finite dimension.

Note that $g = \gamma^*\phi \in \tilde{\mathcal{F}}$. Let $g_s(t) := g(t - s)$. Then we have, for all $s \in \mathbb{R}$, $g_s = \gamma^*\phi_{\gamma(s)} \in \tilde{\mathcal{F}}$. Applying Proposition 16, we see that g is an exponential polynomial. From Proposition 22, we know that $0 \leq g(t) \leq t^2$ and that g is even. Therefore, we must have

$$g(t) = \sum_{i=1}^{N_1} a_i \cos(\alpha_i t) + t \sum_{i=1}^{N_2} b_i \sin(\beta_i t) + t^2 \sum_{i=1}^{N_3} c_i \cos(\gamma_i t),$$

with suitable constants $a_i, b_i, c_i, \alpha_i, \beta_i, \gamma_i \in \mathbb{R}$.

Since $g''(t) \leq 2$ by Proposition 22(c), this simplifies to

$$g(t) = \sum_{i=1}^{N} a_i \cos(\alpha_i t) + ct^2.$$

Differentiating g three times, we obtain

$$\mu''(t) = \frac{1}{2} g'''(t) = \frac{1}{2} \sum_{i=1}^{N} a_i \alpha_i^3 \sin(\alpha_i t).$$

Now, we assume that $\mathrm{Ric}^X < 0$. Hence, Proposition 20(a) yields that $\mu'''(0) = \frac{2\mathrm{Ric}^X}{n(n+2)} < 0$, which implies the existence of $\delta, r_0 > 0$ such that $\mu''(r_0) = -\delta < 0$. Since μ'' is a finite sum of sines, μ'' is almost periodic, and there exists a sequence $r_k \to \infty$ such that $\mu''(r_k) = -\delta$ (see, e.g., [Bohr]). This together with

$$\lim_{r \to \infty} \frac{1}{\mu(r)} = \lim_{r \to \infty} \frac{f(r)}{\int_0^r f(t)dt} = \lim_{r \to \infty} \frac{f'(r)}{f(r)} = h = 0$$

yields $-\mu(r_k)\mu''(r_k) = \delta\mu(r_k) \to \infty$.

But $\mu(r_k)\mu''(r_k) \to \infty$ is in contradiction to Proposition 20(d) and therefore X has to be Ricci flat.

To prove flatness, we consider the Ricatti equation

$$U_v'(r) + U_v^2(r) + R_v(r) = 0,$$

where $U_v(r)$ (a self-adjoint endomorphism on $c_v'(r)^\perp \subset T_{c_v(r)}X$) denotes the second fundamental form of the horosphere through $c_v(r)$, and centered at $c_v(-\infty)$, and $R_v(r)(w) = R(w, c_v'(r))c_v'(r)$ is the Jacobi operator. Note that $\mathrm{tr}\, U_v(r) = (\Delta b_v)(c_v(r)) = h = 0$ and, therefore, also $\mathrm{tr}\, U_v'(r) = (\mathrm{tr}\, U_v)'(r) = 0$. Therefore, the Ricatti equation yields

$$tr U_v^2(r) = -\mathrm{tr}\, R_v(r) = -\mathrm{Ric}^X = 0,$$

i.e., $U_v^2(r) = 0$, which implies that $U_v(r) = 0$. Inserting this back into the Riccati equation, we end up with $R_v(r) = 0$, which shows that (X, g) is flat.

6 The rank of an asymptotically harmonic manifold

Denote by $D(v) = U_v'(0) - S_v'(0)$ the difference of the second fundamental of the stable and unstable horospheres determined by $v \in SX$.

Theorem 25 *[KnPe] Let (X, g) be an asymptotically harmonic manifold such that $\|R\| \leq R_0$ and $\|\nabla R\| \leq R_0'$ with suitable constants $R_0, R_0' > 0$. Then $v \mapsto \det D(v)$ is a constant function on SX.*
Moreover, if X has rank one, there exists $\rho > 0$ such that $D(v) \geq \rho \cdot \mathrm{id}$ for all $v \in SX$.

For harmonic manifolds, this Theorem is a consequence of the relation between $\det D(v)$ and the volume density function observed by the author (see [Kn4]) and stated in Proposition 12. For asymptotically harmonic manifolds this Theorem was proved in [HKS, Cor. 2.1] under the additional condition of strictly negative curvature bounded away from zero. Zimmer [Zi1, Prop. 3.3] provides a proof under the additional assumption of the existence of a compact quotient, using dynamical arguments. The proof of the general case without the assumptions of negative curvature or compact quotient has been very recently given by Peyerimhoff and myself [KnPe]. It requires new subtle estimates for the second fundamental forms of spheres and horospheres. We will discuss the basic ideas.

As a start, it is easy to see that $\det D(v) = \det D(-v)$:

$$\det D(-v) = \det(U(-v) - S(-v)) = \det(-S(v) + U(v)) = \det D(v).$$

Now we work towards the result that $\det D(v)$ is constant on all of SX.

The first step in the proof of Theorem 25 is to show that $\det D(v)$ is constant along the geodesic flow.

Proposition 26 *[HKS, Lemma 2.2] Let (X, g) be asymptotically harmonic. Then for all $v \in SX$, the map $t \mapsto \det(D(\phi^t v))$ is constant.*

Proof For the proof we need besides $D(v)$ the symmetric tensor $H(v) = -\frac{1}{2}(U(v) + S(v))$. Note that U and therefore also S are solutions of the Ricatti equation

$$\frac{d}{dt} U(\phi^t v) + U(\phi^t v)^2 + R_{\phi^t v} = 0.$$

Hence, a straightforward calculation yields for all $v \in SX$

$$(HD + DH)(\phi^t v) = S(\phi^t v)^2 - U(\phi^t v)^2 = \frac{d}{dt} D(\phi^t v). \tag{13}$$

In the case $\det D(\phi^t v) = 0$ for all $t \in \mathbb{R}$, there is nothing to prove. If $\det D(\phi^t v) \neq 0$ for some $t \in \mathbb{R}$, we have

$$
\begin{aligned}
\frac{d}{dt} \log \det D(\phi^t v) &= \frac{1}{\det D(\phi^t v)} \operatorname{tr} \left(\left(\frac{d}{dt} D(\phi^t v) \right) D^{-1}(\phi^t v) \right) \\
&= \frac{1}{\det D(\phi^t v)} \operatorname{tr} \left((HD + DH)(\phi^t v) D^{-1}(\phi^t v) \right) \\
&= \frac{2}{\det D(\phi^t v)} \operatorname{tr} \left(H(\phi^t v) \right) = 0,
\end{aligned}
$$

since $\operatorname{tr} H(w) = -\frac{1}{2}(\operatorname{tr} U(w) + \operatorname{tr} S(w)) = -\frac{1}{2}(\operatorname{tr} U(w) - \operatorname{tr} U(-w)) = 0$. This implies that $t \mapsto \det D(\phi^t v)$ is constant for all $t \in \mathbb{R}$. $\qquad\square$

The next step is to show that $\det D(v)$ is constant along stable and unstable manifolds which have been defined Section 2 by the formulas 9 . It requires the following estimates whose proofs are given in [KnPe].

Proposition 27 *(see [KnPe, Corollary 2.6] Let $\|R_v(t)\| \leq R_0$ for all $v \in SX$ and $t \in \mathbb{R}$ with a constant $R_0 > 0$. Let $\gamma : [0, 1] \to W^s(v)$ be a smooth curve and $\rho > 0$ such that*

$$
D(\phi^t(\gamma(s))) \geq \rho \cdot \mathrm{id}
$$

for all $s \in [0, 1]$ and $t \in \mathbb{R}$. Then there exists a function $b : \mathbb{R} \to (0, \infty)$, only depending on R_0 and ρ, such that we have for all $r > 1$ and all $-\infty < t < r$,

$$
\|S_{\gamma(s),r}(t)\| \leq b(t), \qquad \|U_{\gamma(s),r}(-t)\| \leq b(t). \tag{14}
$$

For $t \geq 0$ we have

$$
b(t) \leq a_2 e^{-\frac{\rho}{2} t}. \tag{15}
$$

Moreover, if $\beta = \pi \gamma$ and $\beta_t = \pi(\phi^t \gamma)$, we have

$$
\|\beta_t'(s)\| \leq b(t) \|\beta'(s)\| \tag{16}
$$

$$
\|(\phi^t \gamma)'(s)\| \leq b(t)\sqrt{1 + R_0} \, \|\beta'(s)\| \tag{17}
$$

for all $s \in [0, 1]$ and $t \in \mathbb{R}$.

Corollary 28 *(see [KnPe, Lemma 3.4]) Let $R_0 > 0$ be a constant such that $\|R\| \leq R_0$. Assume for $v \in SX$ the existence of a constant $\rho > 0$ such that*

$$
D(\phi^t(v)) \geq \rho \, \mathrm{id}
$$

for all $t \in \mathbb{R}$. Then there exists $a \geq 1$, depending only on R_0 and ρ such that

$$
0 < S'_{\phi^t(v)}(0) - S'_{\phi^t(v),r}(0) \leq \frac{a}{r} \quad \text{and} \quad 0 < U'_{\phi^t(v),r}(0) - U'_{\phi^t(v)}(0) \leq \frac{a}{r}
$$

for all $r > 0$ and all $t \in \mathbb{R}$.

Proof Since we have $U'_{w,r}(0) = -S_{-w,r}(0)$ for all $w \in SX$, it suffices to prove the first assertion. The estimates 14 and 15 in Proposition 27 above yields for all $t \geq 0$

$$\|S_w(t)\| \leq a_2 = a_2(R_0, \rho),$$

where $w = \phi^s(v)$ for some $s \in \mathbb{R}$. Recall from Lemma 6 that

$$S'_w(0) - S'_{w,r}(0) = \left(\int_0^r (S_w^* S_w)^{-1}(u) du \right)^{-1}.$$

This implies for all $x \in S_{c_v(s)} X$

$$\langle (S'_w(0) - S'_{w,r}(0)) x, x \rangle \leq \left\| \left(\int_0^r (S_w^* S_w)^{-1}(u) du \right)^{-1} \right\|$$

$$\leq \left(\int_0^r \|(S_w^* S_w)\|^{-1}(u) du \right)^{-1}$$

$$\leq \left(\int_0^r \frac{1}{a_2^2} du \right)^{-1} = \frac{a_2^2}{r},$$

which yields the required estimate. $\qquad\qquad\square$

The following formula for the difference of the second fundamental forms of spheres along curves in horospheres is crucial ingredient in the proof of Theorem 25 and is based on an formula of E. Hopf [Ho, (7.2)] for surfaces.

Proposition 29 *(see [KnPe, Corollary 2.6]) Let $\gamma : [0,1] \to W^s(v)$ be a smooth curve with $\gamma(0) = v$ and $\gamma(1) = \tilde{v}$. Let $e_1(s), \ldots, e_{n-1}(s)$ be an orthonormal frame in $\mathcal{H} = \pi W^s(v)$ along $\beta = \pi\gamma$ which is parallel in $\mathcal{H}$ with the induced connection. Let $e_i(s,t)$ be the parallel translation along the geodesic $c_{\gamma(s)}$. Then in coordinates relative to this frame we obtain*

$$S'_{\tilde{v},r}(0) - S'_{v,r}(0) = \int_0^1 \int_0^r S^*_{\gamma(s),r}(t) \left(\frac{\partial}{\partial s} R_{\gamma(s)}(t) \right) S_{\gamma(s),r}(t) dt\, ds \qquad (18)$$

and

$$U'_{\tilde{v},r}(0) - U'_{v,r}(0) = - \int_0^1 \int_{-r}^0 U^*_{\gamma(s),r}(t) \left(\frac{\partial}{\partial s} R_{\gamma(s)}(t) \right) U_{\gamma(s),r}(t) dt\, ds. \qquad (19)$$

Finally one obtains the following estimate:

Theorem 30 *(see [KnPe, Theorem 2.8]) Let (X, g) be a complete simply connected Riemannian manifold without conjugate points. Assume that $\|R_0\| \leq R_0$*

and $\|\nabla R\| \leq R_0'$ *with suitable constants* $R_0, R_0' > 0$. *Let* $\gamma : [0, 1] \to W^s(v)$ *be a smooth curve and* $\beta = \pi\gamma$. *Assume that* $D(\phi^t(\gamma(s))) \geq \rho \cdot \mathrm{id}$ *for all* $s \in [0, 1]$ *and* $t \in \mathbb{R}$ *and some constant* $\rho > 0$. *Let* $r > 1$. *Then there exists a constant* $C > 0$, *only depending on* R_0, R_0', ρ *and* r, *such that*

$$\|S'_{\gamma(1),r}(0) - S'_{\gamma(0),r}(0)\| \leq C\, \ell(\beta)$$

and

$$\|U'_{\gamma(1),r}(0) - U'_{\gamma(0),r}(0)\| \leq C\, \ell(\beta),$$

where $\ell(\beta)$ *denotes the length of the curve* β.

Now we are able to prove Theorem 25:

Step 1: det D is constant on stable and unstable manifolds: Let's assume that (X, g) has rank one, i.e., we have $\det D(w) > 0$ for some $w \in SX$. It suffices to show that $w \mapsto \det D(w)$ is locally constant on $W^s(v)$. Let $v \in SX$ and $\rho > 0$ such that $\det D(v) = 2\rho$. Since $w \mapsto \det D(w)$ is continuous on SX by Lemma 11, we find an open neighbourhood $U \subset SX$ of v such that $\det D(w) \geq \rho$ for all $w \in U$. Let $v, \tilde{v} \in U \cap W^s(v)$ and $\gamma : [0, 1] \to U \cap W^s(v)$ be a smooth curve with $\gamma(0) = v$ and $\gamma(1) = \tilde{v}$. We need to show that for every $\epsilon > 0$ the estimate $|\det D(v) - \det D(\tilde{v})| < \epsilon$ holds.

Proposition 26 implies

$$|\det D(v) - \det D(\tilde{v})| = |\det D(\phi^t(v)) - \det D(\phi^t(\tilde{v}))|$$

for all $t \in \mathbb{R}$ and Proposition 27 yields

$$\lim_{t \to \infty} d(\phi^t(v), \phi^t(\tilde{v})) = 0, \tag{20}$$

where because of the estimate (15) and (17) the convergence is exponentially fast.

Since by Lemma 8 the operators $D(w) = U(w) - S(w) \geq 0$ are uniformly bounded by $2\sqrt{R_0}$ and the determinant is a smooth function, there is a uniform Lipschitz constant $A > 0$ such that

$$|\det D(w_1) - \det D(w_2)| \leq A\, \|D(w_1) - D(w_2)\|.$$

Therefore, it suffices to show that, for every $\delta > 0$, there exists $t > 0$ such that

$$\|D(\phi^t(v)) - D(\phi^t(\tilde{v}))\| < \delta. \tag{21}$$

Let $D_r(w) = U'_{w,r}(0) - S'_{w,r}(0)$. Corollary 28 implies

$$\|D(\phi^t(w)) - D_r(\phi^t(w))\| \leq \frac{2a}{r}$$

for all $w \in \gamma([0, 1])$ and $t \in \mathbb{R}$. Therefore, we can choose $r > 1$ large enough such that we have

$$\|D(\phi^t(w)) - D_r(\phi^t(w))\| \leq \frac{\delta}{3}$$

for all $w \in \gamma([0, 1])$ and $t \in \mathbb{R}$. This implies that (21) holds if

$$\|D_r(\phi^t(v)) - D_r(\phi^t(\widetilde{v}))\| < \frac{\delta}{3}$$

for some $t > 0$. But this is a direct consequence of (20) and Theorem 30.

This shows that $w \rightarrow \det D(w)$ is locally and therefore also globally constant on $W^s(v)$. Note that $w \rightarrow \det D(w)$ is also constant on $W^u(v)$: Let $w \in W^u(v)$. Then $-w \in W^s(-v)$ and

$$\det D(w) = \det D(-w) = \det D(-v) = \det D(v).$$

Step 2: $\det D$ is constant on SX:

In the case $\det D(v) = 0$ for all $v \in SX$ there is nothing to prove. Therefore, we assume that there exists $v \in SX$ with $\det D(v) \neq 0$.

For $v \in SX$, let

$$W^{0s}(v) = \bigcup_{t \in \mathbb{R}} \phi^t(W^s(v)) = \{- \operatorname{grad} b_v(q) \mid q \in X\},$$

$$W^{0u}(v) = \bigcup_{t \in \mathbb{R}} \phi^t(W^u(v)) = \{\operatorname{grad} b_{-v}(q) \mid q \in X\}.$$

$W^{0s}(v)$ resp. $W^{0s}(v)$ are called the weak stable resp. unstable manifold. Observe that $W^{0u}(v) = -W^{0s}(-v)$ holds for all $v \in SX$.

We define a vector $w \in SX$ to be *asymptotic to* $v \in SX$ if $w \in W^{0s}(v)$. Since X has continuous asymptote, being asymptotic is an equivalence relation (see [Es, Prop. 3]). We write $v \sim w$ for asymtotic vectors $v, w \in SX$. Note that a flow line $\phi^{\mathbb{R}}(v_1)$ can intersect a leaf $W^u(v_2)$ in at most one vector, since the footpoint sets of these leafs are level sets of Busemann functions and $b_v(\pi(\phi^t(w))) = b_v(\pi(w)) - t$ for asymptotic vectors $v, w \in SX$.

Lemma 31 *Let $v, v' \in SX$ with $\det D(v) \neq 0$. Assume that $W^u(v) = W^u(v')$ and $v' \in W^{0s}(v)$. Then $v = v'$.*

Proof $v' \in W^{0s}(v)$ implies that v and v' are asymptotic. Since $W^u(v) = -W^s(-v)$, $W^u(v) = W^u(v')$ implies that $-v$ and $-v'$ are also asymptotic. Therefore, v and v' are bi-asymptotic. We have $v' \notin \phi^{\mathbb{R}}(v)$, since both v and v' lie in the same unstable manifold $W^u(v)$. By [Es, Thm. 1 (iv)], there exists a central Jacobi field along c_v, i.e., $\ker D(v) \neq 0$. But this contradicts to $\det D(v) \neq 0$. $\square$

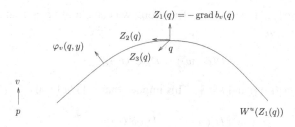

Figure 1 Illustation of the map $\varphi_v : X \times B_\delta(0) \to SX$

The assumption $\|R\| \leq R_0$ implies that the intrinsic sectional curvatures of all horospheres are also uniformly bounded in absolute value, by the Gauss equation. Therefore, there exists $\delta > 0$ such that for all horospheres $\mathcal{H}$ and all $p \in \mathcal{H}$, the intrinsic exponential map $\exp_{p,\mathcal{H}} : T_p\mathcal{H} \to \mathcal{H}$ is a diffeomorphism on the ball $B_{p,\mathcal{H}}(\delta) = \{v \in T_p\mathcal{H} \mid \|v\| < \delta\}$.

Assume that $n = \dim X$. Let $v \in SX$ be a fixed vector with $\det D(v) \neq 0$. Now, we define the following continuous map (see Figure 1)

$$\varphi_v : X \times B_\delta(0) \to SX,$$

where $B_\delta(0) = \{y \in \mathbb{R}^{n-1} \mid \|y\| < \delta\}$: Choose a smooth global orthonormal frame $Z_1 = -\operatorname{grad} b_v, Z_2, \ldots, Z_n$ on X. Define

$$\varphi_v(q, y) = \psi^u_{Z_1(q)} \left[\exp_{q, \pi W^u(Z_1(q))} \left(\sum_{i=2}^n y_i Z_i(q) \right) \right] \in W^u(Z_1(q)),$$

where $\psi^u_w : \pi W^u(w) \to W^u(w)$ is defined by $\psi^u_w(q) = \operatorname{grad} b_{-w}(q)$.

We now show that φ_v is injective: Let $\varphi_v(q, y) = \varphi_v(q', y')$. Then $W^u(Z_1(q)) = W^u(Z_1(q'))$ and

$$Z_1(q') = -\operatorname{grad} b_v(q') \sim v \sim -\operatorname{grad} b_v(q) = Z_1(q),$$

which implies $Z_1(q') \in W^{0s}(Z_1(q))$. We conclude from the previous subsections that $\det D(Z_1(q)) = \det D(v) \neq 0$. Using Lemma 31, we obtain $Z_1(q) = Z_1(q')$, i.e., $q = q'$. The equality $y = y'$ follows now from the injectivity of the exponential maps and ψ^u_w.

Since $\dim X \times B_\delta(0) = 2n - 1 = \dim SX$, we conclude that $U = \varphi_v(X \times B_\delta(0)) \subset SX$ is an open neighborhood of v, by Brouwer's domain invariance. Moreover, $\det D(w) = \det D(v) \neq 0$ for all $w \in U$, using that $\det D$ is constant along unstable manifolds, as well.

Without loss of generality, we assume that there exists a vector $v_0 \in SX$ with $\det D(v_0) = \alpha \neq 0$. Let $SX_\alpha = \{w \in SX \mid \det D(w) = \alpha\}$. By continuity of $w \mapsto \det D_w$, the set $SX_\alpha \subset SX$ is closed. Since $v_0 \in SX_\alpha$, we know that

SX_α is non-empty. The above arguments show for every vector $v \in SX_\alpha$ that the open neighbourhood $\varphi_v(U)$ is contained in SX_α, i.e., SX_α is open. Since SX is connected, we conclude that $SX_\alpha = SX$.

Finally $\|R\| \leq R_0$ implies that X has bounded sectional curvature. Hence by Lemma 8, the second fundamental forms of horospheres are bounded and therefore the eigenvalues of the positive endomorphism $D(v) = U'_v(0) - S'_v(0)$ are uniformly bounded from above. Since $\det D(v) = \mathrm{const} > 0$, we obtain that the smallest eigenvalue of $D(v)$ is uniformly bounded from below by a constant $\rho > 0$. This finishes the proof of Theorem 25.

7 Characterization of hyperbolicity for asymptotically harmonic manifolds

Theorem 32 *[KnPe] Let (X, g) be an asymptotically harmonic manifold such that $|R| \leq R_0$ and $|\nabla R| \leq R'_0$ with suitable constants $R_0, R'_0 > 0$. Let $h \geq 0$ be the mean curvature of its horospheres, i.e. $h = \mathrm{tr}\, U(v)$. Then the following properties are equivalent.*

(a) X has rank one.
(b) X has Anosov geodesic flow $\phi^t : SX \to SX$.
(c) X is Gromov hyperbolic.
(d) X has purely exponential volume growth with growth rate $h_{vol} = h$.

The main tool in proving these equivalences is provided by Theorem 25. This Theorem has been first obtained by the author [Kn4] in the case of *noncompact harmonic manifolds*. Under the assumption that (X, g) is an asymptotically harmonic manifold *with compact quotient*, this equivalence has been derived by Zimmer [Zi1].

Recently, we proved together with Norbert Peyerimhoff that Zimmer's assumption of a compact quotient can be dropped. Note that for harmonic manifolds the curvature tensor and its covariant derivative are bounded ([Be, Props. 6.57 and 6.68]), i,e, the Theorem above is a true generalization of the previous results of the author and Zimmer.

However under the assumption of a compact quotient the results become particularly interesting if we combine it with the rigidity Theorem of [BCG] together with [BFL] and [FL].

Theorem 33 *Let (M, g) be a compact Riemannian manifold such that the geodesic flow is Anosov. Assume that the mean curvature of the horospheres* *is constant. Then (M, g) is isometric to a locally symmetric space (M_0, g_0) of negative curvature.*

Proof From the work of P. Foulon and F. Labourie [FL] follows that the stable and unstable distribution E^s and E^u of an Anosov geodesic flow are C^∞ provided the mean curvature of the horospheres is constant. The results of Y. Benoist, P. Foulon and F. Labourie imply that the geodesic flow on the unit tangent bundle of (M, g) is smoothly conjugate to the geodesic flow on the unit tangent bundle of a locally symmetric space (M_0, g_0) of negative curvature. Furthermore, M, M_0 are homotopy equivalent and the topological entropy as well as the volume of both manifolds (M, g) and (M_0, g_0) coincide. Since by a result of A. Freiré and R. Mañé [FM] the volume entropy and the topological entropy for metrics without conjugate points coincide, the work of G. Besson, G. Courtois and S. Gallot implies that (M, g) and (M_0, g_0) are isometric. $\square$

Hence one obtains:

Corollary 34 *Let (X, g) be an asymptotically harmonic manifold admitting a compact quotient. Then X is a symmetric space of negative curvature provided one of the four equivalent conditions in Theorem 32 is fulfilled.*

From now on, we assume that (X, g) is asymptotically harmonic with $\|R\| \leq R_0$ and $\|\nabla R\| \leq R'_0$. We prove Theorem 32 proving each of the implications separately.

1. Rank one implies Anosov geodesic flow.

Observe first that $h > 0$. Otherwise, $\operatorname{tr} D(v) = \operatorname{tr} U(v) - \operatorname{tr} S(v) = h - h = 0$ which would imply $D(v) = 0$ for all $v \in SX$, since $D(v) \geq 0$. This contradicts $\operatorname{rank}(X) = 1$. Now we assume that $\operatorname{rank}(X) = 1$ and, therefore, $D(v) \geq \rho > 0$, by Theorem 25. By [Bo, Theorem, p. 107] this implies that the geodesic flow is Anosov.

2. Anosov geodesic flow implies Gromov hyperbolicity.

Recall that a geodesic metric space is called *Gromov hyperbolic* if there exists $\delta > 0$ such that every geodesic triangle is δ-thin, i.e., every side of the triangle is contained in the union of the δ-tubular neighborhoods of the other two sides.

Assume now that the geodesic flow $\phi^t : SX \to SX$ is Anosov with respect to the Sasaki metric. For $v \in SX$ consider the normal Jacobi tensor along c_v with $A_v(0) = 0$ and $A'_v(0) = \operatorname{id}$. Then the Anosov property implies (see [Bo, p. 113])

$$\|A_v(t)x\| \geq ce^{\alpha t}\|x\|$$

for $t \geq 1$ with suitable constants $c, \alpha > 0$. Consider two distinct geodesic rays $c_1 : [0, \infty) \to X$ and $c_2 : [0, \infty) \to X$ with $c_1(0) = c_2(0) = p$ and define

$$d_t^p(c_1(t), c_2(t)) := \inf\{L(\gamma) \mid \gamma : [a, b] \to X \setminus B(p, t) \text{ piecewise}$$
$$\text{smooth curve joining } c_1(t) \text{ and } c_2(t)\} \qquad (22)$$

where $B(p, t) = \{q \in X \mid d(p, q) < t\}$. Let $t \geq 1$ and $\gamma : [0, 1] \to X \setminus B(p, t)$ be a curve connecting $c_1(t)$ and $c_2(t)$. Let $v_1 = c_1'(0) \in S_p X$ and $v_2 = c_2'(0) \in S_p X$. Then $\gamma(s) = \exp_p(r(s)v(s))$ with $r : [0, 1] \to [t, \infty)$ and $v : [0, 1] \to S_p X$ and

$$\gamma'(s) = D \exp_p(r(s)v(s)) \left(r'(s)v(s) + r(s)v'(s)\right)$$
$$= r'(s)c_{v(s)}'(r(s)) + A_v(r(s))v'(s).$$

Since $c_{v(s)}'(r) \perp A_v(r)v'(s)$, we conclude that

$$\|\gamma'(s)\| \geq ce^{\alpha r(s)}\|v'(s)\|.$$

This implies that

$$L(\gamma) = \int_0^1 \|\gamma'(s)\| ds \geq ce^{\alpha t} \angle(v_1, v_2),$$

and therefore

$$\liminf_{t \to \infty} \frac{\log d_t^q(c_1(t), c_2(t))}{t} \geq c_0 \alpha \qquad (23)$$

with a suitable constant $c_0 > 0$. This implies, using [BH, Chapter III, Prop. 1.26] that X is Gromov hyperbolic. (Note that the condition there is $\liminf_{t \to \infty} \frac{d_t^q(c_1(t), c_2(t))}{t} = \infty$, which is a priori weaker than (23). In fact, both conditions are equivalent to Gromov hyperbolicity, see [BH, Chapter III, Prop. 1.25].)

3. Gromov hyperbolicity implies purely exponential volume growth and $h = h_{vol}$.

We like to note first that simply connected Riemannian manifolds X without conjugate points which are Gromov hyperbolic spaces *admitting compact quotients* have purely exponential volume growth (see [Coor, Thm. 7.2]). Here we consider the special case of an asymptotically harmonic manifold without the additional assumption that X admits a compact quotient.

Definition 35 Let X be a Riemannian manifold with $h_{vol} = h_{vol}(X) > 0$, where $h_{vol}(X)$ denotes the volume entropy defined in 14. Then X has *purely exponential volume growth* if for every $p \in X$, there exists a constant

$C = C(p) \geq 1$ with

$$\frac{1}{C} e^{h_{vol}r} \leq \text{vol } B_r(p) \leq C e^{h_{vol}r}$$

for all $r \geq 1$.

Remark Note, that corollary 13 implies $h \leq h_{vol}(X)$ and $h = h_{vol}(X)$ holds either under the assumption of the existence of a compact quotient or bounded asymptote.

We begin by proving the following general lemma.

Lemma 36 *Let X be a δ-hyperbolic space without conjugate points and bounded curvature. Then the volume of any geodesic sphere grows exponentially. In particular, we have $h_{vol}(X) > 0$.*

Proof Fix $p \in X$ and geodesic rays $c_1, c_2 : [0, \infty) \to X$ with $c_1(0) = c_2(0)$. As remarked above, Gromov hyperbolicity implies

$$\liminf_{t \to \infty} \frac{\log d_t^q(c_1(t), c_2(t))}{t} \geq c(\delta),$$

where $c(\delta) > 0$ depends only on the Gromov constant δ. In particular, there exists $t_0 > 0$ such that for all $t \geq t_0$

$$d_{S_p(t)}(c_1(t), c_2(t)) \geq e^{tc(\delta)/2},$$

where $d_{S_p(t)}$ is the intrinsic distance in the sphere $S_p(t) \subset X$. Let $\gamma_t : [0, l(t)] \to S_p(t)$ be a minimal geodesic in $S_p(t)$ connnecting $c_1(t)$ and $c_2(t)$. The $1/4$-balls in $S_p(t)$ with centers $\gamma_t(k)$ and $k \in \mathbb{Z} \cap [0, l(t)]$ are pairwise disjoint. Lemma 8 implies that the second fundamental forms of $S_p(t)$ are bounded by a universal constant for all $t \geq t_0 > 0$. Using the Gauss equation, this implies that the curvatures of the spheres $S_p(t)$ are uniformly bounded for $t \geq t_0$, as well. Therefore, the $1/4$-balls in $S_p(t)$ have a uniform lower volume bound $A_0 > 0$. Hence, we have

$$\text{vol}(S_t(p)) \geq A_0(e^{tc(\delta)/2} - 1)$$

for all $t \geq t_0$. This finishes the proof of the Lemma.   □

Recall the following result in [Kn4, Cor. 4.6].

Proposition 37 *Let X be a simply connected δ-hyperbolic manifold without conjugate points. Consider for $v \in S_pX$, $\ell = \delta + 1$ and $r > 0$ the spherical cone in X given by*

$$A_{v,\ell}(r) := \{c_w(t) \mid 0 \leq t \leq r, w \in S_pX, d(c_v(\pm\ell), c_w(\pm\ell)) \leq 1\}.$$

Then, for $\rho = 4\delta + 2$ the set $A_{v,\ell}(r)$ is contained in

$$H_{v,\rho}(r) := \{c_q(t) \mid -\rho/2 \leq t \leq r, \; c_q \text{ is an integral curve of}$$
$$\text{grad } b_{-v} \text{ with } c_q(0) = q \in b_{-v}^{-1}(0) \cap B(p, \rho)\}.$$

This result has the following consequence.

Corollary 38 *Let (X, g) be a Gromov hyperbolic asymptotically harmonic manifold and $p \in X$. Then there exists a constant $C_2(p) > 0$ such that*

$$\text{vol } B_r(p) \leq C_2(p) \int_{-\rho/2}^{r} e^{hs} ds,$$

where ρ is defined as in Proposition 37. In particular, we have

$$h_{vol}(X) \leq h.$$

Proof Let $p \in X$. Choose $l = \delta + 1$. Then we have

$$S_p X = \bigcup_{v \in S_p X} U_{v,\ell}(r),$$

with the open sets

$$U_{v,\ell}(r) = \{w \in S_p X \mid d(c_v(\pm\ell), c_w(\pm\ell)) < 1\}.$$

Since $S_p X$ is compact, we find finitely many vectors $v_1, \ldots, v_k \in S_p X$ with

$$S_p X = \bigcup_{j=1}^{k} U_{v_j,\ell}(r),$$

which implies for $\rho = 4\delta + 2$

$$B_r(p) \subset \bigcup_{j=1}^{k} A_{v_j,\ell}(r) \subset \bigcup_{j=1}^{k} H_{v_j,\rho}(r).$$

Using

$$\text{vol}(H_{v,\rho}(r)) = \int_{-\rho/2}^{r} e^{hs} ds \, \text{vol}_0(b_v^{-1}(0) \cap B_\rho(p))$$

where vol_0 denotes the induced volume on the horosphere $b_v^{-1}(0)$, we conclude

$$\text{vol } B_r(p) \leq \left(\sum_{j=1}^{k} \text{vol}_0(b_{v_j}^{-1}(0) \cap B_\rho(p)) \right) \int_{-\rho/2}^{r} e^{hs} ds.$$

Setting $C_2(p) = \sum_{j=1}^{k} \mathrm{vol}_0(b_{v_j}^{-1}(0) \cap B_\rho(p))$ proves the first part of the Corollary. The inequality $h_{vol}(X) \le h$ follows then from the definition of $h_{vol}(X)$.   □

Now we are able to prove:

Proposition 39 *Let (X, g) be a Gromov hyperbolic asymptotically harmonic space with bounded curvature. Then X has purely exponential volume growth and $h = h_{vol}$.*

Proof By Lemma 36 Gromov hyperbolicity implies $h_{vol}(X) > 0$. From Corollary 13 and Corollary 38 we obtain $h = h_{vol}(X)$. Moreover, we derive from Corollary 38 that

$$\mathrm{vol}\, B_r(p) \le \frac{C_2(p)}{h} e^{hr}.$$

Since Corollary 13 implies for $r \ge 2$ that

$$\frac{\mathrm{vol}\, B_r(p)}{e^{hr}} \ge \frac{\int_{r-1}^{r} \mathrm{vol}\, S_t(p)dt}{e^{hr}} \ge \frac{1}{e^h} \int_{r-1}^{r} \frac{\mathrm{vol}\, S_t(p)}{e^{ht}} dt \ge \frac{\omega_{n-1}}{e^h} \left(\frac{n-1}{2h}\right)^{n-1}$$

we obtain that X has purely exponential volume growth.   □

4. Purely exponential volume growth with $h = h_{vol}$ implies rank one.

Finally, we show the remaining implication of Theorem 32. This closes the chain of implications and finishes the proof that all four properties listed in (a), (b), (c) and (d) are equivalent.

Assume that (X, g) is asymptotically harmonic with purely exponential volume growth and $h = h_{vol}$. Proposition 12 implies

$$\int_{r-1}^{r} e^{ht} \int_{S_p X} \frac{1}{\det(U(v) - S'_{v,t}(0))} d\theta_p(v)dt \le \mathrm{vol}(B_r(p)).$$

Hence,

$$\frac{1}{e^h} \int_{r-1}^{r} \int_{S_p X} \frac{1}{\det(U(v) - S'_{v,t}(0))} d\theta_p(v)dt \le \frac{\mathrm{vol}(B_r(p))}{e^{hr}} \le C(p)$$

for some constant $C(p) > 0$. Assume that $\det(U(v) - S'_{v,t}(0)) \to 0$ for all $v \in S_p X$. Then, monotonicity and Dini's Theorem imply that this convergence is uniform. However, this contradicts the above inequality. Therefore, there exist $v \in S_p X$ with $\det(U(v) - S(v)) \ne 0$ and therefore, (X, g) has rank one.

8 Harmonic manifolds with bounded asymptote

In this section, we show now that for a harmonic manifold of bounded asymptote the rank is constant. In odd dimensions this implies by a result of N.E. Steenrod and J.H.C. Whitehead [SW] that the rank is one. The notion of bounded asymptote has been introduced by J.-H. Eschenburg [Es]. In particular, M has bounded asymptote if M has nonpositive curvature or, more generally, no focal points.

Definition 40 Let M be a manifold without conjugate points. Then M is called of bounded asymptote if there exists a constant $\alpha \geq 1$ such that for all $v \in SX$

$$\|S_v(t)x_t\| \leq \alpha \|x\|$$

for all $t \geq 0$ and parallel vector fields x_t with $x_0 = x \in v^\perp$.

Remark Note that bounded asymptote implies for all $v \in SX$ the estimate

$$\|U_v(t)x_t\| \geq \frac{1}{\alpha}\|x\|$$

for all $t \geq 0$ and parallel vector fields x_t with $x_0 = x \in v^\perp$, as well.

Proposition 41 *Let X be a simply connected n-dimensional manifold of bounded asymptote. Then for all $x \in \ker(U'_v(0) - S'_v(0))$*

$$\frac{1}{\alpha^2 t}\langle x, x\rangle \leq \langle (U'_v(0) - S'_{v,t}(0))x, x\rangle \leq \frac{\alpha^2}{t}\langle x, x\rangle.$$

Let

$$\lambda_1(v, t) \leq \ldots \lambda_k(v, t) \leq \lambda_{k+1}(v, t) \leq \ldots \leq \lambda_{n-1}(v, t)$$

be the eigenvalues of $U'_v(0) - S'_{v,t}(0)$ and

$$\beta_t(v) := \lambda_{k+1}(v, t) \cdot \ldots \cdot \lambda_{n-1}(v, t),$$

then $\beta_t(v)$ is monotonically decreasing and converging to the product of the positive eigenvalues $\beta(v)$ of $U'_v(0) - S'_v(0)$. Furthermore,

$$\frac{\beta(v)}{\alpha^{2k}} \leq \det(U'_v(0) - S'_{v,t}(0))t^k \leq \alpha^{2k}\beta_t(v).$$

Proof Since Lemma 6 implies

$$(U'_v(0) - S'_{v,t}(0)) = \left(\int_0^t \left(U_v^* U_v\right)^{-1}(u)du\right)^{-1}$$

we have for each unit vector $x \in v^\perp$

$$\langle (U_v'(0) - S_{v,t}'(0))x, x \rangle \geq \frac{1}{\max \left\{ \int_0^t \langle (U_v^* U_v)^{-1}(u) y_u, y_u \rangle du \mid \|y\| = 1 \right\}}$$

$$= \frac{1}{\left\| \int_0^t (U_v^* U_v)^{-1}(u) du \right\|}.$$

Using

$$\left\| \int_0^t (U_v^* U_v)^{-1}(u) du \right\| \leq \int_0^t \|(U_v^* U_v)^{-1}(u)\| du \leq \int_0^t \|U_v^{-1}(u)\|^2 du$$

and

$$\|U_v^{-1}(u)\| = \frac{1}{\min\{\|U_v(u)x_u\| \mid \|x\| = 1\}} \leq \alpha,$$

we obtain

$$\frac{1}{\alpha^2 t} \leq \langle (U_v'(0) - S_{v,t}'(0))x, x \rangle.$$

On the other hand, lemma 6 implies

$$(S_v'(0) - S_{v,t}'(0)) = \left(\int_0^t (S_v^* S_v)^{-1}(u) du \right)^{-1}$$

as well. This yields for all unit vectors $x \in v^\perp$:

$$\langle (S_v'(0) - S_{v,t}'(0))x, x \rangle \leq \left\| \left(\int_0^t (S_v^* S_v)^{-1}(u) du \right)^{-1} \right\|$$

$$\leq \left(\int_0^t \|(S_v^* S_v)(u)\|^{-1} du \right)^{-1} = \left(\int_0^t \|S_v(u)\|^{-2} du \right)^{-1}$$

$$\leq \left(\int_0^t \frac{1}{\alpha^2(v)} du \right)^{-1} = \frac{\alpha^2}{t}.$$

Therefore,

$$\langle (S_v'(0) - S_{v,t}'(0))x, x \rangle \leq \frac{\alpha^2}{t}. \tag{24}$$

Putting both inequalities together, we obtain for all $x \in \ker(U_v'(0) - S_v'(0))$ with $\|x\| = 1$:

$$\frac{1}{\alpha^2 t} \le \langle (U_v'(0) - S_{v,t}'(0))x, x \rangle \le \frac{\alpha^2}{t},$$

which implies the first assertion. Let $0 < \lambda_1(v, t) \le \ldots \le \lambda_{n-1}(v, t)$ the eigenvalues of $(U_v'(0) - S_{v,t}'(0))$ and $k = \dim(\ker(U_v'(0) - S_v'(0))$. Then using the above estimates and the minmax characterization of eigenvalues we conclude that

$$\frac{1}{\alpha^2 t} \le \lambda_i(v, t) \le \frac{\alpha^2}{t}$$

for $1 \le i \le k$. The remaining eigenvalues $\lambda_{k+1}(v, t) \le \ldots \le \lambda_{n-1}(v, t)$ of $(U_v'(0) - S_{v,t}'(0))$ are monotonically decreasing in t and converging to the positive eigenvalues of $(U_v'(0) - S_v'(0))$. Hence,

$$\frac{\beta(v)}{\alpha^{2k} t^k} \le \det(U_v'(0) - S_{v,t}'(0)) \le \frac{\alpha^{2k} \beta_t(v)}{t^k}, \tag{25}$$

where $\beta_t(v) = \lambda_{k+1}(v, t) \cdot \ldots \cdot \lambda_{n-1}(v, t)$ and $\beta(v) = \lambda_{k+1}(v) \cdot \ldots \cdot \lambda_{n-1}(v)$ is the product of the positive eigenvalues. $\qquad\Box$

Corollary 42 *Let X be a nonflat, harmonic manifold of bounded asymptote and volume density $f(t)$. Then*

$$\frac{1}{\alpha^{2k} \beta_1(v)} \le \frac{f(t)}{e^{ht} \, t^{\mathrm{rank}(v)-1}} \le \frac{\alpha^{2k}}{\beta(v)}$$

for all $t \ge 1$. In particular, $\mathrm{rank}(v)$ is constant. Furthermore, there exists a constant $\rho > 0$ such that

$$\langle (U_v'(0) - S_v'(0))x, x \rangle \ge \rho \langle x, x \rangle$$

for all $x \in \ker(U_v'(0) - S_v'(0))^\perp$.

Proof Using the estimate 25 above together with 12 we have for $t \ge 1$

$$\frac{\beta(v)}{\alpha^{2k} t^{\mathrm{rank}(v)-1}} \le \frac{e^{ht}}{f(t)} = \det(U_v'(0) - S_{v,t}'(0)) \le \det(U_v'(0) - S_{v,1}'(0))$$

$$\le \frac{\alpha^{2k} \beta_1(v)}{t^{\mathrm{rank}(v)-1}}$$

Comparing this inequality at v, $v' \in SX$ shows that $\mathrm{rank}(v)$ is constant.

Now let v_0 be a fixed and v an arbitrary vector in SX. Then the estimate (25) implies

$$\frac{\beta(v_0)}{\alpha^{2k}} \le \det(U'_{v_0}(0) - S'_{v_0,t}(0))t^k = \det(U'_v(0) - S'_{v,t}(0))t^k$$

$$\le \alpha^{2k}\beta_t(v)$$

for all $t \ge 0$ and, therefore,

$$\frac{\beta(v_0)}{\alpha^{2k}} \le \alpha^{2k}\beta(v),$$

where $\beta(v) = \prod_{i=k+1}^{n-1} \lambda_i(v)$ is the product of the positive eigenvalues of $(U'_v(0) - S'_v(0))$. Since the curvature of X is bounded the eigenvalues of $(U'_v(0) - S'_v(0))$ are bounded by Lemma 8. Therefore, the lower bound of $\beta(v)$ implies that all the positive eigenvalues of $(U'_v(0) - S'_v(0))$ are bounded from below as well. □

Theorem 43 *Let X be a harmonic manifold of bounded asymptote and odd dimension. Then for all $v \in SX$ we have* rank$(v) = 1$ *or X is flat. In the first case the geodesic flow is Anosov.*

Proof Since the rank(v) is constant for each $v \in S_p X$ consider the subspace of $v^\perp \cong T_v S_p X$, given by $\mathcal{L}(v) = \ker(U'_v(0) - S'_v(0))$. Since the rank is constant it defines a continuous distribution. If the dimension n of X is odd, the dimension of the sphere $S_p X$ is even. By a result of N.E. Steenrod and J.H.C. Whitehead [SW] the distribution must be trivial, i.e., dim $\mathcal{L}(v)$ is zero or has dimension $n - 1$. In the latter case $U'_v(0) = S'_v(0)$ and, therefore, $e^{ht} = \det U_v(t) = \det S_v(t) = e^{-ht}$. But then $h = 0$. In particular, X is flat by Theorem 19. If dim $\mathcal{L}(v)$ is zero then rank$(v) = 1$ for all $v \in SX$. In this case corollary 42 yields $(U'_v(0) - S'_v(0)) \ge \rho$ id for a constant $\rho > 0$. Hence, be a theorem of Bolton [Bo] we have that the geodesic flow is Anosov and the assertion is a consequence. □

We immediately obtain:

Corollary 44 *Let X be a nonflat harmonic manifold of bounded asymptote and odd dimension. If X admits a compact quotient X is a symmetric space of rank 1.*

Proof Since the geodesic flow is Anosov the assertion is a consequence of the results in [BCG]. □

9 Harmonic and asymptotically harmonic manifolds without focal points

For even dimension the proof of Theorem 43 does not work. However, it was shown in [Kn4] that under the stronger assumption of no focal points the assertion of this Theorem remains true, i.e.

Theorem 45 *[Kn4] Let* (X, g) *be a harmonic manifold without focal points. Then for all* $v \in SX$ *we have* rank$(v) = 1$ *or* X *is flat. In the first case the geodesic flow is Anosov.*

Using the rank rigidity theorem (see [Ba] resp. [BS] in case of nonpositive sectional curvature and [Wa] for its recent extension to no focal points) in combination with Theorem 32 one obtains a partial generalization to asymptotically harmonic manifolds.

Theorem 46 *[KnPe] Let* (X, g) *be an asymptotically harmonic manifold without focal points and uniform bounds on the curvature tensor and its derivative. Suppose that* (X, g) *admits a quotient of finite volume. Then* (X, g) *is either flat or the geodesic flow on* X *is Anosov.*

Proof Since higher rank symmetric spaces are not asymptotically harmonic the rank rigidity Theorem implies that rank$(X) = 1$ or X is flat. Furthermore, Theorem 32 yields that the geodesic flow is Anosov if rank$(X) = 1$. □

Remark If X admits a compact quotient this result is due to Zimmer [Zi1]. In this case the uniform bounds on the curvature tensor and its derivative are automatically satisfied. Moreover, using entropy rigidity one even obtains that X is a rank one symmetric space of noncompact type provided that X is non-flat.

Unfortunately, the rank rigidity Theorem for manifolds X without focal points is not known without further assumptions on X, like the existence of a quotient of finite volume or homogeneity.

However, the author obtained in [Kn4] for manifolds without focal points the following weaker rigidity result which suffices to prove Theorem 45

Theorem 47 *[Kn4] Let* (X, g) *be a manifold with bounded sectional curvature and no focal points such that* $v \mapsto$ rank(v) *is constant. For each* $v \in SX$ *define* $\mathcal{L}(v) = \ker(U'_v(0) - S'_v(0))$ *and assume that the symmetric positive definite operator* $(U'_v(0)) - (S'_v(0)|_{\mathcal{L}^\perp(v)})$ *is uniformly bounded from below by a positive constant. Then the geodesic flow is Anosov (with respect to the Sasaki metric) unless* rank$(v) = \dim X$ *and therefore* X *is flat.*

Since by Corollary 42 the assumptions of Theorem 47 are fulfilled for harmonic manifolds without focal points we obtain Theorem 45.

We are finishing this survey by providing an outline of the main steps of the proof of Theorem 47. Note that X has no focal points if the second fundamental form of horospheres is positive semi-definite or if for all $v \in SX$ the Busemann functions b_v are convex. In terms of the stable and unstable Jacobi tensors S_v and U_v this is equivalent to $U_v'(0) \geq 0$ and therefore $S_v'(0) = -U_v'(0) \leq 0$ for all $v \in SX$. Most of the properties of manifolds of nonpositive curvature are shared by manifolds without focal points. Since $U_{v,t}'(0) > U_v'(0) \geq 0$ for all $t > 0$ geodesic spheres are convex as well. Furthermore, the flat strip Theorem is true, which asserts that two geodesics c_1 and c_2 bound a flat strip if $d(c_1(t), c_2(t)) \leq b$ for some $b \geq 0$ and all $t \in \mathbb{R}$. This means that there exists an isometric, totally geodesic imbedding $F : [0, a] \times \mathbb{R} \to X$ such that $c_1(t) = F(0, t)$ and $c_2(t) = F(a, t)$ (see [Es] for a proof). In particular, c_1 and c_2 are parallel, i.e., $d(c_1(t), c_2(t)) = a$.

Lemma 48 *[Kn4] Let X be a manifold without focal points. Then the following holds*

$$\mathcal{L}(v) = \ker(U_v'(0)) - (S_v'(0))$$
$$= \{x \in v^\perp \mid x_t \text{ is a parallel Jacobi field along } c_v\}.$$

Furthermore $x_t, S_v(t)x_t, U_v(t)x_t \in \mathcal{L}^\perp(\phi^t v)$ for all $x \in \mathcal{L}^\perp(v) = \{x \in v^\perp \mid x \perp \mathcal{L}(v)\}$. Furthermore, the distributions $\mathcal{L}(v)$ and $\mathcal{L}^\perp(v)$ depend smoothly on $v \in SX$.

Proposition 49 *[Kn4] Let X be a nonflat manifold without focal points and bounded sectional curvature. Assume there exists $w \in SX$ such that for all $v = \phi^t(w)$ the eigenvalues of $(U_v'(0) - S_v'(0))$ restricted to $\mathcal{L}^\perp(v) \subset v^\perp$ are bounded from below by a positive constant constant $\rho > 0$. Then there are constants $a \geq 1$ and $\alpha = \rho/2$ such that*

$$\|S_v(t)x_t\| \leq ae^{-\alpha t}\|x\| \quad \text{and} \quad \|S_v(-t)x_{-t}\| \geq \frac{1}{a}e^{\alpha t}\|x\|$$

as well as

$$\|U_v(t)x_t\| \geq \frac{1}{a}e^{\alpha t}\|x\| \quad \text{and} \quad \|U_v(-t)x_{-t}\| \geq ae^{-\alpha t}\|x\|$$

for all $x \in \mathcal{L}^\perp(v)$ and $t \geq 0$. Furthermore, for $t \geq 0$

$$\|A_v(t)x_t\| = t\|x\| \quad \text{for all } x \in \mathcal{L}(v)$$

and there exists a constant a' such that

$$\|A_v(t)x_t\| \geq a' e^{\alpha t} \|x\| \quad \text{for all } x \in \mathcal{L}^\perp(v)$$

for all $t \geq 1$.

Proof $v \in \{\phi^t(w) \mid t \in \mathbb{R}\}$ consider the equation

$$S_v^*(t) \left(U'_{\phi^t(v)}(0) - S'_{\phi^t(v)}(0) \right) S_v^*(t) = \left(\int_{-\infty}^t (S_v^* S_v)^{-1}(u) du \right)^{-1}$$

proved in Lemma 7. Since the eigenvalues of $((U'_{\phi^t(v)}(0) - S'_{\phi^t(v)}(0))$ restricted to $\mathcal{L}^\perp \phi^t(v) \subset \phi^t(v)^\perp$ are bounded from below by a positive constant ρ, we obtain

$$\rho \leq \left\langle \left(\int_{-\infty}^t (S_v^* S_v)^{-1}(u) du \right)^{-1} \Big|_{\mathcal{L}^\perp(\phi^t v)} S_v^{-1}(t)x, \, S_v^{-1}(t)x \right\rangle$$

$$\leq \left\| \left(\int_{-\infty}^t (S_v^* S_v)^{-1}(u) du \right)^{-1} \Big|_{\mathcal{L}^\perp(\phi^t v)} \right\| \|S_v^{-1}(t)x\|^2$$

for all $x \in \mathcal{L}^\perp(\phi^t v)$ with $\|x\| = 1$. Furthermore, we have

$$\left\| \left(\int_{-\infty}^t (S_v^* S_v)^{-1}(u) du \right)^{-1} \Big|_{\mathcal{L}^\perp(\phi^t v)} \right\|$$

$$= \frac{1}{\min \left\{ \int_{-\infty}^t \langle (S_v^* S_v)^{-1}(u) y_u, \, y_u \rangle du \mid y \in \mathcal{L}^\perp(v), \|y\| = 1 \right\}}.$$

Therefore,

$$\rho \min \left\{ \int_{-\infty}^t \langle S_v^{*-1}(u) y_u, \, S_v^{*-1}(u) y_u \rangle du \mid y \in \mathcal{L}^\perp(v), \|y\| = 1) \right\}$$

$$\leq \|S_v^{-1}(t)x\|^2.$$

Defining

$$\varphi(u) := \min \left\{ \|S_v^{*-1}(u)y\|^2 \mid y \in \mathcal{L}^\perp(\phi^u v), \|y\| = 1 \right\}$$

$$= \min \left\{ \|S_v^{-1}(u)y\|^2 \mid y \in \mathcal{L}^\perp(\phi^u v), \|y\| = 1 \right\},$$

we obtain

$$\rho \int_0^t \varphi(u)du \leq \rho \int_{-\infty}^t \varphi(u)du \leq \varphi(t)$$

and, hence,

$$\rho F(t) \leq F'(t)$$

for $F(t) := \int_0^t \varphi(u)du$. This implies $F(t) \geq F(1)e^{\rho t}$ for all $t \geq 1$ and, therefore, $\varphi(t) = F'(t) \geq \rho F(t) \geq \rho F(1)e^{\rho t}$ for all $t \geq 1$. Since the sectional curvature of X is bounded, $\varphi(t)$ is on $[0, 1]$ bounded away from 0. Hence, there exist a constant $a \geq 1$ such that

$$\|S_v^{-1}(t)y\| \geq \frac{1}{a}e^{\alpha t}\|y\| \tag{26}$$

for $\alpha = \frac{\rho}{2}$ and all $y \in \mathcal{L}^\perp(\phi^t v)$ and $t \geq 0$. Since $S_v(t) : \mathcal{L}^\perp(\phi^t v) \to \mathcal{L}^\perp(\phi^t v)$ is an isomorphism, we obtain for all $x \in \mathcal{L}^\perp(\phi^t v)$ and $t \geq 0$

$$\|S_v(t)x\| \leq ae^{-\alpha t}\|x\|.$$

Note, that for each $u \in \mathbb{R}$, we have

$$S_{\phi^u v}(t)x_t = S_v(t+u)(S_v^{-1}(u)x)_t,$$

where x_t is the parallel translation of $x \in \mathcal{L}^\perp(\phi^u(v))$ along $c_{\phi^u v}(t)$. Hence, for $t = -u \leq 0$ we obtain with (26)

$$\|S_{\phi^u v}(-u)x_{-u}\| = \|(S_v^{-1}(u)x)_{-u}\| = \|(S_v^{-1}(u)x)\| \geq \frac{1}{a}e^{\alpha u}\|x\|.$$

In particular, for $w = \phi^u v$ and $u \geq 0$ the estimate

$$\|S_w(-u)x_{-u}\| \geq \frac{1}{a}e^{\alpha u}\|x\|$$

holds for all $x \in \mathcal{L}^\perp(w)$. Since $U_v(t) = S_{-v}(-t)$ the second estimate of the Proposition follows.

To prove the remaining assertions we recall that

$$A_v(t)x_t = U_v(t) \int_0^t \left(U_v^* U_v\right)^{-1}(s)x_s ds.$$

If $x \in \mathcal{L}(v)$ we have $\left(U_v^* U_v\right)^{-1}(s)x_s = x_s$ and, therefore, $A_v(t)x_t = U_v(t)(tx_t) = tx_t$.

If $x \in \mathcal{L}^{\perp}(v)$ we have $\int_0^t \left(U_v^* U_v\right)^{-1}(s)x_s ds \in \mathcal{L}^{\perp}(\phi^t v)$. Therefore,

$$\|A_v(t)x_t\| \geq ae^{\alpha t} \left\| \int_0^t \left(U_v^* U_v\right)^{-1}(s)x_s ds \right\|.$$

Since Lemma 6 implies

$$((U_v'(0) - S_{v,t}'(0))^{-1}x)_t = \int_0^t (U_v^* U_v)^{-1}(s)x_s ds$$

and since for $t \geq 1$ there exists a constant $b > 0$ such that $\langle (U_v'(0) - S_{v,t}'(0))x, x \rangle \leq b\langle x, x \rangle$ for all $x \in v^{\perp}$ the last estimate follows. $\qquad\square$

Using Proposition 49 we obtain that under the assumption of Theorem 47 the geodesic flow is partially hyperbolic. For the proof consider the distributions

$$E^p(v) = \{(x + \lambda v, 0) \mid x \in \mathcal{L}(v), \lambda \in \mathbb{R}\},$$
$$E^c(v) = \{(x + \lambda v, y) \mid x, y \in \mathcal{L}(v), \lambda \in \mathbb{R}\},$$
$$E^s(v) = \{x, S_v'(0)x\} \mid x \in \mathcal{L}^{\perp}(v)\},$$
$$E^u(v) = \{x, U_v'(0)x\} \mid x \in \mathcal{L}^{\perp}(v)\}.$$

We call E^p the parallel, E^c the central, E^s the stable and E^u the unstable distribution. Obviously $E^p(v) \subset E^c(v)$. Furthermore, the central, stable and the unstable distributions are transversal and

$$\dim E^c(v) = 2\operatorname{rank}(X) - 1, \ \dim E^s(v) = \dim E^u(v) = n - \operatorname{rank} X.$$

Therefore, the sum of the dimension is equal to $2n - 1$ and hence

$$T_v SX = E^c(v) \oplus E^s(v) \oplus E^u(v).$$

Proposition 50 *Under the assumption of Theorem 47, the geodesic flow is partially hyperbolic with respect to the Sasaki metric. More precisely, there are constants $b, c \geq 1$ such that for all $\xi \in E^s(v)$*

$$\|D\phi^t(v)\xi\| \leq b\|\xi\|e^{\alpha t}, \ t \geq 0 \ \text{and} \ \|D\phi^t(v)\xi\| \geq \frac{1}{b}\|\xi\|e^{-\alpha t}, \ t \leq 0$$

and for all $\xi \in E^u(v)$

$$\|D\phi^t(v)\xi\| \geq \frac{1}{b}\|\xi\|e^{\alpha t}, \ t \geq 0 \ \text{and} \ \|D\phi^t(v)\xi\| \leq b\|\xi\|e^{\alpha t}, \ t \leq 0.$$

Furthermore, for all $\xi \in E^c(v)$ and $t \in \mathbb{R}$ we have

$$\|D\phi^t(v)\xi\| \leq c\|\xi\|(|t| + 1).$$

Proof For $\xi = (x, S'_v(0)x) \in E^s(v)$ we obtain

$$\|D\phi^t(v)\xi\| = \|(S_v(t)x_t, S'_v(t)x_t)\| = \sqrt{\|S_v(t)x_t\|^2 + \|S'_v(t)x_t\|^2}$$

$$= \|S_v(t)x_t\|\sqrt{1 + \frac{\|S'_v(t)x_t\|^2}{\|S_v(t)x_t\|^2}}.$$

Moreover, Lemma 7 implies

$$\frac{\|S'_v(t)x_t\|}{\|S_v(t)x_t\|} \leq \|S'_v(t)S_v(t)^{-1}\| = \|S'_{\phi^t v}(0)\| \leq \beta$$

and Proposition 49 yields

$$\|D\phi^t(v)\xi\| \leq ae^{-\alpha t}\|x\|\sqrt{1 + \beta^2} \leq ae^{-\alpha t}\|\xi\|\sqrt{1 + \beta^2}.$$

The remaining assertions are obtained in a similar way. $\qquad\square$

As we will see, all the distributions are integrable. Define

$$P(v) = \{w \in SX \mid d(c_w(t), c_v(t)) \text{ is constant}\}$$

to be the subset of SX consisting of vectors tangent to the parallel geodesics of c_v and denote by $F(v) := \pi(P(v))$ its projection on X.

Proposition 51 *Assume that X fulfills the assumption of Theorem 47. Then the distributions E^p and E^c are integrable and provide flow invariant and smooth foliations. The integral manifolds of E^p are given by $P(v)$. The projection $F(v) := \pi(P(v))$ of each leaf is a k-flat. The integral manifolds of E^c are given by the unit tangent bundles $SF(v)$ of the k-flats. The distributions E^s and E^u are integrable as well and the leaves are the stable and unstable manifolds given by*

$$W^s(v) = \{w \in SX \mid d(c_v(t), c_w(t)) \leq ae^{-ct}d(\pi(v), \pi(w)), t \geq 0\}$$

and

$$W^u(v) = \{w \in SX \mid d(c_v(t), c_w(t)) \leq ae^{-ct}d(\pi(v), \pi(w)), t \leq 0\}$$

for constants $c, a > 0$.

Proof The integrability of E^p follows as in Lemma 2.2 in [BBE], which was given under the assumption of nonpositive curvature and compact quotient, but not necessarily constant rank. By Lemma 48 the distribution $E^p(v) =$

$\{(x + \lambda v, 0) \mid x \in \mathcal{L}(v)\}$ is smooth. Choose a smooth curve $\rho : [0, a] \to SX$ with $\rho(0) = v$ tangent to E^p, i.e. $\frac{d}{ds}\rho(s) = (x(s) + \lambda(s)\rho(s), 0) \in E^p(\rho(s))$, and, therefore, $x(s) \in \mathcal{L}(\gamma(s))$. Consider for each $t \in \mathbb{R}$ the curve $\gamma_t : I \to SX$ with $\gamma_t(s) = c_{\rho(s)}(t)$. Hence,

$$\frac{d}{ds}\gamma_t(s) = J_{\rho(s)}(t),$$

where $J_{\rho(s)}(t)$ is the parallel Jacobi field with $J_{\rho(s)}(0) = x(s) + \lambda(s)\rho(s)$. Consequently, the length of $\gamma_t(I)$ is constant, the distance $d(c_v(t), c_{\rho(s)}(t))$ is bounded and the geodesic $c_{\rho(s)}(t)$ is parallel to c_v. If ξ and η are two smooth vector fields tangent to E^p the commutator $[\xi, \eta]$ is tangent to E^p as well. To prove this consider the flows φ_ξ and φ_η. Then for s

$$f(s^2) = \varphi_\eta^{-s} \circ \varphi_\xi^{-s} \circ \varphi_\eta^{s} \circ \varphi_\xi^{s}(v)$$

is parallel to v and, hence,

$$\frac{d}{ds}\bigg|_{s=0} f(s^2) = [\xi, \eta](v) \in E^p(v).$$

Therefore, each leaf of E^p though v is a k-dimensional submanifold of SX, given by $P(v)$. Now the flat strip Theorem (see [Es]) implies that the projections $F(v) := \pi(P(v))$ are k-flats, i.e., totally geodesic flat spaces isometric to the Euclidean space $\mathbb{R}^k$. Therefore, the unit tangent bundle of a k-flat is flow invariant and as one easily checks tangent to the central distribution E^c.

Since ϕ^t defines a partially hyperbolic flow the stable and unstable distributions E^s and E^u are integrable and tangent to W^s and W^u. $\qquad\square$

Let X be a simply connected manifold without conjugate points. Let $B(q, t)$ be the open ball of radius t about q. For $q_1, q_2 \in X \setminus B(q, t)$ we recall the definition

$$d_t^q(q_1, q_2) = \inf\{L(\gamma) \mid \gamma : [a, b] \to X \setminus B(q, t)$$

$$\text{piecewise smooth curve joining } q_1 \text{ and } q_2\}.$$

given in 22. The following Lemma is crucial in the proof of Theorem 47.

Lemma 52 *Let X be a simply connected manifold without focal points which fulfills the assumption of Theorem 47 and assume that its rank is at least 2. Given $v \in S_q X$ then for each $w \in S_q X$ the following is equivalent*

(a) $w \in S_q F(v)$,

(b) $\overline{\lim_{t\to\infty}} \dfrac{1}{t} d_t^q(c_v(t), c_w(t)) < \infty$,

(c) *there exists* $\rho \geq 0$ *such that* $\overline{\lim_{t\to\infty}} \dfrac{1}{t} d_{t-\rho}^q(c_v(t), c_w(t)) < \infty$.

Proof We can assume that X is not flat. Assume $w \in S_q F(v)$. Consider a shortest curve $x : [0, a] \to S_q F(v)$ such that $x(0) = v, x(a) = w$ and $\|x'(s)\| = 1$. In particular, $x'(s) \in \mathcal{L}(x(s))$. Then $\gamma_t(s) = \exp_q(tx(s))$ connects $c_v(t)$ and $c_w(t)$ and the image is in the complement of $B(q, t)$. Furthermore, $\dot{\gamma}_t(s) = J(t)$ is the Jacobi field along the geodesic $c_{x(s)}$ with $J(0) = 0$ and $J'(0) = x'(s)$. Hence, $J(t) = A_{x(s)}(t)(x'(s))_t$, where $A_{x(s)}$ is the Jacobi tensor along $c_{x(s)}$ with $A_{x(s)}(0) = 0$ and $A'_{x(s)}(0) = \mathrm{id}$. Since $x'(s) \in \mathcal{L}(x(s))$ and $\|x'(s)\| = 1$ Proposition 49 implies

$$\|\dot{\gamma}_t(s)\| = \|A_{x(s)}(t)(x'(s))_t\| = t.$$

Hence, $L(\gamma_t) = ta$ and, therefore, (a) implies (b).

Since for all $\rho \geq 0$ and $q_1, q_2 \in X \setminus B(q, t)$ we have $d_{t-\rho}^q(q_1, q_2) \leq d_t^q(q_1, q_2)$, assertion (c) follows from (b).

Now assume that (c) holds and $w \notin S_q F(v)$. Let $\gamma : [0, 1] \to X$ be a smooth curve such that $\gamma(0) = v, \gamma(1) = w$ and $d(\gamma(s), q) = f(s) \geq t - \rho$. Consider the curve $x : [0, 1] \to S_q X$ such that $\gamma(s) = \exp_q(f(s)x(s))$ and the geodesic variation

$$\varphi(s, u) = \exp_q(ux(s)), \quad \text{where } 0 \leq u \leq f(s).$$

Then

$$\frac{\partial}{\partial s}\varphi(s, u) = J_{x(s)}(u)$$

is the perpendicular Jacobi field along $c_{x(s)}(u)$ with initial conditions $J_{x(s)}(0) = 0$ and $J'_{x(s)}(0) = x'(s)$. Since

$$\frac{\partial}{\partial s}\bigg|_{s=s_0} \gamma(s) = \frac{\partial}{\partial s}\bigg|_{s=s_0} \varphi(s, f(s_0)) + \frac{\partial}{\partial u}\bigg|_{u=f(s_0)} \varphi(s_0, u) f'(s_0)$$
$$= J_{x(s_0)}(f(s_0)) + \dot{c}_{x(s_0)}(f(s_0)) f'(s_0)$$

the estimate

$$\left\|\frac{\partial}{\partial s}\bigg|_{s=s_0} \gamma(s)\right\| \geq \|J_{x(s_0)}(f(s_0))\|$$

holds. As above we have $J_{x(s)}(u) = A_{x(s)}(u)(x'(s))_u$. Decompose $x'(s) = y(s) + z(s)$, where $y(s) \in \mathcal{L}(x(s))$ and $z(s) \in \mathcal{L}^\perp(x(s))$. Since $w \notin S_q F(v)$ there is a constant $b > 0$ such $\int_0^1 \|z(s)\|ds \geq b$. Using Proposition 49, we obtain

$$L(\gamma) = \int_0^1 \|\dot{\gamma}(s)\|ds \geq a'e^{\alpha(t-\rho)} \int_0^1 \|z(s)\|ds \geq a'e^{\alpha(t-\rho)}b$$

for all $t > 0$ in contradiction to (c). $\qquad\qquad\qquad\qquad\qquad\qquad$ □

Proof of Theorem 47 Assume that the rank of X is at least 2. Consider $v \in S_pX$ and $v' \in W^s(v)$ where $v' \neq v$ and $\pi(v') = q$. This implies that $d(c_v(t), c_{v'}(t))$ converges to 0 as t tends to ∞. For each $w \in S_pF(v)$ define $w' = -\operatorname{grad} b_w(q)$. Since X has no focal points $d(c_{w'}(t), c_w(t)) \leq d(p, q)$. In particular, for $\rho = d(p, q)$ we obtain

$$\frac{1}{t}d^p_{t-\rho}(c_{w'}(t), c_{v'}(t))$$

$$\leq \frac{1}{t}\left(d^p_{t-\rho}(c_{w'}(t), c_w(t)) + d^p_{t-\rho}(c_w(t), c_v(t)) + d^p_{t-\rho}(c_v(t), c_{v'}(t))\right)$$

is bounded for $t \geq 0$. Using Lemma 52 this implies $w' \in S_qF(v')$. Furthermore,

$$\theta' := \sphericalangle_q(w', v') = \sphericalangle_p(w, v) =: \theta,$$

where $\theta, \theta' \in [-\pi, \pi]$. To see this we note that c_v and c_w respectively $c_{v'}$ and $c_{w'}$ are lying in the Euclidean spaces $F(v)$ resp. $F(v')$. Therefore, we have

$$d(c_v(t), c_w(t)) = 2t \sin\left(\frac{\theta}{2}\right) \quad \text{and} \quad d(c_{v'}(t), c_{w'}(t)) = 2t \sin\left(\frac{\theta'}{2}\right)$$

and the triangle inequality implies

$$2t\left|\sin\left(\frac{\theta}{2}\right) - \sin\left(\frac{\theta'}{2}\right)\right| = |d(c_v(t), c_w(t) - d(c_{v'}(t), c_{w'}(t))|$$

$$\leq |d(c_v(t), c_{v'}(t)) + d(c_{w'}(t), c_w(t))| \leq A$$

for some constant $A > 0$ and for all $t \geq 0$. Hence $\theta = \theta'$. In particular, for $w = -v$ we obtain

$$w' = (-v)' = -(v')$$

since

$$\sphericalangle_q((-v)', v') = \sphericalangle_p(-v, v) = \pi.$$

Consequently

$$d(c_{v'}(-t), c_v(-t)) = d(c_{-(v')}(t), c_{-v}(t)) = d(c_{(-v)'}(t), c_{-v}(t)) \leq d(p, q)$$

for all $t \geq 0$. Since $d(c_v(t), c_{v'}(t))$ converges to 0 for $t \to \infty$ we have $v = v'$ which is in contradiction to the assumption. Hence the rank of X must be one. $\qquad\qquad\square$

Acknowledgement. The author would like to thank Norbert Peyerimhoff for a careful reading of this article.

Gerhard Knieper
Faculty of Mathematics, Ruhr University Bochum,
44780 Bochum, GERMANY
Email: gerhard.knieper@rub.de

References

[Ba] W. Ballmann. *Nonpositively curved manifolds of higher rank*, Ann. of Math. **122** (2) (1985), 597–609.

[BBE] W. Ballmann, M. Brin, P. Eberlein. *Structure of manifolds of nonpositive curvature I*, Ann. of Math. **122** (2) (1985), 171–203.

[BCG] G. Besson, G. Courtois, S. Gallot. *Entropies et regidités des espaces localement symétriques de courbure strictement négative.*, Geom. Funct. Anal. **5** (5) (1995), 731–799.

[BFL] Y. Benoist, P. Foulon, F. Labourie. *Flots d' Anosov à distributions stable et instable différéntiables*, J. Amer. Math. Soc. **5** (1) (1992), 33–74.

[Be] A.L. Besse. *Manifolds all of whose geodesics are closed*, Ergebnisse der Mathematik und ihrer Grenzgebiete [Results in Mathematics and Related Areas] **93**, Springer Verlag, Berlin-New York, 1978.

[BH] M.R. Bridson, A. Haefliger. *Metric spaces of nonpositive curvature.* Grundlehren der mathematischen Wissenschaften, [Fundamental Principles of Mathematical Sciences], **319**, Springer-Verlag, Berlin, 1999. xxii + 643 pp.

[Bohr] H. Bohr. *Almost Periodic Functions*, Chelsea Publishing Company, New York, 1947.

[Bo] J. Bolton. *Conditions under which a geodesic flow is Anosov*, Inst. Hautes Études Sci. Publ. Math., **65** Math. Ann. **240** (2) (1979), 103–113.

[BS] K. Burns, R. Spatzier. *Manifolds of nonpositive curvature and their buildings*, (1987), 35–59.

[Bu] K. Burns. *The flat strip theorem fails for surfaces with no conjugate points*, Proc. Am. Math. Soc. **115** (1) (1992), 199–206.

[CS] P. Castillon, A. Sambusetti. *On asymptotically harmonic manifolds of negative curvature*, Math. Zeitschrift, published online, March 2014, see also arXiv:1203.2482, 12 March 2012.

[Coor] M. Coornaert. *Mesures de Patterson-Sullivan sur le bord d'un espace hyperbolique au sens de Gromov*, Pacific J. Math. **159** (2) (1993), 241–270.

[DR] E. Damek, F. Ricci. *A class of nonsymmetric harmonic Riemannian spaces*, Bull. Amer. Math. Soc. (N.S.) **27** (1) (1992), 139–142.

[Eb] P. Eberlein. *When ist the geodesic flow of Anosov type? I.*, J. Differential Geometry **8** (1973), 437–463.

[EO] P. Eberlein, B. O'Neill. *Visibility manifolds*, Pacific J. Math. **46** (1973), 45–109.

[Es] J.H. Eschenburg. *Horospheres and the stable part of the geodesic flow*, Math. Zeitschrift **153** (3) (1977), 237–251.

[FM] A. Freiré, R. Mañé. *On the entropy of the geodesic flow in manifolds without conjugate points*, Invent. Math. **69** (3) (1982), 375–392.

[FL] P. Foulon, F. Labourie. *Sur les variétés compactes asymptotiquement harmoniques*, Invent. Math. **109** (1) (1992), 97–111.

[Gr] M. Gromov. *Hyperbolic groups*, In: Essays in group theory, Math. Sci. Res. Inst. Publ., **8**, Springer, New York (1987), 75–263.

[Gre] L. Green. *A theorem by Hopf*, Michigan Math. J. **5** (1) (1958), 31–34.

[He] J. Heber. *On harmonic and asymptotically harmonic homogeneous spaces*, Geom. Funct. Anal. **16** (4) (2006), 869–890.

[HKS] J. Heber, G. Knieper, H. M. Shah. *Asymptotically harmonic spaces in dimension 3*, Proc. Amer. Math. Soc. **135** (3) (2007), 845–849.

[Ho] E. Hopf. *Statistik der Lösungen geodätischer Probleme vom unstabilen Typus. II*, Math. Ann. **117** (1940), 590–608.

[Kl] W. Klingenberg. *Riemannian manifolds with geodesic flow of Anosov type*, Ann. of Math. **99** (2) (1974), 1–13.

[Kn1] G. Knieper. *Mannigfaltigkeiten ohne konjugierte Punkte*, Bonn. Math. Schr. **168** (1986).

[Kn2] G. Knieper. *On the asymptotic geometry of nonpositively curved manifolds*, Geom. Funct. Anal. **7** (4) (1997), 755–782.

[Kn3] G. Knieper. *Hyperbolic Dynamics and Riemannian Geometry*, in Handbook of Dynamical Systems, Vol. 1A 2002, Elsevier Science B., eds. B. Hasselblatt and A. Katok, (2002), 453–545.

[Kn4] G. Knieper. *New results on noncompact harmonic manifolds*, Comment. Math. Helv. **87** (3) (2012), 669–703, see also arXiv:0910.3872, 20 Oct. 2009.

[KnPe] G. Knieper, N.Peyerimhoff. *Geometric properties of rank one asymptotically harmonic manifolds*, J. Differential Geom. **100** (3) (2015), 507-532.

[Le] F. Ledrappier. *Harmonic measures and Bowen-Margulis measures*, Israel J. Math. **71** (3) (1990), 275–287.

[Led] J. Ledger. *Symmetric harmonic spaces*, J. London Math. Soc. **32** (1957), 53–56.

[Li] A. Lichnerowicz. *Sur les espaces riemanniens complèment harmoniques* (French), Bull. Soc. Math. France **72** (1944), 146–168.

[LiW] P. Li, J. Wang. *Mean value inequalities*, Indiana Univ. Math. J. **48** (4) (1999), 1257–1283.

[Ma] R. Mané. *On a theorem of Klingenberg*, Dynamical Systems and Bifurcation Theory, (Rio de Janeiro, 1985), M. Camacho, M. Pacifico and F. Takens eds., Pitman Res. Notes Math. **160** Longman Sci. Tech., Harlow (1987), 319–345.

[Ni] Y. Nikolayevsky. *Two theorems on harmonic manifolds*, Comment. Math. Helv. **80** (1) (2005), 29–50.

[PeSa] N. Peyerimhoff, E. Samiou. *Integral geometric properties of noncompact harmonic spaces*, J. Geom. Anal. **25** (1) (2015), 122-148.

[RSh1] A. Ranjan, H. Shah. *Harmonic manifolds with minimal horospheres*, J. Geom. Anal. **12** (4) (2002), 683–694.

[Ru] H. S. Ruse, : *On the elementary solution of Laplace's equation*, Proc. Edinburgh Math. Soc. **2** (2) (1931), 135–139.

[SS] V. Schroeder, H. Shah. *On 3-dimensional asymptotically harmonic manifolds*, Arch. Math. **90** (2008), 275–278.

[SW] N.E. Steenrod, J.H.C. Whitehead. *Vector fields on the n-Sphere*, Proc. Nat. Acad. Sci. U. S. A. **37** (1951), 58–63.

[Sz] Z.I. Szabó. *The Lichnerowicz Conjecture on Harmonic manifolds*, J. Differential Geom. **31** (1) (1990), 1–28.

[Wal] A.G. Walker. *On Lichnerowicz's conjecture for harmonic 4-spaces*, J. London Math. Soc. **24** (1949), 21–28.

[Wa] J. Watkins. *The Higher Rank Rigidity Theorem for Manifolds With No Focal Points*, Geom. Dedicata, **164** (2013), 319–349.

[Wi1] T.J. Willmore. *Mean value theorems in harmonic Riemannian spaces*, J. London Math. Soc. **25** (1950), 54–57.

[Wi2] T.J. Willmore. *Riemannian geometry*. Oxford Science Publications, The Clarendon Press Oxford University Press, New York, 1993.

[Zi1] A. M. Zimmer. *Compact asymptotically harmonic manifolds*, J. Mod. Dynamics. **6** (3) (2012), 377–403.

[Zi2] A. M. Zimmer. *Boundaries of non-compact harmonic manifolds*, Geom. Dedicata, **168** (2014), 339–357.

6

The Atiyah conjecture

PETER A. LINNELL[1]

Abstract

We will give a survey on the Atiyah conjecture and related questions. In particular we will consider recent progress on approximating L^2-invariants and mod-p analogues.[2]

1 Introduction

Let G be a discrete group and let $c_0(G)$ denote all formal sums $\sum_{x\in G} a_x x$, where $a_x \in \mathbb{C}$ and for every $\epsilon > 0$, the set $\{x \mid |a_x| > \epsilon\}$ is finite. Define $*: c_0(G) \to c_0(G)$ by $\left(\sum_{x\in G} a_x x\right)^* = \bar{a}_x x^{-1}$, where $\bar{a}_x$ denotes the complex conjugate of a_x. Let $\ell^2(G)$ denote the Hilbert space with Hilbert basis G. Then $\ell^2(G)$ is the set of formal sums $\sum_{x\in G} a_x x$ with $a_x \in \mathbb{C}$ and $\sum_{x\in G} |a_x|^2 < \infty$, so it can be viewed as a $\mathbb{C}$-subspace of $c_0(G)$. The usual multiplication (convolution) $\mathbb{C}G \times \mathbb{C}G \to \mathbb{C}G$ extends to a multiplication $\ell^2(G) \times \ell^2(G) \to c_0(G)$ defined by

$$\sum_{x\in G} a_x x \sum_{x\in G} b_x x = \sum_{x\in G} \left(\sum_{y\in G} a_{xy^{-1}} b_y\right) x.$$

This in turn extends to matrix multiplication for matrices with entries in $\ell^2(G)$. Then the group von Neumann algebra $\mathcal{N}(G)$ is

$$\{\alpha \in \ell^2(G) \mid \alpha\beta \in \ell^2(G) \; \forall \beta \in \ell^2(G)\}.$$

Thus $\mathcal{N}(G)$ is a subspace of $\ell^2(G)$ which contains $\mathbb{C}G$ and is closed under addition and multiplication, so $\mathcal{N}(G)$ is a $\mathbb{C}$-algebra. However $\ell^2(G)$ is not closed

[1] Partially support by a grant from the NSA
[2] **Keywords:** Atiyah conjecture, pro-p group, localization, Ore condition.
AMS codes: Primary 20C07, Secondary 16S34, 20C07 22D25, 46L10.

under multiplication in general. On the other hand $\ell^1(G) := \{\sum_{x \in G} a_x x \mid a_x \in \mathbb{C}$ and $\sum_{x \in G} |a_x| < \infty\}$ is closed under multiplication, and so is always a $\mathbb{C}$-algebra. Furthermore $\mathcal{N}(G) = \{\alpha \in \ell^2(G) \mid \beta\alpha \in \ell^2(G) \; \forall\beta \in \ell^2(G)\}$. It follows that $\mathcal{N}(G)$ acts on both the left and right of $\ell^2(G)$, by left and right multiplication respectively. Another description of $\mathcal{N}(G)$ is as follows. Let $\mathcal{B}(H)$ denote the bounded linear operators on the Hilbert space H, which we shall regard as acting on the left of H. Then $\mathbb{C}G$ acts faithfully on $\ell^2(G)$ as bounded linear operators by $\theta(\alpha) = \theta\alpha$ for $\theta \in \mathbb{C}G$ and $\alpha \in \ell^2(G)$. Then $\mathcal{N}(G)$ is the weak closure of $\mathbb{C}G$ in $\mathcal{B}(\ell^2(G))$. A further description of $\mathcal{N}(G)$ is that it is elements of $\mathcal{B}(\ell^2(G))$ which commute with the right action of $\mathbb{C}G$ on $\ell^2(G)$. We also have the reduced group C^*-algebra $C_r^*(G)$ of G: this is the operator norm closure of $\mathbb{C}G$ in $\mathcal{B}(\ell^2(G))$.

Let $\mathcal{U}(G)$ denote the closed densely defined unbounded operators affiliated to $\mathcal{N}(G)$. An example of such an operator is an element $\alpha \in \ell^2(G)$, because such an element acts by left multiplication $\beta \mapsto \alpha\beta$ on the dense subspace $\mathbb{C}G$ of $\ell^2(G)$. This map has closed graph and commutes with the right action of $\mathcal{N}(G)$ on $\ell^2(G)$, and it follows that this well defines an element of $\mathcal{U}(G)$.

We now have a sequence of inclusions

$$\mathbb{C}G \subseteq \ell^1(G) \subseteq C_r^*(G) \subseteq \mathcal{N}(G)$$
$$\subseteq \ell^2(G) \subseteq \ell^2(G)\ell^2(G) \subseteq c_0(G) \cap \mathcal{U}(G).$$

Clearly if G is finite, then we have equality at all stages. Conversely if G is infinite, then one would expect to have strict inequality at all stages, though I don't think this has been proven for all stages. Obviously if G is infinite, then $\mathbb{C}G \neq \ell^1(G)$.

Given $p \in (1, \infty)$, the L^p-conjecture in the case of discrete groups states that if G is an infinite group, then $\ell^p(G)$ is not closed under multiplication. This was solved in the affirmative for $p \geq 2$ in [36, Theorem 3], and in general for not necessarily discrete groups in [37, Theorem 1]. Thus $\ell^2(G)$ is not closed under multiplication if G is infinite, and it follows in this case that $\mathcal{N}(G) \neq \ell^2(G) \neq \ell^2(G)\ell^2(G)$.

For a ring R, the element $s \in R$ is a non-zerodivisor means $rs \neq 0 \neq sr$ whenever $0 \neq r \in R$. Let $S \subset R$ which contains 1 and is closed under multiplication. Then we say that R satisfies the right Ore condition with respect to S if given $r \in R$ and $s \in S$, there exists $r_1 \in R$ and $s_1 \in S$ such that $rs_1 = sr_1$. Similarly R satisfies the left Ore condition with respect to S if given $r \in R$ and $s \in S$, there exists $r_1 \in R$ and $s_1 \in S$ such that $s_1 r = sr_1$. If R satisfies the right Ore condition with respect to S, we can construct the ring of fractions RS^{-1} in the same manner when R is commutative (one can extend this construction to the case when S contains zero divisors, but then one needs the "right reversible condition" [12, p. 168], that is given $r \in R$ and $s \in S$ such that $sr = 0$, then there exists $s_1 \in S$ such that $rs_1 = 0$). Then R embeds naturally in RS^{-1}, every element of S becomes invertible in RS^{-1}, and every element of RS^{-1} can be written in the form rs^{-1} with $r \in R$ and $s \in S$.

Let S denote the non-zerodivisors of $\mathcal{N}(G)$. Then every element of S is invertible in $\mathcal{U}(G)$, and $\mathcal{N}(G)$ satisfies the right and left Ore condition with respect to S. Thus $\mathcal{U}(G) = \mathcal{N}(G)S^{-1}$. Also $\mathcal{N}(G)$ is semihereditary (all finitely generated submodules of a projective module are projective) and $\mathcal{U}(G)$ is von Neumann regular (for every $a \in \mathcal{U}(G)$, there exists $x \in \mathcal{U}(G)$ such that $axa = a$, so in particular every finitely generated submodule of a projective module is a direct summand).

Given a subring R of the ring S, one can consider the division closure $\mathcal{D}(R, S)$ of R in S, that is the smallest subring of S containing R which is closed under taking inverses. Since the intersection of two subrings closed under taking inverses is also closed under taking inverses, it is easy to see that such a subring exists. If K is a subring of $\mathbb{C}$, then we shall write $\mathcal{D}(KG) = \mathcal{D}(KG, \mathcal{U}(G))$. Let $\overline{\mathbb{Q}}$ indicate the ring of all algebraic numbers. We will especially be interested in studying $\mathcal{D}(KG)$ when $K = \mathbb{Z}, \overline{\mathbb{Q}}$ or $\mathbb{C}$, in particular whether this is a division ring. Clearly if G is not torsion free, then KG is not a domain, so in particular $\mathcal{D}(KG)$ is not a division ring. On the other hand if G is torsion free, then it is unknown whether $\mathcal{D}(KG)$ is a division ring. As we shall see, this is closely related to L^2-Betti numbers, and variations of this problem are often referred to as "the Atiyah conjecture".

Let $\Sigma(R, S)$ indicate all matrices with entries in R which become invertible over S. Closely related to the division closure is the rational closure $\mathcal{R}(R, S)$. This consists of entries of inverses of all matrices in $\Sigma(R, S)$, and is a subring of S containing R [8, Proposition 7.1.1 and Theorem 7.1.2]. An important property of the rational closure is that given $p \in M_d(\mathcal{R}(R, S))$, then p is stably associated to a matrix over R, that is there exists $e \in \mathbb{N}$ and invertible matrices $U, V \in M_{d+e}(S)$ such that $U \operatorname{diag}(p, I_e)V \in M_{d+e}(R)$ [7, Proposition 7.1.3]. The division closure is always contained in the rational closure [8, Exercise 7.1.3], but the converse is not true in general. Bergman and Dicks [5] wrote down details of an explicit example for which the rational closure is not contained in the division closure. The example is constructed from the theory of [4, pp. 267–271].

2 Notation and Terminology

We will let $\mathbb{N}$ denote the positive integers $\{1, 2, \dots\}$. All rings will have a 1 and subrings will have the same 1. If $H \le G$ are groups, then $[G : H]$ indicates the index of H in G. Modules will be unital right modules unless otherwise stated. For $d \in \mathbb{N}$, the ring of d by d matrices with entries in the ring R will be denoted $M_d(R)$, and $\mathcal{Z}(R)$ indicates the center of R. The identity matrix of $M_d(R)$ will be denoted I_d. If A and B are square matrices, then $\text{diag}(A, B)$ will denote the square matrix $\begin{pmatrix} A & 0 \\ 0 & B \end{pmatrix}$.

3 Group von Neumann algebras and dimension

Lück [31, Section 6] gave a definition for $\dim_{\mathcal{N}(G)} M$ valid for all $\mathcal{N}(G)$-modules M; this quantity is often called the von Neumann dimension of M. It is always a non-negative real number or ∞. To define this dimension, we first need the trace map $\text{tr}_G : c_0(G) \to \mathbb{C}$ defined by $\text{tr}_G(\sum_{x \in G} a_x x) = a_1$, where 1 indicates the identity of G. This extends to matrices $B = (b_{ij}) \in M_d(c_0(G))$ by setting $\text{tr}_G(B) = \sum_{i=1}^{d} \text{tr}_G(b_{ii})$. Also one can define $(b_{ij})^* = (b_{ji}^*)$. Then an idempotent is an element e such that $e = e^2$, and is a projection if in addition $e = e^*$. Given a finitely generated projective right $\mathcal{N}(G)$-module P, we may choose an idempotent $e \in M_d(\mathcal{N}(G))$ such that $e\mathcal{N}^d \cong P$. In general e is not unique, and can always be chosen to be a projection. Then one defines $\dim_G(P) = \text{tr}_G e$. For a general $\mathcal{N}(G)$-module M, one defines the integer $\dim_G(M)$ to be

$$\sup\{\dim_G(P) \mid P \text{ is a finitely generated projective submodule of } M\}.$$

Then $\dim_G$ has all the properties one would like for a dimension function, in particular

- If P is a projective $\mathcal{N}(G)$-module, $\dim_G(P) = 0$ if and only $P = 0$.
- If $0 \to M \to L \to N \to 0$ is an exact sequence, then $\dim_G(L) = \dim_G(M) + \dim_G(N)$.
- If $\{M_i \mid i \in I\}$ is a cofinal system of submodules of M (i.e. $M = \bigcup_{i \in I} M_i$ and given $i, j \in I$, there exists $k \in I$ such that $M_i, M_j \subseteq M_k$), then $\dim_G(M) = \sup_i \dim_G(M_i)$.
- If G is finite and M is a $\mathcal{N}(G)$-module (which is the same as a $\mathbb{C}G$-module in this situation), then $\dim_G(M) = \dim_{\mathbb{C}}(M)/|G|$.
- $\dim_G(\mathcal{N}(G)) = \dim_G(\mathcal{U}(G)) = 1$.

In the case of a finitely presented $\mathcal{N}(G)$-module M, we may represent M by an exact sequence $\mathcal{N}(G)^k \xrightarrow{A} \mathcal{N}(G)^l \to M \to 0$, where $A \in M_{l \times k}(\mathcal{N}(G))$, and then A induces a continuous map $\ell^2(G)^k \to \ell^2(G)^l$; let K denote the kernel of this map and let e denote the projection of $\ell^2(G)^l$ onto K. Then it turns out that e is a projection in $M_k(\mathcal{N}(G))$, and we have $\dim_G(M) = l - k + \text{tr}_G(e)$.

A refined variant of $\dim_G$ is the center-valued dimension function $\dim_G^u$, which depends on the center-valued trace tr_G^u. This is the uniquely defined $\mathbb{C}$-linear map $\text{tr}_G^u \colon \mathcal{N}(G) \to \mathcal{Z}(\mathcal{N}(G))$ such that for all $a, b \in \mathcal{N}(G)$ and $c \in \mathcal{Z}(\mathcal{N}(G))$, we have

- $\text{tr}_G^u(ab) = \text{tr}_G^u(ba)$;
- $\text{tr}_G^u(c) = c$;
- $\text{tr}_G^u(a) \in (\mathcal{Z}(\mathcal{N}(G)))^+$ for all $a \in (\mathcal{N}(G))^+$, where $^+$ indicates the positive operators.

As with tr_G, this can be extended to $M_d(\mathcal{N}(G))$ by defining $\text{tr}_G^u(A) = \sum_{i=1}^d \text{tr}_G^u(a_{ii})$ for $A = (a_{ij})$. That such a trace exists is established e.g. in [19, Chapter 8]. If G is an abelian group and M is a finitely generated projective $\mathcal{N}(G)$-module, then we may write $M \cong e\, M_d(\mathcal{N}(G))$ for some projection $e = (e_{ij})$, and then $\dim_G^u(M) = \sum_{i=1}^d e_{ii}$.

For a general $\mathcal{N}(G)$-module M, we may have $\dim_G(M) = 0$ even if $M \neq 0$. On the other hand $\dim_G(M)$ still carries useful information about M, so the following result is especially interesting.

Theorem 1 *Let $H \lhd G$ be groups with H infinite amenable, and let M be a $\mathbb{C}G$-module. Then $\dim_G\big(\text{Tor}_n^{\mathbb{C}G}(M, \mathcal{N}(G))\big) = 0$ for all $n \in \mathbb{N}$.*

See [31, Theorem 7.2] for a proof and further results. Theorem 1 tells us that $\mathcal{N}(G)$ is dimension flat over $\mathbb{C}G$ if G is amenable. The converse of this statement is an open problem [31, Conjecture 6.48]. One can also consider when $\mathcal{N}(G)$ is flat over $\mathbb{C}G$ [31, Conjecture 6.49].

Conjecture 2 *A group G is locally virtually cyclic if and only if $\mathcal{N}(G)$ is flat as a $\mathbb{C}G$-module.*

This is [32, Conjecture (B)], where interesting progress on this problem can be found.

4 Formulation of the Atiyah conjecture

In [1], Atiyah introduced L^2-Betti numbers for manifolds with cocompact free G-action for a discrete group G. He asked [1, p. 72] about the possible values these numbers can take, in particular whether they could be irrational. The L^2-Betti numbers are defined using the L^2-chain complex. This chain complex is a sequence of groups of the form $\ell^2(G)^d$, and with differentials given by convolution with matrices over $\mathbb{Z}G$. This means that one could give a more algebraic formulation of Atiyah's questions [31, Section 10.1.1].

Let K be a subring of $\mathbb{C}$ and let $d \in \mathbb{N}$. Given $A \in \mathrm{M}_d(KG)$, we have an induced map of right $\mathcal{N}(G)$-modules given by $u \mapsto Au \colon \mathcal{N}(G)^d \to \mathcal{N}(G)^d$ for $u \in \mathcal{N}(G)^d$, which we shall also denote by A. We shall write $\mathrm{lcm}(G)$ for the least common multiple of the orders of the finite subgroups of G; if these orders are unbounded, then we define $\mathrm{lcm}(G) = \infty$.

Definition 3 Let G be a group with a bound on the orders of its finite subgroups. Then we say that G satisfies the strong Atiyah conjecture over K, or KG satisfies the strong Atiyah conjecture, if for every $d \in \mathbb{N}$ and every $A \in \mathrm{M}_d(KG)$ we have $\dim_G(\ker A) \in \frac{1}{\mathrm{lcm}(G)}\mathbb{Z}$.

An equivalent formulation of Definition 3 is

Definition 4 (cf. [31, Lemma 10.7]) Let G be a group with a bound on the orders of its finite subgroups. Then we say that G satisfies the strong Atiyah over K, or KG satisfies the strong Atiyah conjecture, if $\dim_G(M \otimes_{KG} \mathcal{N}(G)) \in \frac{1}{\mathrm{lcm}(G)}\mathbb{Z}$ for every finitely presented KG-module M.

We note that if F is the field of fractions of K, then the strong Atiyah conjecture holds over K if and only if it holds over F. Indeed if $A \in \mathrm{M}_d(KG)$, then $A \in \mathrm{M}_d(FG)$, which proves the "if" part. On the other hand if $A \in \mathrm{M}_d(FG)$, then there exists $k \in K \setminus 0$ such that $kA \in \mathrm{M}_d(KG)$, and the "only if" part follows because $\dim_G(\ker kA) = \dim_G(\ker A)$.

If H is a finite subgroup of G, let p_H denote the projection $\sum_{h \in H} h/|H| \in \mathbb{Q}H$. Then $\dim_G(\ker p_h) = 1/|H| \in \frac{1}{|H|}\mathbb{Z}$, and it follows that the additive subgroup of $\mathbb{Z}$ generated by $\dim_G(\ker A)$ in Definition 3 at least contains $\frac{1}{\mathrm{lcm}(G)}\mathbb{Z}$; the strong Atiyah conjecture implies that this subgroup is exactly $\frac{1}{\mathrm{lcm}(G)}\mathbb{Z}$.

The strong Atiyah conjecture is related to various conjectures in ring theory, in particular the Kaplansky zero divisor conjecture, that is if k is a field and G is a torsion-free group, then kG is a domain. In fact we have

Proposition 1 *Let G be a torsion-free group. Then KG satisfies the strong Atiyah conjecture if and only if $\mathcal{D}(KG)$ is a division ring.*

Proof Since G is torsion free, $\mathrm{lcm}(G) \in \mathbb{Z}$. First suppose that $D := \mathcal{D}(KG)$ is a division ring. Let $A \in \mathrm{M}_n(KG)$ and write A_n, A_u and A_d for the corresponding maps $\mathcal{N}(G)^n \to \mathcal{N}(G)^n$, $\mathcal{U}(G)^n \to \mathcal{U}(G)^n$ and $D^n \to D^n$. Since D is a division ring, we may write $\ker A_d \cong D^k$ for some $k \in \mathbb{Z}$. Then $\ker A_u \cong \mathcal{U}^k$ and $\dim_G(\ker A_n) = \dim_G(\ker A_u) = k$ and hence $\dim_G(\ker A_n) \in \mathbb{Z}$. This proves the "if" part.

Now suppose the strong Atiyah conjecture is true. If D is not a division ring, then there exists $d \in D \setminus 0$ such that d is not invertible in $\mathcal{U}(G)$. Thus if

$\delta \colon \mathcal{U}(G) \to \mathcal{U}(G)$ is the map $x \mapsto dx$, we see that $0 < \ker \delta < 1$. Since $d \in D$, so $d \in \mathcal{R}(KG)$ and hence there exists $n \in \mathbb{N}$ such that $\mathrm{diag}(d, I_{n-1}) = UAV$, where U, V are invertible matrices in $\mathcal{U}(G)^n$. Then $\dim_G(\ker \delta) = \dim_G(\ker A)$ and we see that $\dim_G(\ker A) \notin \mathbb{Z}$, which proves the "only if" part. $\qquad\square$

Remark 5 If G is torsion free, then it is clear that the strong Atiyah conjecture is inherited by subgroups of G. On the other hand if G is not torsion free, then it is unknown whether the strong Atiyah conjecture is inherited by subgroups, though for subgroups H with $\mathrm{lcm}(H) = \mathrm{lcm}(G)$, the strong Atiyah conjecture is inherited by H. For example if $\mathrm{lcm}(H) = 1$ (so H is torsion free) and $\mathrm{lcm}(G) \neq 1$ (so G is not torsion free), then we need to prove that $\mathcal{D}(KH)$ is a division ring even though $\mathcal{D}(KG)$ is not a division ring, by Proposition 1.

The finite conjugate subgroup $\Delta(G)$ is the set of elements of G which have only finitely many conjugates, that is $\{g \in G \mid [G : C_G(g)] < \infty\}$, where $C_G(x)$ denotes the centralizer of g in G. The subset of $\Delta(G)$ consisting of elements of finite order is denoted by $\Delta^+(G)$. Then $\Delta(G)$ and $\Delta^+(G)$ are normal subgroups of G, and $\mathcal{Z}(KG) \subseteq K\Delta(G)$, $\mathcal{Z}(\mathcal{N}(G)) \subseteq \mathcal{N}(\Delta(G))$). Obviously if $\mathrm{lcm}(G) < \infty$, then $\Delta^+(G)$ is finite and $|\Delta^+(G)| \mid \mathrm{lcm}(G)$. One can extend Proposition 1 to

Proposition 2 *Let G be a group with $\mathrm{lcm}(G) < \infty$ and $\Delta^+(G) = 1$, Then KG satisfies the strong Atiyah conjecture if and only if $\mathcal{D}(KG) \cong M_{\mathrm{lcm}(G)}(D)$ for some division ring D.*

More generally, one has

Proposition 3 *Let G be a group with $\mathrm{lcm}(G) < \infty$ and let K be a subfield of $\mathbb{C}$. If KG satisfies the strong Atiyah conjecture, then $\mathcal{D}(KG)$ is a semisimple Artinian ring.*

If $\Delta^+(G) \neq 1$, then the strong Atiyah conjecture doesn't give enough information to describe the semisimple Artinian ring $\mathcal{D}(G)$. A semisimple Artinian ring is a finite direct sum of matrix rings over division rings, and we would like to know the precise size of these matrix rings. Here we need the center-valued Atiyah conjecture, which is defined as follows.

Definition 6 Let G be a group with $\mathrm{lcm}(G) < \infty$, and let K be a subring of $\mathbb{C}$ which is closed under complex conjugation. Let $L(G)$ denote the additive subgroup of $\mathcal{Z}(\mathcal{N}(G))$ generated by $\mathrm{tr}_G^u(P) \in \mathcal{Z}(\mathbb{C}G)$, where P runs through projections $P \in M_d(KH)$, with $H \leq G$ and $d \in \mathbb{N}$. Then we say that G satisfies the center-valued Atiyah conjecture over K, or KG satisfies the center-valued Atiyah conjecture if for every $d \in \mathbb{N}$ and every $A \in M_d(KG)$, we have $\dim_G^u(\ker A) \in L(G)$.

We note that in the case $\Delta^+(G) = 1$, the center-valued Atiyah conjecture is identical to the strong Atiyah conjecture. Then we have the following extension of Proposition 2.

Proposition 4 *Let G be a group with $\mathrm{lcm}(G) < \infty$, and let K be a subring of $\mathbb{C}$ which is closed under complex conjugation. Then KG satisfies the center-valued Atiyah conjecture if and only if the natural map*

$$P \mapsto P \otimes_{KH} D(KG): \bigoplus_{H \leq G \, | \, |H| < \infty} K_0(KH) \to K_0(D(KG))$$

is onto and $D(KG)$ is semisimple Artinian.

Given the center-valued Atiyah conjecture, one can also give a precise formula of the sizes of the matrix rings involved in $D(KG)$; this and the proof of Proposition 4 is done in [21, Definition 3.7, Theorem 3.8].

Let $\Delta^+ = \Delta^+(G)$ and let $p_1, \ldots, p_m$ denote the primitive central idempotents of $K\Delta^+$. Define $P_i = \sum_{j \text{ s.t. } \exists g \in G: g p_i g^{-1} = p_j} p_j$. Then the P_i are precisely the primitive central idempotents of $K\Delta^+$ and we have that $P_i D(KG) P_i$ is an $L_i \times L_i$ matrix ring over some division ring D_i, where $L_i \in \mathbb{N}$. Thus $D(KG)$ is the finite direct sum of matrix rings $M_{L_i}(D_i)$. The formula for the L_i is as follows. Consider all irreducible sub-projections $Q_\alpha \in KH_\alpha$ of P_i (i.e. those satisfying $Q_\alpha P_i = Q_\alpha$), where H_α runs through the finite subgroups of G containing Δ^+. Then

$$L_i = \frac{\dim_{\mathbb{C}}(P_i \mathbb{C}\Delta^+)\,\mathrm{lcm}(G)}{\gcd\big(\dim_{\mathbb{C}}(P_i \mathbb{C}\Delta^+)\,\mathrm{lcm}(G), \dim_C(Q_\alpha \mathbb{C} H_\alpha)|\Delta^+|\,\mathrm{lcm}(G)/|H_\alpha| \mid \alpha\big)}.$$

What happens when $\mathrm{lcm}(G) = \infty$? At one time, it seemed plausible that the conjecture should be that $\dim_G(\ker A) \in \langle 1/|H| \mid H \leq G, \ |H| < \infty \rangle$. However building on results of Grigorchuk and Żuk [15], it was shown in [14] that the Lamplighter group $\mathbb{Z}/2\mathbb{Z} \wr \mathbb{Z}$, a group in which the orders of the finite subgroups are powers of 2, is a counterexample to this, in particular there exist $A \in M_1(\mathbb{Q}G)$ with $\dim_G(\ker A) = 1/3$. It took another decade before Tim Austin [2] produced examples of groups G and matrices A with $\dim_G(\ker A)$ transcendental. Further results in this direction were subsequently obtained in [13, 35], in particular [35, Theorem 11.1] shows that for every $r \in \mathbb{R}$ with $r \geq 0$, there is a finitely generated group G, a $d \in \mathbb{N}$ and $A \in M_d(\mathbb{Q}G)$ such that $\dim_G(\ker A) = r$.

5 Early results on the Atiyah conjecture

Let C denote the smallest class of groups which

(1) contains all free groups;
(2) is closed under directed unions;
(3) satisfies $G \in C$ whenever $H \lhd G$, $H \in C$ and G/H is elementary amenable.

Then [28, Theorem 1.5] proves

Theorem 7 *If $G \in C$ and the finite subgroups of G have bounded order, then the strong Atiyah conjecture for G over $\mathbb{C}$ is true.*

A closely related conjecture is

Conjecture 8 *Let G be a torsion-free group, $0 \neq \alpha \in \mathbb{C}G$ and $0 \neq \beta \in \ell^2(G)$, then $\alpha\beta \neq 0$.*

This is really the strong Atiyah conjecture for torsion-groups in the case $d = 1$ of Definition 3, so could be thought of as the 1×1 torsion-free Atiyah conjecture. It is also implies the Kaplansky zero divisor conjecture over $\mathbb{C}$. The following result is [27, Theorem 2].

Theorem 9 *Let $H \lhd G$ be groups such that G/H is left orderable and Conjecture 8 is true for H. Then Conjecture 8 is true for G.*

Thus, for example, by combining Theorems 7 and 9, we find that Conjecture 8 is true in the case G has a normal subgroup $H \in C$ such that G/H is left orderable.

6 Approximation Techniques

The origin of these results is in Lück's paper [30]. Let $d \in \mathbb{N}$, let $A \in \mathrm{M}_d(\mathbb{Q}G)$, and let A also denote the map left multiplication by $A \colon \mathcal{N}(G)^d \to \mathcal{N}(G)^d$. Suppose $G = G_0 \geqslant G_1 \geqslant \cdots$ is a descending sequence of normal subgroups of G with $|G/G_i| < \infty$ for all i and $\bigcap_i G_i = 1$ and for $i \in \mathbb{N}$, let A_i denote the induced map by left multiplication $\mathbb{Q}[G/G_i]^d \to \mathbb{Q}[G/G_i]^d$. Then Lück proved

$$\dim_G(\ker A) = \lim_{i \to \infty} |G/G_n|^{-1} \dim_{\mathbb{Q}}(\ker A_i).$$

This was generalized as follows in [9] (see also [10, Errata]). Let $\mathcal{G}$ denote the smallest nonempty class of groups such that

(1) If $H \lhd G$, $H \in \mathcal{G}$ and G/H is amenable, then $G \in \mathcal{G}$.
(2) $\mathcal{G}$ is closed under taking subgroups
(3) If $G = \lim_{i \in I} G_i$ is the direct or inverse limit of a directed system of groups with $G_i \in \mathcal{G}$, then $G \in \mathcal{G}$.

Then [9, Theorem 1.6] we have

Theorem 10 *Let $d \in \mathbb{N}$, let $A \in M_d(\overline{\mathbb{Q}}G)$, and let A also denote the map left multiplication by $A\colon \mathcal{N}(G)^d \to \mathcal{N}(G)^d$. Suppose $G = G_0 > G_1 > \cdots$ is a descending sequence of normal subgroups of G with $G/G_i \in \mathcal{G}$ for all i and $\bigcap_i G_i = 1$ and for $i \in \mathbb{N}$, let A_i denote the induced map left multiplication $\mathcal{N}[G/G_i]^d \to \mathcal{N}[G/G_i]^d$.*

$$\dim_G(\ker A) = \lim_{i \to \infty} \dim_{G/G_i}(\ker A_i). \tag{1}$$

It would be nice to prove Theorem 10 with $\overline{\mathbb{Q}}G$ replaced by $\mathbb{C}G$. However all known strategies for proving Theorem 10 depend heavily on the property that the matrices involved are defined over the field of algebraic numbers. Theorem 10 is particularly attractive in the case that all the G/G_i are torsion free and satisfy the strong Atiyah conjecture over $\overline{\mathbb{Q}}$, because then equation 1 tells us that $\dim_G(\ker A)$ is the limit of a sequence of integers and is therefore an integer. Thus we can prove the strong Atiyah conjecture over $\overline{\mathbb{Q}}$ for a large class of groups as follows.

Let $\mathcal{D}$ denote the smallest nonempty class of groups such that

(1) If G is torsion free, A is elementary amenable, and we have an epimorphism $p\colon G \to A$ such that $p^{-1}(E) \in \mathcal{D}$ for every finite subgroup E of A, then $G \in \mathcal{D}$.
(2) $\mathcal{D}$ is subgroup closed.
(3) Let $G_i \in \mathcal{D}$ be a directed system of groups and G its (direct or inverse) limit. Then $G \in \mathcal{D}$.

Clearly $\mathcal{D}$ contains all elementary amenable groups, and also all free groups, because free groups are residually torsion-free nilpotent. Then [9, Theorem 1.4] (see also [10, Errata])

Theorem 11 *Let $G \in \mathcal{D}$. Then $\overline{\mathbb{Q}}G$ satisfies the strong Atiyah conjecture.*

One can interpret the approximation theorems in terms of modules. For example, Theorem 10 becomes

Theorem 12 *Let G be a group and let M be a finitely presented $\overline{\mathbb{Q}}G$-module. If $G = G_0 > G_1 > \cdots$ is a descending sequence of normal subgroups of G*

with $G/G_i \in \mathcal{G}$ for all i and $\bigcap_i G_i = 1$, then

$$\dim_G(M \otimes_{\overline{\mathbb{Q}}G} \mathcal{N}(G)) = \lim_{i \to \infty} \dim_{G/G_i}(M \otimes_{\overline{\mathbb{Q}}G} \mathcal{N}(G/G_i)).$$

Proof Since M is finitely presented, there is an exact sequence

$$0 \longrightarrow \ker A \longrightarrow \overline{\mathbb{Q}}G^d \xrightarrow{A} \overline{\mathbb{Q}}G^d \longrightarrow M \longrightarrow 0,$$

where $d \in \mathbb{N}$, $A \in M_d(\overline{\mathbb{Q}}G)$, and $A \colon \overline{\mathbb{Q}}G^d \to \overline{\mathbb{Q}}G^d$ denotes the map left multiplication by A. Also for $i \in \mathbb{N}$, let $A \colon \mathcal{N}(G)^d \to \mathcal{N}(G)^d$ and $A_i \colon \mathcal{N}[G/G_i]^d \to \mathcal{N}[G/G_i]^d$ denote the corresponding induced maps. Now $M \cong \operatorname{coker} A$, and $\operatorname{coker}(\alpha \otimes 1) \cong (\operatorname{coker} \alpha) \otimes_R S$ for a general map $R \to S$ by right exactness of tensor products. Therefore we have exact sequences

$$0 \longrightarrow \ker A \longrightarrow \mathcal{N}(G)^d \xrightarrow{A} \mathcal{N}(G)^d \longrightarrow M \otimes_{\overline{\mathbb{Q}}G} \mathcal{N}(G) \longrightarrow 0$$

$$0 \longrightarrow \ker A_i \longrightarrow \mathcal{N}(G/G_i)^d \xrightarrow{A} \mathcal{N}(G/G_i)^d \longrightarrow M \otimes_{\overline{\mathbb{Q}}G} \mathcal{N}(G/G_i) \longrightarrow 0.$$

The result now follows from Theorem 10 and that $\dim_G$ has the usual properties of a dimension function. $\qquad\qquad\qquad\qquad\qquad\qquad\qquad\qquad\qquad\qquad \square$

The center-valued analogue of Theorem 10 was proved by Anselm Knebusch in [20, Theorem 3.2] (see also [10, Errata]), namely the same result with $\dim_G(\ker A)$ replaced with $\dim_G^u(\ker A)$. Also in that paper, Knebusch proved the same result with the class $\mathcal{G}$ replaced with the class of sofic groups.

7 Finite extensions

The theme here is given $H \lhd G$ groups such that G/H is finite and H satisfies the strong Atiyah conjecture (over some field), then G/H also satisfies the strong Atiyah conjecture. To prove this seems way out of reach at this time, and could in any case well be false, but it turns out that it is true in many interesting situations. First we make the following definition.

Definition 13 Let G be a group and let p be a prime. Let $\hat{G}$ denote the profinite completion of G and $\hat{G}^p$ denote the pro-p completion of G. Thus $\hat{G}$ and $\hat{G}^p$ are compact totally disconnected groups. Then for a topological $\hat{G}$ or $\hat{G}^p$-module, the cohomology groups $H^n(\hat{G}, M)$ and $H^n(\hat{G}^p, M)$ will always denote the Galois cohomology. We have natural maps $G \to \hat{G}$ and $G \to \hat{G}^p$; the former map is an injection if and only if G is a residually finite group, and the latter is an injection if and only if G is a residually finite p-group. We say that G is cohomologically p-complete if the natural map $H^*(\hat{G}^p, \mathbb{F}_p) \to H^*(G, \mathbb{F}_p)$ is an isomorphism, and G is cohomologically complete if G is cohomologically

p-complete for all primes p. Finally in the terminology of [38, Section 2] and [39, p. 16], we say that G is *good* if the natural map $H^*(\hat{G}, M) \to H^*(G, M)$ is an isomorphism for all finite G-modules M (i.e. $|M| < \infty$).

In addition to cohomological p-completeness, another key property required is "enough torsion-free quotients".

Definition 14 Let G be a group. We say that G has *enough torsion-free quotients* (respectively *enough torsion-free nilpotent quotients*) [26, Definition 4.53] if given a prime p and a normal subgroup H such that G/H is a finite p-group, then there exists $N \lhd G$, $N \subseteq H$ such that G/N is a torsion-free elementary amenable group (respectively torsion-free nilpotent group). Also G has the *factorization* property [38, Section 3] if given a normal subgroup H of finite index in G, then there exists $N \subseteq H$ such that G/N is a torsion-free elementary amenable group.

Let us illustrate the methods used in the special case G is torsion free. So let H be a normal subgroup of finite index in G. We suppose that H has type FP, that is the trivial $\mathbb{Z}H$-module $\mathbb{Z}$ has a resolution of finite length with finitely generated projective $\mathbb{Z}H$-modules, that H is cohomologically complete, and that H has enough torsion-free quotients. Now let p be a prime and let P/H be a subgroup of order p in G/H. Then a spectral sequence argument shows that P is cohomologically p-complete. Also H has type FP and because G is torsion free, we see that P is also of type FP by a theorem of Serre [6, Proposition 6.6], and we deduce that $H^n(P, \mathbb{F}_p) = 0$ for n greater than the cohomological dimension of H. Therefore $H^n(\hat{P}, \mathbb{F}_p) = 0$ for all n greater than the cohomological dimension of H and we conclude that $\hat{P}^p$ is also torsion free, in particular the extension $\hat{H}^p \hookrightarrow \hat{P}^p \twoheadrightarrow \hat{P}^p/\hat{H}^p$ does not split. This means that there is $L \lhd P$ such that $L \subset H$, P/L is a finite p-group, and the extension $L \hookrightarrow P \hookrightarrow P/L$ does not split. Now H has enough torsion-free quotients, from which we deduce that there exists a normal subgroup N of P contained in L such that P/N is torsion-free elementary amenable. By doing this for all such subgroups P, and then intersecting the conjugates of all such N obtained (a finite intersection), we obtain $M \lhd G$ with $M \subseteq H$ and G/M torsion free. It now follows from Theorem 11 that the strong Atiyah conjecture is true for $\overline{\mathbb{Q}}G$ if H is residually torsion-free elementary amenable.

Generalizations of this method are considered in [26], for example we have the following results.

Theorem 15 *[26]* *Let* $H \lhd G$ *with* G/H *elementary amenable and* $\operatorname{lcm}(G/H) < \infty$. *Suppose* H *has a finite classifying space, is cohomologically complete and has enough torsion-free quotients. If* H *satisfies the strong*

Atiyah conjecture over $\overline{\mathbb{Q}}$, *then* G *satisfies the strong Atiyah conjecture over* $\overline{\mathbb{Q}}$.

Theorem 16 *[26, Corollary 5.41] Let* H *denote Artin's pure braid group (usually denoted* P_n*) and suppose* H *is a normal subgroup of the group* G *such that* G/H *is elementary amenable and* $\mathrm{lcm}(G) < \infty$. *Then the strong Atiyah conjecture for* G *over* $\overline{\mathbb{Q}}$ *is true. Furthermore if* G/H *is finite and* G *is torsion free, then there exists a descending sequence of subgroups* $H > H_1 > H_2 > \cdots$ *such that* $H_i \lhd G$ *for all* i, $\bigcap_{i \in \mathbb{N}} H_i = 1$, *and* G/H_i *is torsion-free virtually nilpotent for all* $i \in \mathbb{N}$.

More generally, Theorem 16 remains true if the pure braid group is replaced by the braid groups associated to the Coxeter groups C_n, G_2 or $I_2(p)$, and the classical braid group is replaced by the full braid groups associated to the Coxeter groups C_n, G_2 or $I_2(p)$. Building on these techniques, the following result was proven in [25, Theorem 2].

Theorem 17 *Let* H *be a right-angled Artin group and assume that* H *is a normal subgroup of the group* G *such that* G/H *is elementary amenable and* $\mathrm{lcm}(G/H) < \infty$. *Then* G *satisfies the strong Atiyah conjecture over* $\overline{\mathbb{Q}}$.

However it could not prove Theorem 17 with H being instead a right-angled Coxeter group in complete generality. In [38], a clever variant of the above methods was used, namely replacing "cohomological completeness" with "good", and "enough torsion-free quotients" with "factorization property". Though the factorization property is harder to verify than the enough torsion-free quotients property, the good property is easier to verify than the cohomological completeness property. Actually the factorization property has a nicer closure property, namely that if H is a subgroup of finite index in the group G which has the factorization property, then H also has the factorization property [38, Lemma 3.1], which is not true for the enough torsion-free quotients property. Using this new method, [38] shows that Theorem 17 remains true for H a right-angled Coxeter group. Moreover [38, Theorem 1.1] proves

Theorem 18 *Let* $H \lhd G$ *such that* G/H *elementary amenable and* $\mathrm{lcm}(G/H) < \infty$. *Suppose* H *has a finite classifying space, is good, and has the factorization property. If* H *satisfies the strong Atiyah conjecture over* $\overline{\mathbb{Q}}$, *then* G *satisfies the strong Atiyah conjecture over* $\overline{\mathbb{Q}}$.

Using this, [38, Theorem 1.2] shows

Theorem 19 *Let $H \lhd G$ be groups such that H is a cocompact special group and G/H is elementary amenable with $\mathrm{lcm}(G) < \infty$. Then G satisfies the strong Atiyah conjecture over $\overline{\mathbb{Q}}$.*

Let us explain the concept of a cocompact special group. Special cube complexes were introduced in [16, Section 3]. A cube complex K is a polyhedral complex where each cell is isometric to the Euclidean cube $[0, 1]^n$. A cube complex is non-positively curved if the link of each vertex is a flag simplicial complex. A non-positively curved cube complex K is then said to be special if the hyperplanes of K avoid certain configurations, which are defined precisely in [16, Definition 3.2]. In [16, Section 3], it is then shown that the fundamental group $\pi_1(K)$ of a non-positively curved cube complex injects into a right-angled Artin group if and only if K is special. Following the terminology of [38, Section 6], a group is special if it is the fundamental group of a special cube complex, and is cocompact special if that complex is compact. Since right-angled Artin groups satisfy the strong Atiyah conjecture by Theorem 17, it follows from Remark 5 that special groups satisfy the strong Atiyah conjecture, as observed in [38, Theorem 6.1]. Furthermore if G is cocompact special, then G has a finite classifying space, is good, and has the factorization property. Thus Theorem 19 follows from Theorem 18.

8 Pro-p groups of finite rank

Let H be a profinite group and let $\Phi(H)$ denote the Frattini subgroup of H, that is the intersection of maximal open subgroups of H. In a profinite group, the open subgroups are precisely the closed subgroups of finite index. In general, not every subgroup of finite index is open. On the other hand, if H is finitely generated (as a profinite group, i.e. there exists a finite subset S of H such that the closure of the subgroup of H generated by S is H), then every subgroup of finite index in H is open [33, Theorem 1.1]. More generally, define $\Phi^i(H)$ for $i \geq 0$ by $\Phi^0(H) = H$ and $\Phi^{i+1} = \Phi(\Phi^i(H))$. Finally let $d(H)$ denote the minimum number of elements required to generate H and for $n \in \mathbb{Z}$, let H^n denote the closure of $\langle h^n \mid h \in H \rangle$. Then $d(H) = d(H/\Phi(H))$. Also define the rank $\mathrm{rk}(H)$ to be $\max\{d(K) \mid K \leqslant H\}$, so the rank of H is the smallest integer r such that every subgroup of H can be generated by r elements; this number is ∞ if no such integer r exists. In the case H is a pro-p group (where p is a prime), $H/\Phi(H)$ is an abelian group of exponent p, and H is finitely generated if and only if $H/\Phi(H)$ is a finite elementary abelian p-group.

We will be particularly interested in pro-p groups of finite rank.

Definition 20 Let G be a pro-p group of finite rank. Then G is uniform if and only if G/G^p (G/G^4 if $p = 2$) is abelian and $|G/\Phi(G)| = |\Phi^i(G)/\Phi^{i+1}(G)|$ for all $i \geq 0$.

Let $\mathbb{Z}_p$ denote the p-adic integers and let $d \in \mathbb{N}$.

Example 21

(i) $\mathbb{Z}_p^d$ is a pro-p group of rank d.
(ii) Let $n \in \mathbb{N}$ and let $G_n = \{A \in \mathrm{GL}_d(\mathbb{Z}_p) \mid A \equiv I \mod p^n\}$ (a congruence subgroup). If $p \neq 2$, then G_1 is a uniform pro-p group and $\Phi^i(G_1) = G_{i+1}$ for all $i \in \mathbb{N}$. If $p = 2$, then G_2 is a uniform pro-2 group.

Remark 22 Let G be a topological group. Then G is a compact p-adic analytic (or Lie) group if and only if G has an open normal subgroup which is a pro-p group of finite rank.

Uniform pro-p groups have particularly nice completed group rings. If G is a group and k is a commutative ring, then we let ε_G denote the augmentation map, that is $\varepsilon_G \sum_{g \in G} a_g g = \sum_{g \in G} a_g$. Then ε_G is a k-homomorphism and we set $\omega(G) = \ker \varepsilon_G$, so $\omega(G)$ is the k-linear span of $\{g - 1 \mid g \in G\}$. Now let G be a pro-p group and let k be a finite field of characteristic p. We give k the discrete topology and define the completed group algebra $k[[G]]$ to be $\varprojlim_{N \lhd G} k[G/N]$, where N runs through the open normal subgroups of G. Then ε_G extends to an epimorphism $k[[G]] \to k$, which we shall also call ε_G. Let $\overline{\omega}(G)$ denote the closure of $\omega(G)$ in $k[[G]]$. Then $\ker \varepsilon_G \colon k[[G]] \to k = \overline{\omega}(G)$. Furthermore if H is an open normal subgroup of G, then we have a natural homomorphism $\varepsilon_{G,H} \colon k[[G]] \to k[G/H]$, and then $\ker \varepsilon_{G,H} = \overline{\omega}(H)G$ [41, Proposition 7.1.2].

Now suppose G is a uniform pro-p group. Since G/H is a finite p-group, $\omega(G/H)^n = 0$ for some $n \in \mathbb{N}$ and we deduce that $\overline{\omega}(G)^n \subseteq \ker_{G,H}$. On the other hand we have that for each $n \in \mathbb{N}$, there exists an open normal subgroup H of G such that $H \subseteq \overline{\omega}(G)^n$ cf. [11, Lemma 3.3]. Thus $k[[G]] \cong \varprojlim_{n \in \mathbb{N}} \overline{\omega}(G)^n$. It turns out that the completed group rings of uniform pro-p groups behave rather like group rings of abelian groups. If G is a free abelian group of rank d, then the associated graded ring $\bigoplus_{n \in \mathbb{N}} \omega(G)^{n-1}/\omega(G)^n$ is a polynomial ring $k[x_1, \ldots, x_d]$ in d variables. In fact we have the following Theorem due to Lazard [23], see [41, Theorem 8.7.10]:

Theorem 23 *Let G be a uniform pro-p group of rank d and let k be a finite field of characteristic p. Then the associated graded ring $\bigoplus_{n\in\mathbb{N}} \omega(G)^{n-1}/\omega(G)^n$ is a polynomial ring $k[x_1, \ldots, x_d]$ in d variables.*

From this it is easy to deduce that if G is a uniform pro-p group, then $k[[G]]$ is a noetherian domain. In fact this result remains true over arbitrary fields k of characteristic p. Here we need the theory of linearly compact modules over a Hausdorff linearly topological ring [40, Chapter VII]. In this more general situation, we still define $k[[G]] = \varprojlim_{N\lhd G} k[G/N]$, where N runs through the open normal subgroups of G, and give each $k[G/N]$ the discrete topology. Also, we will let $\overline{\omega}(G)$ denote the kernel of the natural map $k[[G]] \to k$, which extends the definition for when k is a finite field. Since each $k[G/N]$ is artinian, these rings are linearly compact [40, 28.14] and we deduce from [40, 28.15] (essentially Tychonoff's theorem) that $k[[G]]$ is linearly compact. Then we have (cf. [11, Theorem 4.3])

Theorem 24 *Let G be a uniform pro-p group of rank d and let k be an arbitrary field of characteristic p.*

(a) *The associated graded ring $\bigoplus_{n\in\mathbb{N}} \omega(G)^{n-1}/\omega(G)^n$ is a polynomial ring $k[x_1, \ldots, x_d]$ in d variables.*

(b) *$k[[G]]$ is a noetherian domain.*

Now let G be an arbitrary pro-p group of finite rank. Then G has uniform open characteristic subgroup [41, Theorem 8.5.3]. Using Theorem 24 together with the fact that G has finite cohomological dimension [41, Theorem 11.6.9], we can deduce (cf. [11, Theorem 6.1])

Theorem 25 *Let k be a field of characteristic p and let G be a torsion-free pro-p group of finite rank. Then $k[[G]]$ is a noetherian domain.*

We can use the ingredients of the proof of Theorem 25 to obtain (cf. [11, Proposition 4 and following remark]):

Theorem 26 *Let R be a complete discrete valuation ring with maximal ideal qR such that R/qR is a field of characteristic p, and let G be a pro-p group of finite rank. Then $R[[G]]$ is a noetherian domain.*

The information obtained in Theorem 23 can be applied to the Atiyah conjecture by using Lück's approximation theorem (Theorem 10). Let $0 \neq \alpha \in k[x_1, \ldots, x_d]$ and for $n \in \mathbb{N}$, let $k[x_1, \ldots, x_d]_n$ denote the truncated polynomial ring consisting of polynomials of total degree at most n. Let $\alpha_n \colon k[x_1, \ldots, x_d]_n \to k[x_1, \ldots, x_d]_n$ denote the map induced by left multiplication by α. Then it's not difficult to show that

$\lim_{n\to\infty}(\dim_k \ker \alpha_n)/\dim_k(k[x_1, \ldots, x_d]) = 0$. A similar argument can be applied to the associated graded rings of completed group algebras, and we get (cf. [11, Lemma 5.2]):

Theorem 27 *Let k be a field of characteristic p and let G be a uniform pro-p group. Let $0 \neq \alpha \in k[[G]]$ and for $n \in \mathbb{N}$, let $G_n = \Phi^n(G)$ and $\alpha_n: k[G/G_n] \to k[G/G_n]$ denote the map induced by left multiplication by α in $k[[G]]$. Then $\lim_{n\to\infty}(\dim_k \ker \alpha_n)/|G/G_n| = 0$.*

Then as in Theorem 25, we can exploit the fact that a pro-p group of finite rank has a uniform open characteristic subgroup to obtain (cf. [11, Lemma 6.3])

Theorem 28 *Let k be a field of characteristic p, let G be a pro-p group of finite rank, and let $0 \neq \alpha \in k[[G]]$. Let H be a uniform open characteristic subgroup of G and for $n \in \mathbb{N}$, let $H_n = \Phi^n(H)$ and $\alpha_n: k[G/H_n] \to k[G/H_n]$ denote the map induced by left multiplication by α in $k[[G]]$. Then $\lim_{n\to\infty}(\dim_k \ker \alpha_n)/|G/H_n| = 0$.*

Now let R be a discrete valuation ring with maximal ideal qR, let K be the field of fractions of R, and let $k = R/qR$. If M is an R-module, we will define $\dim_R M := \dim_K M \otimes_R K$. In the case M is a submodule of a finitely generated R-module, then M is free of unique rank because R is a PID, and this rank is $\dim_R M$.

Suppose M is a finitely generated free R-module, say $M \cong R^d$ and $x_1, \ldots, x_e \in M$ such that the images $\bar{x}_1, \ldots, \bar{x}_e$ in M/qM are linearly independent over k, where $0 \leq e \leq d$. Then we may choose $y_1, \ldots, y_{d-e} \in M$ such that the images $\bar{y}_1, \ldots, \bar{y}_{d-e}$ in M/qM, together with the $\bar{x}_i$ form a k-basis for M/qM. By Nakayama's lemma, $\{x_1, \ldots, x_e, y_1, \ldots, y_{d-e}\}$ generates M and it follows that $\{x_1, \ldots, x_e\}$ generates a free R-submodule of M. We deduce that if $\alpha: R^d \to R^d$ is an R-module map and $\bar{\alpha}$ is the map $(R/qR)^d \to (R/qR)^d$ is the map induced by α, then $\dim_k \operatorname{im} \bar{\alpha} \leq \dim_R \operatorname{im} \alpha$, consequently

$$\dim_R \ker \alpha \leq \dim_k \ker \bar{\alpha}. \tag{2}$$

We can use (2) to deduce from Theorem 28 (cf. [11, Lemma 6.3])

Theorem 29 *Let R be a complete discrete valuation ring with maximal ideal qR such that R/qR is a field of characteristic p, let G be a torsion-free pro-p group of finite rank, and let $0 \neq \alpha \in R[[G]]$. Let H be a uniform open characteristic subgroup of G and for $n \in \mathbb{N}$, let $H_n = \Phi^n(H)$ and $\alpha_n: R[G/H_n] \to R[G/H_n]$ denote the map induced by left multiplication by α in $R[[G]]$. Then $\lim_{n\to\infty}(\dim_k \ker \alpha_n)/|G/H_n| = 0$.*

Now we can use Theorem 26, that $R[[G]]$ is a noetherian domain for R a complete discrete valuation ring, to go to matrices over $R[[G]]$ [11, Lemma 7.1].

Theorem 30 *Let R be a complete discrete valuation ring with maximal ideal qR such that R/qR is a field of characteristic p, let G be a torsion-free pro-p group of finite rank, let H be a uniform open characteristic subgroup of G, and for $n \in \mathbb{N}$, let $H_n = \Phi^n(H)$. For $d \in \mathbb{N}$ and $A \in \mathrm{M}_d(R[[G]])$, let $A_n \colon R[G/H_n]^d \to R[G/H_n]^d$ denote the right $R[G/H_n]$-module map induced by left multiplication by A. Then $\lim_{n\to\infty}(\dim_R \ker A_n)/|G/H_n| \in \mathbb{Z}$.*

We can now prove the strong Atiyah conjecture over $\overline{\mathbb{Q}}$ for torsion-free pro-p groups finite rank [11, Theorem 1.1]:

Theorem 31 *Let G be a torsion-free pro-p group of finite rank. Then G satisfies the strong Atiyah conjecture over $\overline{\mathbb{Q}}$.*

Proof Let $d \in \mathbb{N}$ and let $A \in \mathrm{M}_d(\overline{\mathbb{Q}}G)$. Let H be a uniform open characteristic subgroup of G and for $n \in \mathbb{N}$, set $H_n = \Phi_n(H)$. We need to prove $\dim_G(\ker A) \in \mathbb{Z}$. According to Theorem 10 we want to show that $\lim_{n\to\infty} \dim_{\overline{\mathbb{Q}}}(\ker A_n)/|G/H_n| \in \mathbb{Z}$. Since A has only finitely many entries, we may assume that $A \in \mathrm{M}_d(KG)$, where K is an algebraic number field. Let $p \in \mathbb{N}$ be a prime, and extend the p-adic evaluation of $\mathbb{Q}$ to K and complete to obtain $\hat{K}$. Let R be the complete discrete valuation ring determined by the extended valuation. Then we may choose $r \in R \setminus 0$ such that $rA \in \mathrm{M}_d(R)$. Since $\ker A = \ker rA$, we may assume that $A \in \mathrm{M}_d(R)$. Then $\dim_{\overline{\mathbb{Q}}}(\ker A_n) = \dim_R(\ker A_n)$. Application of Theorem 29 finishes the proof. $\square$

At this time, it is still an open question to as whether Theorem 31 remains true for an arbitrary pro-p group of finite rank.

In [29], the following is proved (here G^n indicates the direct product of n copies of G).

Proposition 5 *Let G be a group. Suppose for $n \in \mathbb{N}$, the division closure of $\overline{\mathbb{Q}}[G^n]$ in $\mathcal{U}[G^n]$ is a division ring. Then for every $n \in \mathbb{N}$, the division closure of $\mathbb{C}[G^n]$ in $\mathcal{U}[G^n]$ is also a division ring.*

Now if G is a torsion-free pro-p group of finite rank, then so is G^n for all $n \in \mathbb{N}$. Thus by combining Theorem 31 and Proposition 5, we obtain:

Theorem 32 *Let G be a torsion-free pro-p of finite rank. Then G satisfies the strong Atiyah conjecture over $\mathbb{C}$.*

9 Approximation over arbitrary fields

The purpose of this section is to consider what can be done with Theorem 10 when $\overline{\mathbb{Q}}$ is replaced with a field of nonzero characteristic. First we consider amenable groups.

Let k be a division ring, let G be an elementary amenable group with $\mathrm{lcm}(G) < \infty$, and let $k*G$ be a crossed product [34, Section 1.1]. First suppose $\Delta^+(G) = 1$. Then [22, Proposition 4.2] shows that $k*G$ has a right and left artinian quotient ring. This means that if S is the set of non-zerodivisors of $k*G$, then S satisfies the left and right Ore condition, and $S^{-1}k*G$ is a $d \times d$ matrix over a division ring D for some $d \in \mathbb{N}$. For M a right $k*G$-module, we define $\dim_{k*G}^{\mathrm{Ore}} M = \frac{1}{d}\dim_D M \otimes_{k*G} D$. Then $\dim_{k*G}^{\mathrm{Ore}}$ has all the nice properties one would like of a dimension function, in particular:

- $\dim_{k*G}^{\mathrm{Ore}} k*G = 1$
- If $0 \to L \to M \to N \to 0$ is an exact sequence of $k*G$-modules, then $\dim_{k*G}^{\mathrm{Ore}} M = \dim_{k*G}^{\mathrm{Ore}} L + \dim_{k*G}^{\mathrm{Ore}} N$.

Now suppose that G has a subgroup H of finite index such that $\Delta^+(H) = 1$. Then we define

$$\dim_{k*G}^{\mathrm{Ore}} M := [G : H]^{-1} \dim_{k*H}^{\mathrm{Ore}} M. \tag{3}$$

It is not difficult to show that this is well defined, i.e. is independent of the choice of H, and still has the nice properties of a dimension function, such as above. Then the following was proved in [24].

Theorem 33 *Let G be an elementary amenable group with $\mathrm{lcm}(G) < \infty$, let $G = G_0 \supseteq G_1 \supseteq G_2 \supseteq \cdots$ be a descending sequence of normal subgroups of finite index such that $\bigcap_{i\in\mathbb{N}} G_i = 1$, let k be a division ring, and let M be a finitely presented kG-module. Then*

$$\dim_{kG}^{\mathrm{Ore}} M = \lim_{n\to\infty} |G/G_n|^{-1} \dim_{kG}^{\mathrm{Ore}}(M \otimes_{kG_n} k).$$

For nonamenable groups, we need to put some further restriction on the sequence of normal subgroups of finite index. So let p be a prime and let G be a group with a descending sequence of normal subgroups $G = G_1 \supseteq G_2 \supseteq \cdots$ such that $\bigcap_{i\in\mathbb{N}} G_i = 1$, G/G_i is a finite p-group, and $\sup_i \mathrm{rk}_i(G/G_i) < \infty$. Then $\hat{G} := \varprojlim_i G/G_i$ is a pro-p group of finite rank, so G embeds naturally in a pro-p group of finite rank. Let $\mathbb{F}_p$ denote the field with p-elements, and let $R = \mathbb{F}_p$ or $\mathbb{Z}_p$. Then G has an open uniform characteristic subgroup H, and then $R[[H]]$ is a noetherian domain by Theorem 26, so as in (3), we can define $\dim_{R[[G]]}^{\mathrm{Ore}} M = |G/H|^{-1} \dim_{R[[H]]}^{\mathrm{Ore}} M$ for an $R[[G]]$-module M. Also, the dimension of G is $\mathrm{rk}(H)$; this does not depend on the choice of H. Then the following theorem of Harris [17, 18] is proven in [3, Theorem 2.1] using the above techniques.

Theorem 34 *Let $R = \mathbb{Z}_p$ or $\mathbb{F}_p$, let G be a pro-p group of finite rank and dimension d, and let M be a finitely generated $R[[G]]$-module. Let $G = G_0 \supseteq G_1 \supseteq \cdots$ be a descending sequence of normal subgroups of finite index in G such that $\bigcap_i G_i = 1$. Then*

$$\dim_R(M \hat{\otimes}_{R[[G_i]]} R) = |G/G_i| \dim_{R[[G]]}(M) + O\big(|G/G_i|^{1-1/d}\big).$$

Here $\hat{\otimes}$ denotes the completed tensor product. In the case $R = \mathbb{F}_p$, we have $M \hat{\otimes}_{R[[G_i]]} R = M / M \overline{\omega}(G_i)$.

In [3, Theorem 1.1], this is applied to give the asymptotic growth of Betti numbers for covers of a connected compact CW-complex.

P. A. Linnell
Department of Mathematics, Virginia Tech,
Blacksburg, VA 24061-0123 U.S.A.
E-mail: plinnell@math.vt.edu

References

[1] M. F. Atiyah. Elliptic operators, discrete groups and von Neumann algebras. In *Colloque "Analyse et Topologie" en l'Honneur de Henri Cartan (Orsay, 1974)*, pages 43–72. Astérisque, No. 32–33. Soc. Math. France, Paris, 1976.

[2] Tim Austin. Rational group ring elements with kernels having irrational dimension. *Proc. London Math. Soc., to appear*, 2009. available from http://arxiv.org/abs/0909.2360.

[3] Nicholas Bergeron, Peter Linnell, Wolfgang Lück, and Roman Sauer. On the growth of Betti numbers in p-adic analytic towers. *Groups Geom. Dyn., to appear*, 2013. available from http://arxiv.org/abs/1204.3298.

[4] George M. Bergman. Some examples in PI ring theory. *Israel J. Math.*, 18:257–277, 1974.

[5] George M. Bergman and Warren Dicks. Some examples in PI theory, errata. available from http://math.berkeley.edu/~gbergman/papers/updates/3pi_eg.html, 2013.

[6] Kenneth S. Brown. *Cohomology of groups*, volume 87 of *Graduate Texts in Mathematics*. Springer-Verlag, New York, 1994. Corrected reprint of the 1982 original.

[7] P. M. Cohn. *Free rings and their relations*, volume 19 of *London Mathematical Society Monographs*. Academic Press Inc. [Harcourt Brace Jovanovich Publishers], London, second edition, 1985.

[8] P. M. Cohn. *Free ideal rings and localization in general rings*, volume 3 of *New Mathematical Monographs*. Cambridge University Press, Cambridge, 2006.

[9] Józef Dodziuk, Peter Linnell, Varghese Mathai, Thomas Schick, and Stuart Yates. Approximating L^2-invariants and the Atiyah conjecture. *Comm. Pure Appl. Math.*, 56(7):839–873, 2003. Dedicated to the memory of Jürgen K. Moser.

[10] Józef Dodziuk, Peter Linnell, Varghese Mathai, Thomas Schick, and Stuart Yates. Approximating L^2-invariants and the Atiyah conjecture. available from http://arxiv.org/abs/math/0107049, 2011.

[11] Daniel R. Farkas and Peter A. Linnell. Congruence subgroups and the Atiyah conjecture. In *Groups, rings and algebras*, volume 420 of *Contemp. Math.*, pages 89–102. Amer. Math. Soc., Providence, RI, 2006.

[12] K. R. Goodearl and R. B. Warfield, Jr. *An introduction to noncommutative Noetherian rings*, volume 61 of *London Mathematical Society Student Texts*. Cambridge University Press, Cambridge, second edition, 2004.

[13] Łukasz Grabowski. On the Atiyah problem for the lamplighter groups. available from http://arxiv.org/abs/1009.0229, 2010.

[14] Rostislav I. Grigorchuk, Peter Linnell, Thomas Schick, and Andrzej Żuk. On a question of Atiyah. *C. R. Acad. Sci. Paris Sér. I Math.*, 331(9):663–668, 2000.

[15] Rostislav I. Grigorchuk and Andrzej Żuk. The lamplighter group as a group generated by a 2-state automaton, and its spectrum. *Geom. Dedicata*, 87(1-3):209–244, 2001.

[16] Frédéric Haglund and Daniel T. Wise. Special cube complexes. *Geom. Funct. Anal.*, 17(5):1551–1620, 2008.

[17] Michael Harris. *p*-adic representations arising from descent on abelian varieties. *Compositio Math.*, 39(2):177–245, 1979.

[18] Michael Harris. Correction to: "*p*-adic representations arising from descent on abelian varieties" [Compositio Math. **39** (1979), no. 2, 177–245; MR0546966 (80j:14035)]. *Compositio Math.*, 121(1):105–108, 2000.

[19] Richard V. Kadison and John R. Ringrose. *Fundamentals of the theory of operator algebras. Vol. II*, volume 16 of *Graduate Studies in Mathematics*. American Mathematical Society, Providence, RI, 1997. Advanced theory, Corrected reprint of the 1986 original.

[20] Anselm Knebusch. Approximation of center-valued Betti-numbers. *Houston J. Math.*, 37(1):161–179, 2011.

[21] Anselm Knebusch, Peter A. Linnell, and Thomas Schick. On the center-valued Atiyah conjecture for L^2-betti numbers. in preparation, 2013.

[22] P. H. Kropholler, P. A. Linnell, and J. A. Moody. Applications of a new K-theoretic theorem to soluble group rings. *Proc. Amer. Math. Soc.*, 104(3):675–684, 1988.

[23] Michel Lazard. Groupes analytiques *p*-adiques. *Inst. Hautes Études Sci. Publ. Math.*, (26):389–603, 1965.

[24] Peter Linnell, Wolfgang Lück, and Roman Sauer. The limit of $\mathbb{F}_p$-Betti numbers of a tower of finite covers with amenable fundamental groups. *Proc. Amer. Math. Soc.*, 139(2):421–434, 2011.

[25] Peter Linnell, Boris Okun, and Thomas Schick. The strong Atiyah conjecture for right-angled Artin and Coxeter groups. *Geom. Dedicata*, 158:261–266, 2012.

[26] Peter Linnell and Thomas Schick. Finite group extensions and the Atiyah conjecture. *J. Amer. Math. Soc.*, 20(4):1003–1051, 2007.

[27] Peter A. Linnell. Zero divisors and $L^2(G)$. *C. R. Acad. Sci. Paris Sér. I Math.*, 315(1):49–53, 1992.

[28] Peter A. Linnell. Division rings and group von Neumann algebras. *Forum Math.*, 5(6):561–576, 1993.

[29] Peter A. Linnell and Thomas Schick. The Atiyah conjecture over $\mathbb{C}$. in preparation, 2013.

[30] W. Lück. Approximating L^2-invariants by their finite-dimensional analogues. *Geom. Funct. Anal.*, 4(4):455–481, 1994.

[31] Wolfgang Lück. *L^2-invariants: theory and applications to geometry and K-theory*, volume 44 of *Ergebnisse der Mathematik und ihrer Grenzgebiete. 3. Folge. A Series of Modern Surveys in Mathematics [Results in Mathematics and Related Areas. 3rd Series. A Series of Modern Surveys in Mathematics]*. Springer-Verlag, Berlin, 2002.

[32] Wade Mattox. *Homology of group von Neumann algebras*. PhD thesis, Virginia Tech, 2012.

[33] Nikolay Nikolov and Dan Segal. On finitely generated profinite groups. I. Strong completeness and uniform bounds. *Ann. of Math. (2)*, 165(1):171–238, 2007.

[34] Donald S. Passman. *Infinite crossed products*, volume 135 of *Pure and Applied Mathematics*. Academic Press Inc., Boston, MA, 1989.

[35] Mikäel Pichot, Thomas Schick, and Andrzej Zùk. On the Atiyah problem for the lamplighter groups. available from http://arxiv.org/abs/1005.1147, 2010.

[36] M. Rajagopalan. On the L^p-space of a locally compact group. *Colloq. Math.*, 10:49–52, 1963.

[37] Sadahiro Saeki. The L^p-conjecture and Young's inequality. *Illinois J. Math.*, 34(3):614–627, 1990.

[38] Kevin Schreve. The strong Atiyah problem for compact special groups. *Math. Ann.*, 2014.

[39] Jean-Pierre Serre. *Galois cohomology*. Springer Monographs in Mathematics. Springer-Verlag, Berlin, english edition, 2002. Translated from the French by Patrick Ion and revised by the author.

[40] Seth Warner. *Topological rings*, volume 178 of *North-Holland Mathematics Studies*. North-Holland Publishing Co., Amsterdam, 1993.

[41] John S. Wilson. *Profinite groups*, volume 19 of *London Mathematical Society Monographs. New Series*. The Clarendon Press Oxford University Press, New York, 1998.

7

Cannon-Thurston Maps for Surface Groups: An Exposition of Amalgamation Geometry and Split Geometry

MAHAN MJ[1]

Abstract

This is an expository paper, aimed at giving a leisurely account of some model geometries associated to surface Kleinian groups. We describe the notion of manifolds of amalgamation geometry and its generalization, split geometry. We show that the limit set of any surface group of split geometry is locally connected, by constructing a natural Cannon-Thurston map.[2]

1 Introduction

In [Mj14a] we prove the existence of Cannon-Thurston maps for arbitrary surface Kleinian groups without accidental parabolics. The proof proceeds by constructing a coarse model geometry, called *split geometry*, satisfied by all associated hyperbolic 3-manifolds. In this paper we give an expository account of the model geometries that go into [Mj14a].

1.1 Questions, Conjectures and Statement of Results

In Section 6 of [CT85], Cannon and Thurston raise the following problem:

Question 1 Suppose a closed surface group $\pi_1(S)$ acts freely and properly discontinuously on $\mathbb{H}^3$ by isometries. Does the inclusion $\tilde{i} : \tilde{S} \to \mathbb{H}^3$ extend continuously to the boundary?

[1] Research partly supported by CEFIPRA Indo-French Research Grant 4301-1.
An earlier version of this paper [Mj05] had been written as the first of a two-part paper proving the existence of Cannon-Thurston Maps for surface Kleinian groups.
[2] **AMS Codes:** 57M50, 57M05.

The authors of [CT85] point out that for a simply degenerate group, this is equivalent to asking if the limit set is locally connected.

In [McM01], McMullen makes the following more general conjecture:

Conjecture 2 *For any hyperbolic 3-manifold N with finitely generated fundamental group, there exists a continuous, $\pi_1(N)$-equivariant map*

$$F : \partial \pi_1(N) \to \Lambda \subset S_\infty^2$$

where the boundary $\partial \pi_1(N)$ is constructed by scaling the metric on the Cayley graph of $\pi_1(N)$ by the conformal factor of $d(e, x)^{-2}$, then taking the metric completion. (cf. Floyd [Flo80])

The author raised the following question in his thesis [Mit97] (see also [Bes04]):

Question 3 Let G be a hyperbolic group in the sense of Gromov acting freely and properly discontinuously by isometries on a hyperbolic metric space X. Does the inclusion of the Cayley graph $i : \Gamma_G \to X$ extend continuously to the (Gromov) compactifications?

A similar question may be asked for relatively hyperbolic groups (in the sense of Gromov [Gro85] and Farb [Far98]). The question for relatively hyperbolic groups unifies all the above questions and conjectures.

In this paper we describe a model geometry we call *amalgamation geometry* which is, in a way, a considerable generalization of the notion of *i-bounded geometry* introduced in [Mj11]. We then further generalize it by weakening the hypothesis to the notion of *split geometry*. For ease of exposition we shall restrict ourselves to surfaces without punctures (parabolics). The main theorem of this paper is the following.

Theorem 50: Let $\rho : \pi_1(S) \to PSL_2(C)$ be a faithful representation of a surface group without punctures, and without accidental parabolics. Let $M = \mathbb{H}^3/\rho(\pi_1(S))$ be of split geometry. Let i be an embedding of S in M that induces a homotopy equivalence. Then the embedding $\tilde{i} : \tilde{S} \to \tilde{M} = \mathbb{H}^3$ extends continuously to a map $\hat{i} : \mathbb{D}^2 \to \mathbb{D}^3$. Further, the limit set of $\rho(\pi_1(S))$ is locally connected.

1.2 History

In [Abi76], Abikoff (1976) gave an approach to showing that limit sets of simply degenerate surface Kleinian groups were never locally connected. Thurston and Kerckhoff found a gap in this approach in about 1980.

The first major positive result was proved by Cannon and Thurston [CT85] [CT07] for hyperbolic 3-manifolds fibering over the circle with fiber a closed surface group.

This was generalized by Minsky who proved the Cannon-Thurston result for bounded geometry Kleinian closed surface groups [Min94].

An alternate approach was given by the author in [Mit98b] proving the Cannon-Thurston result for hyperbolic 3-manifolds of bounded geometry without parabolics and with freely indecomposable fundamental group. A different approach based on Minsky's work was given by Klarreich [Kla99]. In [Mit98a] we gave some further extensions to the domain of Gromov-hyperbolic groups (see also [MP11]).

Bowditch [Bow07] [Bow02] proved the Cannon-Thurston result for punctured surface Kleinian groups of bounded geometry. In [Mj09] we extended Bowditch's results to 3 manifolds of bounded geometry whose cores are incompressible away from cusps (see also [Mj10a]).

McMullen [McM01] proved the Cannon-Thurston result for punctured torus groups, using Minsky's model for these groups [Min99]. In [Mj11] we identified a large-scale coarse geometric structure involved in the Minsky model for punctured torus groups (and called it **i-bounded geometry**). We gave a proof for models of *i-bounded geometry*.

In this paper, we define *amalgamation geometry* and prove the Cannon-Thurston result for models of *amalgamation geometry*. We then weaken this assumption to what we call *split geometry* and prove the Cannon-Thurston property for such geometries. In [Mj14a] we show that the Minsky model for general simply or totally degenerate surface groups [Min10] [BCM12] gives rise to a model of *split geometry*. This shows that all surface groups have the Cannon-Thurston property and hence have locally connected limit sets. In [Mj10b] we extend the results of [Mj14a] to all finitely generated Kleinian groups. In [DM10], [Mj14b] we describe point-preimages of Cannon-Thurston maps.

2 Preliminaries and Amalgamation Geometry

2.1 Hyperbolic Metric Spaces

We start off with some preliminaries about hyperbolic metric spaces in the sense of Gromov [Gro85]. For details, see [CDA90], [GdlH90]. Let (X, d) be a hyperbolic metric space. The **Gromov boundary** of X, denoted by ∂X, is the collection of equivalence classes of geodesic rays $r : [0, \infty) \to \Gamma$

with $r(0) = x_0$ for some fixed $x_0 \in X$, where rays r_1 and r_2 are equivalent if $sup\{d(r_1(t), r_2(t))\} < \infty$. Let $\widehat{X} = X \cup \partial X$ denote the natural compactification of X topologized the usual way(cf.[GdlH90] pg. 124).

Definition 4 A subset Z of X is said to be k-**quasiconvex** if any geodesic joining points of Z lies in a k-neighborhood of Z. A subset Z is **quasiconvex** if it is k-quasiconvex for some k. (For simply connected real hyperbolic manifolds this is equivalent to saying that the convex hull of the set Z lies in a bounded neighborhood of Z. We shall have occasion to use this alternate characterization.)

A map f from one metric space (Y, d_Y) into another metric space (Z, d_Z) is said to be a (K, ϵ)-**quasi-isometric embedding** if

$$\frac{1}{K}(d_Y(y_1, y_2)) - \epsilon \le d_Z(f(y_1), f(y_2)) \le K d_Y(y_1, y_2) + \epsilon$$

If f is a quasi-isometric embedding, and every point of Z lies at a uniformly bounded distance from some $f(y)$ then f is said to be a **quasi-isometry**. A (K, ϵ)-quasi-isometric embedding that is a quasi-isometry will be called a (K, ϵ)-quasi-isometry.

A (K, ϵ)-**quasigeodesic** is a (K, ϵ)-quasi-isometric embedding of a closed interval in $\mathbb{R}$. A (K, K)-quasigeodesic will also be called a K-quasigeodesic.

Let (X, d_X) be a hyperbolic metric space and Y be a subspace that is hyperbolic with the inherited path metric d_Y. By adjoining the Gromov boundaries ∂X and ∂Y to X and Y, one obtains their compactifications $\widehat{X}$ and $\widehat{Y}$ respectively.

Let $i : Y \to X$ denote inclusion.

Definition 5 Let X and Y be hyperbolic metric spaces and $i : Y \to X$ be an embedding. A **Cannon-Thurston map** $\hat{i}$ from $\widehat{Y}$ to $\widehat{X}$ is a continuous extension of i.

The following lemma (Lemma 2.1 of [Mit98a]) says that a Cannon-Thurston map exists if for all $M > 0$ and $y \in Y$, there exists $N > 0$ such that if λ lies outside an N ball around y in Y then any geodesic in X joining the end-points of λ lies outside the M ball around $i(y)$ in X. For convenience of use later on, we state this somewhat differently.

Lemma 6 *A Cannon-Thurston map from $\widehat{Y}$ to $\widehat{X}$ exists if the following condition is satisfied:*

Given $y_0 \in Y$, there exists a non-negative function $M(N)$, such that $M(N) \to \infty$ as $N \to \infty$ and for all geodesic segments λ lying outside an

N-ball around $y_0 \in Y$ any geodesic segment in Γ_G joining the end-points of $i(\lambda)$ lies outside the $M(N)$-ball around $i(y_0) \in X$.

The above result can be interpreted as saying that a Cannon-Thurston map exists if the space of geodesic segments in Y embeds properly in the space of geodesic segments in X.

2.2 Amalgamation Geometry

We start with a hyperbolic surface S without punctures. The hyperbolic structure is arbitrary, but it is important that a choice be made.

The Amalgamated Building Block

For the construction of an amalgamated block B, I will denote the closed interval $[0, 3]$. We will describe a geometry on $S \times I$. B has a **geometric core** K with bounded geometry boundary and a preferred geodesic $\gamma (= \gamma_B)$ of bounded length.

There will exist ϵ_0, ϵ_1, D (independent of the block B) such that the following hold:

1) B is identified with $S \times I$.
2) B has a **geometric core** K identified with $S \times [1, 2]$. (K, in its intrinsic path metric, may be thought of, for convenience, as a convex hyperbolic manifold with boundary consisting of pleated surfaces. But we will have occasion to use geometries that are only quasi-isometric to such geometries when lifted to universal covers. As of now, we do not impose any further restriction on the geometry of K.)
3) γ is homotopic to a simple closed curve on $S \times \{i\}$ for any $i \in I$.
4) γ is small, i.e. the length of γ is bounded above by ϵ_0.
5) The intrinsic metric on $S \times i$ (for $i = 1, 2$) has bounded geometry, i.e. any closed geodesic on $S \times \{i\}$ has length bounded below by ϵ_1. Further, the diameter of $S \times \{i\}$ is bounded above by D. (The latter restriction would have followed naturally had we assumed that the curvature of $S \times \{i\}$ is hyperbolic or at least pinched negative.)
6) There exists a regular neighbourhood $N_k(\gamma) \subset K$ of γ which is homeomorphic to a solid torus, such that $N_k(\gamma) \cap S \times \{i\}$ is homeomorphic to an open annulus for $i = 1, 2$. We shall have occasion to denote $N_k(\gamma)$ by T_γ and call it the Margulis tube corresponding to γ.
7) $S \times [0, 1]$ and $S \times [1, 2]$ are given the product structures corresponding to the bounded geometry structures on $S \times \{i\}$, for $i = 1, 2$ respectively.

We next describe the geometry of the *geometric core K*. $K - T_\gamma$ has one or two components according as γ does not or does separate S. These components shall be called **amalgamation components** of K. Let K_1 denote such an *amalgamation component*. Then a lift $\widetilde{K_1}$ of K_1 to $\widetilde{K}$ is bounded by lifts $\widetilde{T}_\gamma$ of T_γ. The union of such a lift $\widetilde{K_1}$ along with the lifts $\widetilde{T}_\gamma$ that bound it will be called an **amalgamation component** of $\widetilde{K}$.

Note that two amalgamation components of $\widetilde{K}$, if they intersect, shall do so along a lift $\widetilde{T}_\gamma$ of T_γ. In this case, they shall be referred to as **adjacent amalgamation components**.

In addition to the above structure of B, we require in addition that there exists $C > 0$ (independent of B) such that

- Each amalgamation component of $\widetilde{K}$ is C-quasiconvex in the intrinsic metric on $\widetilde{K}$.

Note 1: Quasiconvexity of an amalgamation component follows from the fact that any geometric subgroup of infinite index in a surface group is quasiconvex in the latter. The restriction above is therefore to ensure uniform quasiconvexity. We shall strengthen this restriction further when we describe the geometry of $\widetilde{M}$, where M is a 3-manifold built up of blocks of *amalgamation geometry* and those of bounded geometry by gluing them end to end. We shall require that each amalgamation component is **uniformly quasiconvex in $\widetilde{M}$** rather than just in $\widetilde{K}$.

Note 2: So far, the restrictions on K are quite mild. There are really two restrictions. One is the existence of a bounded length simple closed geodesic whose regular neighborhood intersects the bounding surfaces of K in annuli. The second restriction is that the two boundary surfaces of K have *bounded geometry*.

The copy of $S \times I$ thus obtained, with the restrictions above, will be called a **building block of amalgamated geometry** or an **amalgamation geometry building block**, or simply an **amalgamation block**.

Thick Block

Fix constants D, ϵ and let $\mu = [p, q]$ be an ϵ-thick Teichmuller geodesic of length less than D. μ is ϵ-thick means that for any $x \in \mu$ and any closed geodesic η in the hyperbolic surface S_x over x, the length of η is greater than ϵ. Now let B denote the universal curve over μ reparametrized such that the length of μ is covered in unit time. Thus $B = S \times [0, 1]$ topologically.

B is given the path metric and is called a **thick building block**.

Note that after acting by an element of the mapping class group, we might as well assume that μ lies in some given compact region of Teichmuller space. This is because the marking on $S \times \{0\}$ is not important, but rather its position relative to $S \times \{1\}$ Further, since we shall be constructing models only up to quasi-isometry, we might as well assume that $S \times \{0\}$ and $S \times \{1\}$ *lie in the orbit* under the mapping class group of some fixed base surface. Hence μ can be further simplified to be a Teichmuller geodesic joining a pair (p, q) amongst a finite set of points in the orbit of a fixed hyperbolic surface S.

The Model Manifold

Note that the boundary of an amalgamation block B_i consists of $S \times \{0, 3\}$ and the intrinsic path metric on each such $S \times \{0\}$ or $S \times \{3\}$ is of bounded geometry. Also, the boundary of a thick block B consists of $S \times \{0, 1\}$, where S_0, S_1 lie in some given bounded region of Teichmuller space. The intrinsic path metrics on each such $S \times \{0\}$ or $S \times \{1\}$ is the path metric on S.

The model manifold of **amalgamation geometry** is obtained from $S \times J$ (where J is a sub-interval of $\mathbb{R}$, which may be semi-infinite or bi-infinite. In the former case, we choose the usual normalization $J = [0, \infty)$) by first choosing a sequence of blocks B_i (thick or amalgamated) and corresponding intervals $I_i = [0, 1]$ or $[0, 3]$ according as B_i is thick or amalgamated. The metric on $S \times I_i$ is then declared to be that on the building block B_i. Implicitly, we are requiring that the surfaces along which gluing occurs have the same metric. Thus we have,

Definition 7 A manifold M homeomorphic to $S \times J$, where $J = [0, \infty)$ or $J = (-\infty, \infty)$, is said to be a model of **amalgamation geometry** if

1) there is a fiber preserving homeomorphism from M to $\widetilde{S} \times J$ that lifts to a quasi-isometry of universal covers,
2) there exists a sequence I_i of intervals (with disjoint interiors) and blocks B_i where the metric on $S \times I_i$ is the same as that on some building block B_i,
3) $\bigcup_i I_i = J$,
4) There exists $C > 0$ such that for all amalgamated blocks B and geometric cores $K \subset B$, all amalgamation components of $\widetilde{K}$ are C-quasiconvex in $\widetilde{M}$.

The last restriction (4) above is a global restriction on the geometry of amalgamation components, not just a local one (i.e. quasiconvexity in $\widetilde{M}$ rather than $\widetilde{B}$ is required.)

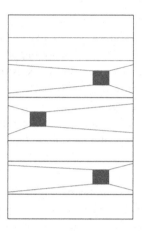

Figure 1 *Model of amalgamated geometry (schematic)*

The figure below illustrates schematically what the model looks like. Filled squares correspond to solid tori along which amalgamation occurs. The adjoining piece(s) denote amalgamation blocks of K. The blocks which have no filled squares are the *thick blocks* and those with filled squares are the *amalgamated blocks*.

Definition 8 A manifold M homeomorphic to $S \times J$, where $J = [0, \infty)$ or $J = (-\infty, \infty)$, is said to have **amalgamated geometry** if there exists $K, \epsilon > 0$ and a model manifold M_1 of *amalgamation geometry* such that

1) there exists a homeomorphism ϕ from M to M_1. This induces from the block decomposition of M_1 a block decomposition of M.
2) We require in addition that the induced homeomorphism $\tilde{\phi}$ between universal covers of blocks is a (K, ϵ) quasi-isometry.

We shall usually suppress the homeomorphism ϕ and take M itself to be a model manifold of *amalgamation geometry*.

Note: We shall later have occasion to introduce a different model, called the **graph model**.

3 Relative Hyperbolicity

In this section, we shall recall first certain notions of relative hyperbolicity due to Farb [Far98], Klarreich [Kla99] and the author [Mj11]. Using these, we shall derive certain Lemmas that will be useful in studying the geometry of the universal covers of building blocks.

3.1 Electric Geometry

We start with a surface S (assumed hyperbolic for the time being) of (K, ϵ) bounded geometry, i.e. S has diameter bounded by K and injectivity radius bounded below by ϵ. Let σ be a simple closed geodesic on S. Replace σ by a copy of $\sigma \times [0, 1]$, by cutting open along σ and gluing in a copy of $\sigma \times [0, 1] = A_\sigma$. (This is like 'grafting' but we shall not have much use for this similarity in this paper.) Let S_G denote the grafted surface. $S_G - A_\sigma$ has one or two components according as σ does not or does separate S. Call these **amalgamation component(s)** of S We shall denote amalgamation components as S_A. We construct a pseudometric on S_G, by declaring the metric on each amalgamation component to be zero and to be the product metric on A_σ. Thus we define:

- the length of any path that lies in the interior of an amalgamation component to be zero,
- the length of any path that lies in A_σ to be its (Euclidean) length in the path metric on A_σ,
- the length of any other path to be the sum of lengths of pieces of the above two kinds.

This allows us to define distances by taking the infimum of lengths of paths joining pairs of points and gives us a path pseudometric, which we call the **electric metric** on S_G. The electric metric also allows us to define geodesics. Let us call S_G equipped with the above pseudometric (S_{Gel}, d_{Gel}) (to be distinguished from a 'dual' construction of an electric metric S_{el} used in [Mj11], where the geodesic σ, rather than its complementary component(s) is electrocuted.)

Important Note: We may and shall regard S as a graph of groups with vertex group(s) the subgroup(s) corresponding to amalgamation component(s) and edge group Z, the fundamental group of A_σ. Then $\widetilde{S}$ equipped with the lift of the above pseudometric is quasi-isometric to the tree corresponding to the splitting on which $\pi_1(S)$ acts.

We shall be interested in the universal cover $\widetilde{S_{Gel}}$ of S_{Gel}. Paths in S_{Gel} and $\widetilde{S_{Gel}}$ will be called electric paths (following Farb [Far98]). Geodesics and quasigeodesics in the electric metric will be called electric geodesics and electric quasigeodesics respectively.

Definition 9 A path $\gamma : I \to Y$ in a path metric space Y is a K-quasigeodesic if we have

$$L(\beta) \leq KL(A) + K$$

for any subsegment $\beta = \gamma|[a, b]$ and any rectifiable path $A : [a, b] \to Y$ with the same endpoints.

γ is said to be an electric K, ϵ-quasigeodesic in $\widetilde{S_{Gel}}$ **without backtracking** if γ is an electric K-quasigeodesic in $\widetilde{S_{Gel}}$ and γ does not return to any any lift $\widetilde{S_A} \subset \widetilde{S_{Gel}}$ (of an amalgamation component $S_A \subset S$) after leaving it.

We collect together certain facts about the electric metric that Farb proves in [Far98]. $N_R(Z)$ will denote the R-neighborhood about the subset Z in the hyperbolic metric d_X. $N_R^e(Z)$ will denote the R-neighborhood about the subset Z in the electric metric d_e.

Lemma 10 *(Lemma 4.5 and Proposition 4.6 of [Far98])*

(1) *Electric quasi-geodesics electrically track hyperbolic geodesics: Given $P > 0$, there exists $K > 0$ with the following property: For some $\widetilde{S_{Gel}}$, let β be any electric P-quasigeodesic without backtracking from x to y, and let γ be the hyperbolic geodesic from x to y. Then $\beta \subset N_K^e(\gamma)$.*

(2) *Hyperbolicity: There exists δ such that each $\widetilde{S_{Gel}}$ is δ-hyperbolic, independent of the curve σ whose lifts are electrocuted.*

Note: As pointed out before, S_{Gel} is quasi-isometric to a tree and is therefore hyperbolic. The above assertion holds in far greater generality than stated. We discuss this below.

We consider a hyperbolic metric space (X, d_X) and a collection $\mathcal{H}$ of *(uniformly) C-quasiconvex uniformly separated subsets*, i.e. there exists $D > 0$ such that for $H_1, H_2 \in \mathcal{H}$, $d_X(H_1, H_2) \geq D$. In this situation X is hyperbolic relative to the collection $\mathcal{H}$. Let d_e denote the hyperbolic metric. The result in this form is due to Klarreich [Kla99]. We give the general version of Farb's theorem below and refer to [Far98] and Klarreich [Kla99] for proofs.

Lemma 11 *(See Lemma 4.5 and Proposition 4.6 of [Far98] and Theorem 5.3 of Klarreich [Kla99]) Given δ, C, D there exists Δ such that if (X, d_X) is a δ-hyperbolic metric space with a collection $\mathcal{H}$ of C-quasiconvex D-separated sets, and (X, d_e) then*

(1) *Electric quasi-geodesics electrically track hyperbolic geodesics: Given $P > 0$, there exists $K > 0$ with the following property: Let β be any electric*

P-quasigeodesic from x to y, and let γ be the hyperbolic geodesic from x to y. Then $\beta \subset N_K^e(\gamma)$.

(2) *γ lies in a hyperbolic (d_X) K-neighborhood of $N_0(\beta)$, where $N_0(\beta)$ denotes the zero neighborhood of β in the* electric metric.

(3) *Hyperbolicity: (X, d_e) is Δ-hyperbolic.*

A special kind of *geodesic without backtracking* will be necessary for universal covers $\widetilde{S_{Gel}}$ of surfaces with some electric metric. Let σ, A_σ be as before.

Let λ_e be an electric geodesic in some $(\widetilde{S_{Gel}}, d_{Gel})$. Then, each segment of λ_e between two lifts $\widetilde{A_\sigma}$ of A_σ (i.e. lying inside a lift of an amalgamation component) is required to be perpendicular to the bounding geodesics. We shall refer to these segments of λ_e as **amalgamation segments** because they lie inside lifts of the amalgamation components.

Let a, b be the points at which λ_e enters and leaves a lift $\widetilde{A_\sigma}$ of A_σ. If a, b lie on the same side, i.e. on a lift of either $\sigma \times \{0\}$ or $\sigma \times \{1\}$, then we join a, b by the geodesic joining them. If they lie on opposite sides of $\widetilde{A_\sigma}$, then assume, for convenience, that a lies on a lift of $\sigma \times \{0\}$ and b lies on a lift of $\sigma \times \{1\}$. Then we join a to b by a union of 2 geodesic segments $[a, c]$ and $[d, b]$ lying along $\widetilde{\sigma} \times \{0\}$ and $\widetilde{\sigma} \times \{1\}$ respectively (for some lift $\widetilde{A_\sigma}$), along with a 'horizontal' segment $[c, d]$, where $[c, d] \subset \widetilde{A_\sigma}$ projects to a segment of the form $\{x\} \times [0, 1] \subset \sigma \times [0, 1]$. We further require that the sum of the lengths $d(a, c)$ and $d(d, b)$ is the minimum possible. The union of the three segments $[a, c], [c, d], [d, b]$ shall be denoted by $[a, b]_{int}$ and shall be referred to as an **interpolating segment**. See figure below.

The union of the *amalgamation segments* along with the *interpolating segments* gives rise to a preferred representative of a quasigeodesic without backtracking joining the end-points of λ_{Gel}. Such a representative of the class of λ_{Gel} shall be called the **canonical representative** of λ_{Gel}. Further, the underlying set of the canonical representative in the *hyperbolic metric* shall be called the **electro-ambient representative** λ_q of λ_e. Since λ_q turns out to be a hyperbolic quasigeodesic (Lemma 13 below), we shall also call it an **electro-ambient quasigeodesic**. See Figure 3 below.

Remark We note first that if we collapse each lift of A_σ along the $I(= [0, 1])$-fibers, (and thus obtain a geodesic that is a lift of σ), then λ_{Gel} becomes an electric geodesic λ_{el} in the universal cover $\widetilde{S_{el}}$ of S_{el}. Here S_{el} denotes the space obtained by electrocuting the geodesic σ (See Section 3.1 of [Mj11].

Let $c : S_G \to S$ be the map that *collapses I-fibers*, i.e. it maps the annulus $A_\sigma = \sigma \times I$ to the geodesic σ by taking (x, t) to x. The lift $\widetilde{c} : \widetilde{S_G} \to \widetilde{S}$ collapses each lift of A_σ along the $I(= [0, 1])$-fibers to a geodesic that is a lift of

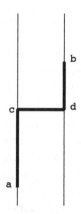

Figure 2 *Interpolating segment*

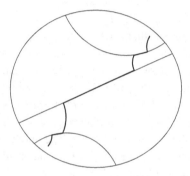

Figure 3 *Electro-ambient quasigeodesic*

σ). Also it takes λ_{Gel} to an electric geodesic λ_{el} in the universal cover $\widetilde{S}_{el}$ of S_{el} (that λ_{el} is an electric geodesic in $\widetilde{S}_{el}$ follows easily, say from normal forms). These were precisely the electro-ambient quasigeodesics in the space $\widetilde{S}_{el}$ (See Section 3.1 of [Mj11] for definitions).

Remark The electro-ambient geodesics in the sense of [Mj11] and those in the present paper differ slightly. The difference is due to the grafting annulus A_{σ} that we use here in place of σ. What is interesting is that whether we electrocute σ (to obtain S_{el}) or its complementary components (to obtain S_{Gel}), we obtain very nearly the same electro-ambient geodesics. In fact modulo c, they *are* the same.

We now recall a lemma from [Mj11]. To distinguish from the electric metrics d_{el} (resp. d_{Gel}) on $\widetilde{S}_{el}$ (resp. $\widetilde{S}_{Gel}$), we shall denote the hyperbolic metrics on $\widetilde{S}_{el}$ as well as $\widetilde{S}_{Gel}$ by d.

Lemma 12 *(See Lemma 3.7 of [Mj11]) There exists (K, ϵ) such that each electro-ambient representative λ_{el} of an electric geodesic in $\widetilde{S}_{el}$ is a (K, ϵ) hyperbolic quasigeodesic in $(\widetilde{S}_{el}, d)$.*

Since $\widetilde{c}$ is clearly a quasi-isometry, it follows easily that:

Lemma 13 *There exists (K, ϵ) such that each electro-ambient representative λ_{Gel} of an electric geodesic in $\widetilde{S}_{Gel}$ is a (K, ϵ) hyperbolic quasigeodesic in $(\widetilde{S}_{Gel}, d)$.*

In the above form, *electro-ambient quasigeodesics* are considered only in the context of surfaces, closed geodesics on them and their complementary (*amalgamation*) components. A considerable generalization of this was obtained in [Mj11], which will be necessary while considering the global geometry of $\widetilde{M}$ (rather than the geometry of $\widetilde{B}$, for an amalgamated building block B).

Definition 14 Given a collection $\mathcal{H}$ of C-quasiconvex, D-separated sets and a number ϵ we shall say that a geodesic (resp. quasigeodesic) γ is a geodesic (resp. quasigeodesic) **without backtracking** with respect to ϵ neighborhoods if γ does not return to $N_\epsilon(H)$ after leaving it, for any $H \in \mathcal{H}$. A geodesic (resp. quasigeodesic) γ is a geodesic (resp. quasigeodesic) **without backtracking** if it is a geodesic (resp. quasigeodesic) without backtracking with respect to ϵ neighborhoods for some $\epsilon \geq 0$.

Note: For strictly convex sets, $\epsilon = 0$ suffices, whereas for convex sets any $\epsilon > 0$ is enough.

Let X be a δ-hyperbolic metric space, and $\mathcal{H}$ a family of C-quasiconvex, D-separated, collection of subsets. Then by Lemma 11, X_{el} obtained by electro-cuting the subsets in $\mathcal{H}$ is a $\Delta = \Delta(\delta, C, D)$-hyperbolic metric space. Now, let $\alpha = [a, b]$ be a hyperbolic geodesic in X and β be an electric P-quasigeodesic without backtracking joining a, b. Replace each maximal subsegment, (with end-points p, q, say) starting from the left of β lying within some $H \in \mathcal{H}$ by a hyperbolic geodesic $[p, q]$. The resulting **connected** path β_q is called an *electro-ambient representative* in X.

In [Mj11] we noted that β_q *need not be a hyperbolic quasigeodesic*. However, we did adapt Proposition 4.3 of Klarreich [Kla99] to obtain the following:

Lemma 15 *(See Proposition 4.3 of [Kla99], also see Lemma 3.10 of [Mj11]) Given δ, C, D, P there exists C_3 such that the following holds: Let (X, d) be a δ-hyperbolic metric space and $\mathcal{H}$ a family of C-quasiconvex, D-separated collection of quasiconvex subsets. Let (X, d_e) denote the electric* *space obtained by electrocuting elements of $\mathcal{H}$. Then, if α, β_q denote respectively a hyperbolic geodesic and an electro-ambient P-quasigeodesic with the same end-points, then α lies in a (hyperbolic) C_3 neighborhood of β_q.*

Note: The above lemma will be needed while considering geodesics in $\widetilde{M}$.

3.2 Electric isometries

Recall that S_G is a grafted surface obtained from a (fixed) hyperbolic metric by grafting an annulus A_σ in place of a geodesic σ.

Now let ϕ be any diffeomorphism of S_G that fixes A_σ pointwise and (in case $(S_G - A_\sigma)$ has two components) preserves each amalgamation component as a set, i.e. ϕ sends each amalgamation component to itself. Such a ϕ will be called a **component preserving diffeomorphism**. Then in the electrocuted surface S_{Gel}, any electric geodesic has length equal to the number of times it crosses A_σ. It follows that ϕ is an isometry of S_{Gel}. (See Lemma 3.12 of [Mj11] for an analogous result in S_{el}.) We state this below.

Lemma 16 *Let ϕ denote a component preserving diffeomorphism of S_G. Then ϕ induces an isometry of (S_{Gel}, d_{Gel}).*

Everything in the above can be lifted to the universal cover $\widetilde{S_{Gel}}$. We let $\widetilde{\phi}$ denote the lift of ϕ to $\widetilde{S_{Gel}}$. This gives

Lemma 17 *Let $\widetilde{\phi}$ denote a lift of a component preserving diffeomorphism ϕ to $(\widetilde{S_{Gel}}, d_{Gel})$. Then $\widetilde{\phi}$ induces an isometry of $(\widetilde{S_{Gel}}, d_{Gel})$.*

3.3 Nearest-point Projections

We need the following basic lemmas from [Mit98b] and [Mj11].

The following lemma says nearest point projections in a δ-hyperbolic metric space do not increase distances much.

Lemma 18 *(Lemma 3.1 of [Mit98b]) Let (Y, d) be a δ-hyperbolic metric space and let $\mu \subset Y$ be a C-quasiconvex subset, e.g. a geodesic segment. Let $\pi : Y \to \mu$ map $y \in Y$ to a point on μ nearest to y. Then $d(\pi(x), \pi(y)) \leq C_3 d(x, y)$ for all $x, y \in Y$ where C_3 depends only on δ, C.*

The next lemma says that quasi-isometries and nearest-point projections on hyperbolic metric spaces 'almost commute'.

Lemma 19 *(Lemma 3.5 of [Mit98b]) Suppose (Y_1, d_1) and (Y_2, d_2) are δ-hyperbolic. Let μ_1 be some geodesic segment in Y_1 joining a, b and let p be*

any vertex of Y_1. Also let q be a vertex on μ_1 such that $d_1(p, q) \le d_2(p, x)$ for $x \in \mu_1$. Let ϕ be a (K, ϵ) - quasi-isometric embedding from Y_1 to Y_2. Let μ_2 be a geodesic segment in Y_2 joining $\phi(a)$ to $\phi(b)$. Let r be a point on μ_2 such that $d_2(\phi(p), r) \le d_2(\phi(p), x)$ for $x \in \mu_2$. Then $d_2(r, \phi(q)) \le C_4$ for some constant C_4 depending only on K, ϵ and δ.

For our purposes we shall need the above lemma for quasi-isometries from $\widetilde{S_a}$ to $\widetilde{S_b}$ for two different hyperbolic structures on the same surface. We shall also need it for electrocuted surfaces.

Yet another property that we shall require for nearest point projections is that nearest point projections in the electric metric and in the 'almost hyperbolic' metric (coming as a lift of the metric on S_G) almost agree. Let $\widetilde{S_G} = Y$ be the universal cover of a surface with the grafted metric. Equip Y with the path metric d as usual. Then Y is quasi-isometric to the hyperbolic plane. Recall that d_{Gel} denotes the electric metric on Y obtained by electrocuting the lifts of complementary components. Now, let $\mu = [a, b]$ be a geodesic on (Y, d) and let μ_q denote the electro-ambient quasigeodesic joining a, b (See Lemma 12). Let π denote the nearest point projection in (Y, d). Tentatively, let π_e denote the nearest point projection in (Y, d_{Gel}). Note that π_e is not well-defined. It is defined up to a bounded amount of discrepancy in the electric metric d_e. But we would like to make π_e well-defined up to a bounded amount of discrepancy in the metric d.

Definition 20 Let $y \in Y$ and let μ_q be an electro-ambient representative of an electric geodesic μ_{Gel} in (Y, d_{Gel}). Then $\pi_e(y) = z \in \mu_q$ if the ordered pair $\{d_{Gel}(y, \pi_e(y)), d(y, \pi_e(y))\}$ is minimized at z.

The proof of the following lemma shows that this gives us a definition of π_e which is ambiguous by a finite amount of discrepancy not only in the electric metric but also in the hyperbolic metric d.

Lemma 21 *There exists $C > 0$ such that the following holds. Let μ be a hyperbolic geodesic joining a, b. Let μ_{Gel} be an electric geodesic joining a, b. Also let μ_q be the electro-ambient representative of μ_{Gel}. Let π_h denote the nearest point projection of Y onto μ. $d(\pi_h(y), \pi_e(y))$ is uniformly bounded.*

Proof: This lemma is similar to Lemma 3.16 of [Mj11], but its proof is somewhat different. For the purposes of this lemma we shall refer to the metric on $\widetilde{S_G}$ as the hyperbolic metric whereas it is in fact only quasi-isometric to it.

$[u, v]$ and $[u, v]_q$ will denote respectively the hyperbolic geodesic and the electro-ambient quasigeodesic joining u, v. Since $[u, v]_q$ is a quasigeodesic by Lemma 12, it suffices to show that for any y, its hyperbolic and electric projections $\pi_h(y)$, $\pi_e(y)$ almost agree.

First note that any hyperbolic geodesic η in $\widetilde{S_G}$ is also an electric geodesic. This follows from the fact that $(\widetilde{S_G}, d_{Gel})$ maps to a tree T (arising from the splitting along σ) with the pullback of every vertex a set of diameter zero in the pseudometric d_{Gel}. Now if a path in $\widetilde{S_G}$ projects to a path in T that is not a geodesic, then it must backtrack. Hence, it must leave an amalgamating component and return to it. Such a path can clearly not be a hyperbolic geodesic in $\widetilde{S_G}$ (since each amalgamating component is convex).

Next, it follows that hyperbolic projections automatically minimize electric distances. Else as in the preceding paragraph, $[y, \pi_h(y)]$ would have to cut a lift of $\tilde{\sigma} = \tilde{\sigma}_1$ that separates $[u, v]_q$. Further, $[y, \pi_h(y)]$ cannot return to $\tilde{\sigma}_1$ after leaving it. Let z be the first point at which $[y, \pi_h(y)]$ meets $\tilde{\sigma}_1$. Also let w be the point on $[u, v]_q \cap \tilde{\sigma}_1$ that is nearest to z. Since amalgamation segments of $[u, v]_q$ meeting $\tilde{\sigma}_1$ are perpendicular to the latter, it follows that $d(w, z) < d(w, \pi_h(y))$ and therefore $d(y, z) < d(y, \pi_h(y))$ contradicting the definition of $\pi_h(y)$. Hence hyperbolic projections automatically minimize electric distances.

Further, it follows by repeating the argument in the first paragraph that $[y, \pi_h(y)]$ and $[y, \pi_e(y)]$ pass through the same set of amalgamation components in the same order; in particular they cut across the same set of lifts of $\tilde{\sigma}$. Let $\tilde{\sigma}_2$ be the last such lift. Then $\tilde{\sigma}_2$ forms the boundary of an amalgamation component $\widetilde{S_A}$ whose intersection with $[u, v]_q$ is of the form $[a, b] \cup [b, c] \cup [c, d]$, where $[a, b] \subset \tilde{\sigma}_3$ and $[c, d] \subset \tilde{\sigma}_4$ are subsegments of two lifts of σ and $[b, c]$ is perpendicular to these two. Then the nearest-point projection of $\tilde{\sigma}_2$ onto each of $[a, b]$, $[b, c]$, $[c, d]$ has uniformly bounded diameter. Hence the nearest point projection of $\tilde{\sigma}_2$ onto the hyperbolic geodesic $[a, d] \subset \widetilde{S_A}$ has uniformly bounded diameter. The result follows. $\qquad\square$

3.4 Coboundedness and Consequences

In this Section, we collect together a few more results that strengthen Lemmas 10 and 11.

Definition 22 A collection $\mathcal{H}$ of uniformly C-quasiconvex sets in a δ-hyperbolic metric space X is said to be **mutually D-cobounded** if for all $H_i, H_j \in \mathcal{H}$, $\pi_i(H_j)$ has diameter less than D, where π_i denotes a nearest point projection of X onto H_i. A collection is **mutually cobounded** if it is mutually D-cobounded for some D.

Lemma 23 *Suppose X is a δ-hyperbolic metric space with a collection $\mathcal{H}$ of C-quasiconvex K-separated D-mutually cobounded subsets. There exists $\epsilon_0 = \epsilon_0(C, K, D, \delta)$ such that the following holds:*

Let β be an electric P-quasigeodesic without backtracking and γ a hyperbolic geodesic, both joining x, y. Then, given $\epsilon \geq \epsilon_0$ there exists $D = D(P, \epsilon)$ such that

(1) *Similar Intersection Patterns 1: if precisely one of $\{\beta, \gamma\}$ meets an ϵ-neighborhood $N_\epsilon(H_1)$ of an electrocuted quasiconvex set $H_1 \in \mathcal{H}$, then the length (measured in the intrinsic path-metric on $N_\epsilon(H_1)$) from the entry point to the exit point is at most D.*

(2) *Similar Intersection Patterns 2: if both $\{\beta, \gamma\}$ meet some $N_\epsilon(H_1)$ then the length (measured in the intrinsic path-metric on $N_\epsilon(H_1)$) from the entry point of β to that of γ is at most D; similarly for exit points.*

Summarizing, we have:

- If X is a hyperbolic metric space and $\mathcal{H}$ a collection of uniformly quasiconvex mutually cobounded separated subsets, then X is hyperbolic relative to the collection $\mathcal{H}$ and satisfies *Bounded Penetration*, i.e. hyperbolic geodesics and electric quasigeodesics have similar intersection patterns in the sense of Lemma 23.

The relevance of co-boundedness comes from the following lemma which is essentially due to Farb [Far98].

Lemma 24 *Let M^h be a hyperbolic manifold of i-bounded geometry, with Margulis tubes $T_i \in \mathcal{T}$ and horoballs $H_j \in \mathcal{H}$. Then the lifts $\widetilde{T_i}$ and $\widetilde{H_j}$ are mutually co-bounded.*

The proof given in [Far98] is for a collection of separated horospheres, but the same proof works for neighborhoods of geodesics and horospheres as well.

A closely related theorem was proved by McMullen (Theorem 8.1 of [McM01]).

As usual, $N_R(Z)$ will denote the R-neighborhood of the set Z.

Let $\mathcal{H}$ be a locally finite collection of horoballs in a convex subset X of $\mathbb{H}^n$ (where the intersection of a horoball, which meets ∂X in a point, with X is called a horoball in X).

Definition 25 *The ϵ-neighborhood of a bi-infinite geodesic in $\mathbb{H}^n$ will be called a ϵ-**thickened geodesic**, or simply a thickened geodesic when the specific ϵ is not important.*

Theorem 26 *[McM01] Let $\gamma : I \to X \setminus \bigcup \mathcal{H}$ be an ambient K-quasigeodesic (for X a convex subset of $\mathbb{H}^n$) and let $\mathcal{H}$ denote a uniformly separated collection of horoballs and thickened geodesics. Let η be the hyperbolic geodesic with the same endpoints as γ. Let $\mathcal{H}(\eta)$ be the union of all the horoballs and thickened geodesics in $\mathcal{H}$ meeting η. Then $\eta \cup \mathcal{H}(\eta)$ is (uniformly) quasiconvex and $\gamma(I) \subset N_R(\eta \cup \mathcal{H}(\eta))$, where R depends only on K.*

4 Universal Covers of Building Blocks and Electric Geometry

4.1 Graph Model of Building Blocks

Amalgamation Blocks

Given a geodesic segment $\lambda \subset \widetilde{S}$ and a basic *amalgamation building block* B, let $\lambda = [a, b] \subset \widetilde{S} \times \{0\}$ be a geodesic segment, where $\widetilde{S} \times \{0\} \subset \widetilde{B}$.

We shall now build a graph model for $\widetilde{B}$ which will be quasi-isometric to an electrocuted version of the original model, where amalgamation components of the geometric core K are electrocuted.

$\widetilde{S} \times \{0\}$ and $\widetilde{S} \times \{3\}$ are equipped with hyperbolic metrics. $\widetilde{S} \times \{1\}$ and $\widetilde{S} \times \{2\}$ are grafted surfaces with electric metric obtained by electrocuting the amalgamation components. This constructs 4 'sheets' of $\widetilde{S}$ comprising the 'horizontal skeleton' of the 'graph model' of $\widetilde{B}$. Now for the vertical strands. On each vertical element of the form $x \times [0, 1]$ and $x \times [2, 3]$ put the Euclidean metric.

To do this precisely, one needs to take a bit more care and perform the construction in the universal cover. For each amalgamation component of $\widetilde{K}$ (recall that such a component is a lift of an amalgamation component of K to the universal cover along with bounding lifts $\widetilde{T}_\sigma$ of the Margulis tubes). For each such component $\widetilde{K}_i$ we construct $\widetilde{K}_i \times [1, 2]$, so that any two copies $\widetilde{K}_i \times [1, 2]$ and $\widetilde{K}_j \times [1, 2]$ intersect (if at all they do) only along the original bounding lifts $\widetilde{T}_\sigma$ of the Margulis tubes. Next electrocute each copy of $\widetilde{K}_i \times [1, 2]$.

This construction is very closely related to the 'coning' construction introduced by Farb in [Far98].

The resulting copy of $\widetilde{B}$ will be called the **graph model of an amalgamation block**.

Next, we give an I-bundle structure to K that preserves the grafting annulus. Thus $A_\sigma \times [1, 2]$ has a structure of a Margulis tube. Let ϕ denote a map from $S \times \{1\}$ to $S \times \{2\}$ mapping $(x, 1)$ to $(x, 2)$. Clearly there is a bound l_B on the length in K of $x \times [1, 2]$ as x ranges over $S \times \{1\}$. That is to say that the core K has a bounded thickness. This bound depends on the block B we are considering.

Let $\widetilde{\phi}$ denote the lift of ϕ to $\widetilde{K}$ Then $\widetilde{\phi}$ is a (k, ϵ)-quasi-isometry where k, ϵ depend on the block B.

Thick Block

For a thick block $B = \widetilde{S} \times [0, 1]$, recall that B is the universal curve over a 'thick' Teichmuller geodesic $\eta_{Teich} = [c, d]$ of length less than some fixed $D > 0$. Each $S \times \{x\}$ is identified with the hyperbolic surface over $(c + \frac{x}{d-c})$ (assuming that the Teichmuller geodesic is parametrized by arc-length).

Here $S \times \{0\}$ is identified with the hyperbolic surface corresponding to a, $S \times \{1\}$ is identified with the hyperbolic surface corresponding to b and each (x, a) is joined to (x, b) by a segment of length 1.

The resulting model of $\widetilde{B}$ is called a **graph model of a thick block**.

Metrics on graph models are called **graph metrics**.

Admissible Paths

Admissible paths consist of the following:

1) Horizontal segments along some $\widetilde{S} \times \{i\}$ for $i = \{0, 1, 2, 3\}$ (amalgamated blocks) or $i = \{0, 1\}$ (thick blocks).
2) Vertical segments $x \times [0, 1]$ or $x \times [2, 3]$ for amalgamated blocks or $x \times [0, 1]$ for thick blocks.
3) Vertical segments of length $\leq l_B$ joining $x \times \{1\}$ to $x \times \{2\}$ for amalgamated blocks.

4.2 Construction of Quasiconvex Sets for Building Blocks

Let M be a manifold homeomorphic to $S \times J$ (where $J[0, \infty)$ or $\mathbb{R}$) obtained by assembling thick and amalgamation blocks one on top of another, giving a model manifold of amalgamation geometry. In the next section, we will construct a set B_λ containing $\lambda (\subset \widetilde{S} \times \{0\} \subset \widetilde{M})$ and a retraction Π_λ of $\widetilde{M}$ onto it. Π_λ will have the property that it does not stretch distances much. This will show that B_λ is quasi-isometrically embedded in $\widetilde{M}$.

In this subsection, we describe the construction of B_λ restricted to a building block B.

Construction of $B_\lambda(B)$ - Thick Block

Let the thick block be the universal curve over a Teichmuller geodesic $[\alpha, \beta]$. Let S_α denote the hyperbolic surface over α and S_β denote the hyperbolic surface over β.

First, let $\lambda = [a, b]$ be a geodesic segment in $\widetilde{S}$. Let λ_{B0} denote $\lambda \times \{0\}$.

Next, let ψ be the lift of the 'identity' map from $\widetilde{S}_\alpha$ to $\widetilde{S}_\beta$. Let Ψ denote the induced map on geodesics and let $\Psi(\lambda)$ denote the hyperbolic geodesic joining $\psi(a)$, $\psi(b)$. Let λ_{B1} denote $\Psi(\lambda) \times \{1\}$.

For the universal cover $\widetilde{B}$ of the thick block B, define:

$$B_\lambda(B) = \bigcup_{i=0,1} \lambda_{Bi}$$

Definition 27 Each $\widetilde{S} \times i$ for $i = 0, 1$ will be called a **horizontal sheet** of $\widetilde{B}$ when B is a thick block.

Construction of $B_\lambda(B)$ - Amalgamation Block

First, recall that $\lambda = [a, b]$ is a geodesic segment in $\widetilde{S}$. Let λ_{B0} denote $\lambda \times \{0\}$.

Next, let λ_{Gel} denote the electric geodesic joining a, b in the electric pseudo-metric on $\widetilde{S}$ obtained by electrocuting lifts of σ. Let λ_{B1} denote $\lambda_{Gel} \times \{1\}$.

Third, recall that $\widetilde{\phi}$ is the lift of a component preserving diffeomorphism ϕ to $\widetilde{S}$ equipped with the electric metric d_{Gel}. Let $\widetilde{\Phi}$ denote the induced map on geodesics, i.e. if $\mu = [x, y] \subset (\widetilde{S}, d_{Gel})$, then $\widetilde{\Phi}(\mu) = [\phi(x), \phi(y)]$ is the geodesic joining $\phi(x)$, $\phi(y)$. Let λ_{B2} denote $\Phi(\lambda_{Gel}) \times \{2\}$.

Fourthly, let $\Phi(\lambda)$ denote the hyperbolic geodesic joining $\phi(a)$, $\phi(b)$. Let λ_{B3} denote $\Phi(\lambda) \times \{3\}$.

For the universal cover $\widetilde{B}$ of the thin block B, define:

$$B_\lambda(B) = \bigcup_{i=0,\cdots,3} \lambda_{Bi}$$

Definition 28 Each $\widetilde{S} \times i$ for $i = 0 \cdots 3$ will be called a **horizontal sheet** of $\widetilde{B}$ when B is a thick block.

Construction of $\Pi_{\lambda,B}$ - Thick Block

On $\widetilde{S} \times \{0\}$, let Π_{B0} denote nearest point projection onto λ_{B0} in the path metric on $\widetilde{S} \times \{0\}$.

On $\widetilde{S} \times \{1\}$, let Π_{B1} denote nearest point projection onto λ_{B1} in the path metric on $\widetilde{S} \times \{1\}$.

For the universal cover $\widetilde{B}$ of the thick block B, define:

$$\Pi_{\lambda,B}(x) = \Pi_{Bi}(x), x \in \widetilde{S} \times \{i\}, i = 0, 1.$$

Construction of $\Pi_{\lambda,B}$ - Amalgamation Block

On $\widetilde{S} \times \{0\}$, let Π_{B0} denote nearest point projection onto λ_{B0}. Here the nearest point projection is taken in the path metric on $\widetilde{S} \times \{0\}$ which is a hyperbolic metric space.

On $\widetilde{S} \times \{1\}$, let Π_{B1} denote the nearest point projection onto λ_{B1}. Here the nearest point projection is taken in the sense of the definition preceding Lemma 21, i.e. minimizing the ordered pair (d_{Gel}, d_{hyp}) (where d_{Gel}, d_{hyp} refer to electric and hyperbolic metrics respectively.)

On $\widetilde{S} \times \{2\}$, let Π_{B2} denote the nearest point projection onto λ_{B2}. Here, again the nearest point projection is taken in the sense of the definition preceding Lemma 21.

Again, on $\widetilde{S} \times \{3\}$, let Π_{B3} denote nearest point projection onto λ_{B3}. Here the nearest point projection is taken in the path metric on $\widetilde{S} \times \{3\}$ which is a hyperbolic metric space.

For the universal cover $\widetilde{B}$ of the thin block B, define:

$$\Pi_{\lambda,B}(x) = \Pi_{Bi}(x), x \in \widetilde{S} \times \{i\}, i = 0, \cdots, 3.$$

$\Pi_{\lambda,B}$ is a retract - Thick Block

The proof for a thick block is exactly as in [Mit98b] and [Mj11]. We omit it here.

Lemma 29 *(Lemma 4.1 of [Mj11] There exists $C > 0$ such that the following holds:*
Let $x, y \in \widetilde{S} \times \{0, 1\} \subset \widetilde{B}$ for some thick block B. Then $d(\Pi_{\lambda,B}(x), \Pi_{\lambda,B}(y)) \leq Cd(x, y)$.

$\Pi_{\lambda,B}$ is a retract - Amalgamation Block

The main ingredient in this case is Lemma 21.

Lemma 30 *There exists $C > 0$ such that the following holds:*
Let $x, y \in \widetilde{S} \times \{0, 1, 2, 3\} \subset \widetilde{B}$ for some amalgamated block B. Then $d_{Gel}(\Pi_{\lambda,B}(x), \Pi_{\lambda,B}(y)) \leq Cd_{Gel}(x, y)$.

Proof: It is enough to show this for the following cases:

1) $x, y \in \widetilde{S} \times \{0\}$ OR $x, y \in \widetilde{S} \times \{3\}$.
2) $x = (p, 0)$ and $y = (p, 1)$ for some p.
3) x, y both lie in the geometric core K.
4) $x = (p, 2)$ and $y = (p, 3)$ for some p.

Case 1: This follows from Lemma 18.

Mahan Mj

Case 2 and Case 4: These follow from Lemma 21 which says that the hyperbolic and electric projections of p onto the hyperbolic geodesic $[a, b]$ and the electro-ambient geodesic $[a, b]_{ea}$ respectively 'almost agree'. If π_h and π_e denote the hyperbolic and electric projections, then there exists $C_1 > 0$ such that

$$d_{Gel}(\pi_h(p), \pi_e(p)) \leq C_1.$$

Hence

$$d_{Gel}(\Pi_{\lambda, B}(p, i)), \Pi_{\lambda, B}((p, i+1))) \leq C_1 + 1, \quad \text{for } i = 0, 2.$$

Case 3: This follows from the fact that K in the graph model with the electric metric is essentially the tree coming from the splitting. Further, by the properties of π_e, each amalgamation component projects down to a set of diameter zero. Hence

$$d_{Gel}(\Pi_{\lambda, B}(p), \Pi_{\lambda, B}(q)) \leq C_1 + 1.$$

Choosing C as the maximum of these constants, we are through. $\qquad\square$

5 Construction of Quasiconvex Sets and Quasigeodesics

5.1 Construction of B_λ and Π_λ

Given a manifold M of amalgamated geometry, we know that M is homeomorphic to $S \times J$ for $J = [0, \infty)$ or $(-\infty, \infty)$. By definition of amalgamated geometry, there exists a sequence I_i of intervals and blocks B_i where the metric on $S \times I_i$ coincides with that on some building block B_i. Denote:

- $B_{\mu, B_i} = B_{i\mu}$
- $\Pi_{\mu, B_i} = \Pi_{i\mu}$

Now for a block $B = S \times I$ (thick or amalgamated), a natural map Φ_B may be defined taking $\mu = B_{\mu, B} \cap \widetilde{S} \times \{0\}$ to a geodesic $B_{\mu, B} \cap \widetilde{S} \times \{k\} = \Phi_B(\mu)$ where $k = 1$ or 3 according as B is thick or amalgamated. Let the map Φ_{B_i} be denoted as Φ_i for $i \geq 0$. For $i < 0$ we shall modify this by defining Φ_i to be the map that takes $\mu = B_{\mu, B_i} \cap \widetilde{S} \times \{k\}$ to a geodesic $B_{\mu, B_i} \cap \widetilde{S} \times \{0\} = \Phi_i(\mu)$ where $k = 1$ or 3 according as B is thick or amalgamated.

We start with a reference block B_0 and a reference geodesic segment $\lambda = \lambda_0$ on the 'lower surface' $\widetilde{S} \times \{0\}$. Now inductively define:

- $\lambda_{i+1} = \Phi_i(\lambda_i)$ for $i \geq 0$
- $\lambda_{i-1} = \Phi_i(\lambda_i)$ for $i \leq 0$

- $B_{i\lambda} = B_{\lambda_i}(B_i)$
- $\Pi_{i\lambda} = \Pi_{\lambda_i, B_i}$
- $B_\lambda = \bigcup_i B_{i\lambda}$
- $\Pi_\lambda = \bigcup_i \Pi_{i\lambda}$

Recall that each $\widetilde{S} \times i$ for $i = 0 \cdots m$ is called a **horizontal sheet** of $\widetilde{B}$, where $m = 1$ or 3 according as B is thick or amalgamated. We will restrict our attention to the union of the horizontal sheets $\widetilde{M_H}$ of $\widetilde{M}$ with the metric induced from the graph model.

Clearly, $B_\lambda \subset \widetilde{M_H} \subset \widetilde{M}$, and Π_λ is defined from $\widetilde{M_H}$ to B_λ. Since $\widetilde{M_H}$ is a 'coarse net' in $\widetilde{M}$ (equipped with the *graph model metric*), we will be able to get all the coarse information we need by restricting ourselves to $\widetilde{M_H}$.

By Lemmas 29 and 30, we obtain the fact that each $\Pi_{i\lambda}$ is a retract. Hence assembling all these retracts together, we have the following basic theorem:

Theorem 31 *There exists $C > 0$ such that for any geodesic $\lambda = \lambda_0 \subset \widetilde{S} \times \{0\} \subset \widetilde{B_0}$, the retraction $\Pi_\lambda : \widetilde{M_H} \to B_\lambda$ satisfies:*

$$\text{Then } d_{Gel}(\Pi_{\lambda, B}(x), \Pi_{\lambda, B}(y)) \leq C d_{Gel}(x, y) + C.$$

Note 1: For Theorem 31 above, note that all that we really require is that the universal cover $\widetilde{S}$ be a hyperbolic metric space. There is *no restriction on $\widetilde{M_H}$*. In fact, Theorem 31 would hold for general stacks of hyperbolic metric spaces with blocks of amalgamated geometry.

Note 2: M_H has been built up out of **graph models of thick and amalgamated blocks** and have sheets that are electrocuted along geodesics.

We want to make *Note 1* above explicit. We first modify the definition of amalgamation geometry as follows, retaining only local quasiconvexity.

Definition: A manifold M homeomorphic to $S \times J$, where $J = [0, \infty)$ or $J = (-\infty, \infty)$, is said to be a model of **weak amalgamation geometry** if

1) there is a fiber preserving homeomorphism from M to $\widetilde{S} \times J$ that lifts to a quasi-isometry of universal covers.
2) there exists a sequence I_i of intervals (with disjoint interiors) and blocks B_i where the metric on $S \times I_i$ is the same as that on some building block B_i. Each block is either thick or has amalgamation geometry.
3) $\bigcup_i I_i = J$.
4) There exists $C_0 > 0$ such that for all amalgamated blocks B_i and geometric cores $C \subset B_i$, all amalgamation components of $\widetilde{C}$ are C_0-quasiconvex in $\widetilde{B_i}$.

Then as a consequence of *the proof* of Theorem 31, we have the following Corollary.

Corollary 32 *Let M be a model manifold of* **weak amalgamation geometry**. *There exists C > 0 such that the following holds:*

Given any geodesic $\lambda \subset \widetilde{S} \times \{0\}$, *let* B_λ, Π_λ *be as before. Then for* $\lambda = \lambda_0 \subset \widetilde{S} \times \{0\} \subset \widetilde{B_0}$, *the retraction* $\Pi_\lambda : \widetilde{M_H} \rightarrow B_\lambda$ *satisfies:*

Then $d_{Gel}(\Pi_{\lambda,B}(x), \Pi_{\lambda,B}(y)) \leq C d_{Gel}(x, y) + C.$

In fact, all that follows in this section may just as well be done for model manifolds of *weak amalgamation geometry*. We shall make this explicit again at the end of this entire section.

Bur before we proceed, we would like to deduce one further Corollary of Theorem 31, which shall be useful towards the end of the paper. Instead of constructing vertical hyperbolic ladders B_λ for finite geodesic segments, first note that λ might as well be bi-infinite. Next, we would like to construct such a B_λ **equivariantly** under the action of $\mathbb{Z}$. That is to say, we would like to construct a vertical annulus in the manifold M homeomorphic to $S \times \mathbb{R}$.

To do this, we start with a simple closed geodesic σ on $S \times \{0\}$. Instead of performing the construction in the universal cover, homotop σ into $S \times \{i\}$ for each level i. Let σ_i denote the shortest electro-ambient geodesic in the *free homotopy class* of $\sigma \times \{i\}$ in the path pseudometric on $S \times \{i\}$. Now let B_σ denote the set $B_\sigma = \bigcup_i \widetilde{\sigma}_i$. Then the proof of Theorem 31 ensures the quasi-convexity of B_σ in the *graph-metric*. Finally, since B_σ has been constructed to be equivariant under the action of the surface group, its quotient in M is an embedded 'quasi-annulus' $A_{P\sigma}$ which partitions the manifold locally. We use the term 'quasi-annulus' because $A_{P\sigma}$ is a collection of disjoint circles at different levels. We finally conclude:

Corollary 33 *Let M be a model manifold of* **weak amalgamation geometry**. *There exists C > 0 such that the following holds:*

Given any simple closed geodesic $\sigma \subset S \times \{0\}$, *let* B_σ *be as above. Then its quotient, the embedded quasi-annulus* $A_{P\sigma}$ *above is C-quasiconvex in M with the graph metric.*

Another Corollary will be used later. Suppose $\Sigma = \Sigma \times \{0\}$ be a subsurface of $S \times \{0\}$ with geodesic boundary components $\sigma^1 \cdots \sigma^k$. Let Σ_i be the subsurface of $S \times \{i\}$ that is bounded by $\sigma_i^1 \cdots \sigma_i^k$. Let $B_\Sigma = \bigcup_i \Sigma_i$.

Corollary 34 *Let M be a model manifold of* **weak amalgamation geometry**. *There exists C > 0 such that the following holds:*

Given any subsurface $\Sigma \subset S \times \{0\}$ *with geodesic boundary components, let* B_Σ *be as above. Then* B_Σ *is C-quasiconvex in M with the graph metric.*

5.2 Heights of Blocks

Recall that each geometric core $C \subset B$ is identified with $S \times I$ where each fibre $\{x\} \times I$ has length $\leq l_C$ for some l_C, called the *thickness* of the block B. If $C \subset B_i$ for one of the above blocks B_i, we shall denote l_C as l_i.

Instead of considering all the horizontal sheets, we would now like to consider only the **boundary horizontal sheets**, i.e. for a thick block we consider $\widetilde{S} \times \{0, 1\}$ and for a thin block we consider $\widetilde{S} \times \{0, 3\}$. The union of all boundary horizontal sheets will be denoted by M_{BH}.

Observation 1: $\widetilde{M_{BH}}$ is a 'coarse net' in $\widetilde{M}$ in the **graph model**, but not in the **model of amalgamated geometry**.

In the graph model, any point can be connected by a vertical segment of length ≤ 2 to one of the boundary horizontal sheets.

However, in the model of amalgamated geometry, there are points within amalgamation components which are at a distance of the order of l_i from the boundary horizontal sheets. Since l_i is arbitrary, $\widetilde{M_{BH}}$ is no longer a 'coarse net' in $\widetilde{M}$ equipped with the model of *amalgamated geometry*.

Observation 2: $\widetilde{M_H}$ is defined only in the **graph model**, but not in the model of amalgamated geometry.

Observation 3: The electric metric on the model of amalgamated geometry on $\widetilde{M}$ obtained by electrocuting amalgamation components is quasi-isometric to the graph model of $\widetilde{M}$.

Bounded Height of Thick Block

Let $\mu \subset \widetilde{S} \times \{0\}\widetilde{B}_i$ be a geodesic in a (thick or amalgamated) block. Then there exists a (K_i, ϵ_i)- quasi-isometry ψ_i ($= \phi_i$ for thick blocks) from $\widetilde{S} \times \{0\}$ to $\widetilde{S} \times \{1\}$ and Ψ_i is the induced map on geodesics. Hence, for any $x \in \mu$, $\psi_i(x)$ lies within some bounded distance C_i of $\Psi_i(\mu)$. But x is connected to $\psi_i(x)$ by

Case 1 - Thick Blocks: a vertical segment of length 1
Case 2 - Amalgamated Blocks: the union of

1) two vertical segments of length 1 between $\widetilde{S} \times \{i\}$ and $\widetilde{S} \times \{i + 1\}$ for $i = 0, 2$.

2) a horizontal segment of length bounded by (some uniform) C' (cf. Lemma 12) connecting $(x, 1)$ to a point on the electro-ambient geodesic $B_\lambda(B) \cap \widetilde{S} \times \{1\}$.

3) a vertical segment of electric length zero in the **graph model** connecting $(x, 1)$ to $(x, 2)$. Such a path has to travel *through an amalgamated block* in the model of **amalgamated geometry** and has length less than l_i, where l_i is the thickness of the ith block B_i.

4) a horizontal segment of length less than C' (Lemma 12) connecting $(\phi_i(x), 3)$ to a point on the hyperbolic geodesic $B_\lambda(B) \cap \widetilde{S} \times \{3\}$.

Thus x can be connected to a point on $x' \in \Psi_i(\mu)$ by a path of length less than $g(i) = 2 + 2C' + l_i$. Recall that λ_i is the geodesic on the lower horizontal surface of the block $\widetilde{B}_i$. The same can be done for blocks $\widetilde{B_{i-1}}$ and *going down* from λ_i to λ_{i-1}. What we have thus shown is:

Lemma 35 *There exists a function g $: \mathbb{Z} \to \mathbb{N}$ such that for any block B_i (resp. B_{i-1}), and $x \in \lambda_i$, there exists $x' \in \lambda_{i+1}$ (resp. λ_{i-1}) for $i \geq 0$ (resp. $i \leq 0$), satisfying:*

$$d(x, x') \leq g(i).$$

5.3 Admissible Paths

We want to define a collection of B_λ-**elementary admissible paths** lying in a bounded neighborhood of B_λ. B_λ is not connected. Hence, it does not make much sense to speak of the path-metric on B_λ. To remedy this we introduce a 'thickening' (cf. [Gro93]) of B_λ which is path-connected and where the paths are controlled. A B_λ-**admissible path** will be a composition of B_λ-elementary admissible paths.

Recall that admissible paths in the graph model of bounded geometry consist of the following :

1) Horizontal segments along some $\widetilde{S} \times \{i\}$ for $i = \{0, 1, 2, 3\}$ (amalgamated blocks) or $i = \{0, 1\}$ (thick blocks).

2) Vertical segments $x \times [0, 1]$ or $x \times [2, 3]$ for amalgamated blocks, where $x \in \widetilde{S}$.

3) Hyperbolic geodesic segments of length $\leq l_B$ in $K \subset B$ joining $x \times \{1\}$ to $x \times \{2\}$ for amalgamated blocks.

4) Vertical segments of length 1 joining $x \times \{0\}$ to $x \times \{1\}$ for thick blocks.

We shall choose a subclass of these admissible paths to define B_λ-elementary admissible paths.

B_λ-elementary admissible paths in the thick block

Let $B = S \times [i, i + 1]$ be a thick block, where each (x, i) is connected by a vertical segment of length 1 to $(x, i + 1)$. Let ϕ be the map that takes (x, i) to $(x, i + 1)$. Also Φ is the map on geodesics induced by ϕ. Let $B_\lambda \cap \widetilde{B} = \lambda_i \cup \lambda_{i+1}$ where λ_i lies on $\widetilde{S} \times \{i\}$ and λ_{i+1} lies on $\widetilde{S} \times \{i + 1\}$. π_j, for $j = i, i + 1$ denote nearest-point projections of $\widetilde{S} \times \{j\}$ onto λ_j. Next, since ϕ is a quasi-isometry, there exists $C > 0$ such that for all $(x, i) \in \lambda_i$, $(x, i + 1)$ lies in a C-neighborhood of $\Phi(\lambda_i) = \lambda_{i+1}$. The same holds for ϕ^{-1} and points in λ_{i+1}, where ϕ^{-1} denotes the *quasi-isometric inverse* of ϕ from $\widetilde{S} \times \{i + 1\}$ to $\widetilde{S} \times \{i\}$. The B_λ-**elementary admissible paths** in $\widetilde{B}$ consist of the following:

1) Horizontal geodesic subsegments of λ_j, $j = \{i, i + 1\}$.
2) Vertical segments of length 1 joining $x \times \{0\}$ to $x \times \{1\}$.
3) Horizontal geodesic segments lying in a C-neighborhood of λ_j, $j = i, i + 1$.

B_λ-elementary admissible paths in the amalgamated block

Let $B = S \times [i, i + 3]$ be an amalgamated block, where each $(x, i + 1)$ is connected by a geodesic segment of zero electric length and hyperbolic length $\leq C(B)$ (due to bounded thickness of B) to $(\phi(x), i + 2)$ (Here ϕ can be thought of as the map from $\widetilde{S} \times \{i + 1\}$ to $.\widetilde{S} \times \{i + 2\}$ that is the identity on the first component. Also Φ is the map on canonical representatives of electric geodesics induced by ϕ. Let $B_\lambda \cap \widetilde{B} = \bigcup_{j=i\cdots i+3} \lambda_j$ where λ_j lies on $\widetilde{S} \times \{j\}$. π_j denotes nearest-point projection of $\widetilde{S} \times \{j\}$ onto λ_j (in the appropriate sense - hyperbolic for $j = i, i + 3$ and electric for $j = i + 1, i + 2$). Next, since ϕ is an electric isometry, but a hyperbolic quasi-isometry, there exists $C > 0$ (uniform constant) and $K = K(B)$ such that for all $(x, i) \in \lambda_i$, $(\phi(x), i + 1)$ lies in an (electric) C-neighborhood and a hyperbolic K-neighborhood of $\Phi(\lambda_{i+1}) = \lambda_{i+2}$. The same holds for ϕ^{-1} and points in λ_{i+2}, where ϕ^{-1} denotes the *quasi-isometric inverse* of ϕ from $\widetilde{S} \times \{i + 2\}$ to $\widetilde{S} \times \{i + 1\}$.

Again, since λ_{i+1} and λ_{i+2} are electro-ambient quasigeodesics, we further note that there exists $C > 0$ (assuming the same C for convenience) such that for all $(x, i) \in \lambda_i$, $(x, i + 1)$ lies in a (hyperbolic) C-neighborhood of λ_{i+1}. Similarly for all $(x, i + 2) \in \lambda_{i+2}$, $(x, i + 3)$ lies in a (hyperbolic) C-neighborhood of λ_{i+3}. The same holds if we go 'down' from λ_{i+1} to λ_i or from λ_{i+3} to λ_{i+2}. The B_λ-**elementary admissible paths** in $\widetilde{B}$ consist of the following:

1) Horizontal subsegments of λ_j, $j = \{i, \cdots i + 3\}$.
2) Vertical segments of length 1 joining $x \times \{j\}$ to $x \times \{j + 1\}$, for $j = i, i + 2$.
3) Horizontal geodesic segments lying in a *hyperbolic C*-neighborhood of λ_j, $j = i, \cdots i + 3$.
4) Horizontal hyperbolic segments of *electric length* $\leq C$ and *hyperbolic length* $\leq K(B)$ joining points of the form $(\phi(x), i + 2)$ to a point on λ_{i+2} for $(x, i + 1) \in \lambda_{i+1}$.
5) Horizontal hyperbolic segments of *electric length* $\leq C$ and *hyperbolic length* $\leq K(B)$ joining points of the form $(\phi^{-1}(x), i + 1)$ to a point on λ_{i+1} for $(x, i + 2) \in \lambda_{i+2}$.

Definition: A B_λ-admissible path is a union of B_λ-elementary admissible paths.

The following lemma follows from the above definition and Lemma 35.

Lemma 36 *There exists a function* $g : \mathbb{Z} \to \mathbb{N}$ *such that for any block* B_i, *and* x *lying on a* B_λ-*admissible path in* $\widetilde{B}_i$, *there exist* $y \in \lambda_j$ *and* $z \in \lambda_k$ *where* $\lambda_j \subset B_\lambda$ *and* $\lambda_k \subset B_\lambda$ *lie on the two boundary horizontal sheets, satisfying:*

$$d(x, y) \leq g(i),$$
$$d(x, z) \leq g(i).$$

Let $h(i) = \Sigma_{j=0\cdots i} g(j)$ be the sum of the values of $g(j)$ as j ranges from 0 to i (with the assumption that increments are by $+1$ for $i \geq 0$ and by -1 for $i \leq 0$). Then we have from Lemma 36 above,

Corollary 37 *There exists a function* $h : \mathbb{Z} \to \mathbb{N}$ *such that for any block* B_i, *and* x *lying on a* B_λ-*admissible path in* $\widetilde{B}_i$, *there exist* $y \in \lambda_0 = \lambda$ *such that:*

$$d(x, y) \leq h(i).$$

Important Note: In the above Lemma 36 and Corollary 37, it is important to note that the distance d is **hyperbolic**, not electric. This is because the number l_i occurring in elementary paths of type 5 and 6 is a hyperbolic length depending only on i (in B_i).

Next suppose that λ lies outside $B_N(p)$, the N-ball about a fixed reference point p on the boundary horizontal surface $\widetilde{S} \times \{0\} \subset \widetilde{B}_0$. Then by Corollary 37, any x lying on a B_λ-admissible path in $\widetilde{B}_i$ satisfies

$$d(x, p) \geq N - h(i).$$

Also, since the electric, and hence hyperbolic 'thickness' (the shortest distance between its boundary horizontal sheets) is ≥ 1, we get,

$$d(x, p) \geq |i|.$$

Assume for convenience that $i \geq 0$ (a similar argument works, reversing signs for $i < 0$). Then,

$$d(x, p) \geq \min\{i, N - h(i)\}.$$

Let $h_1(i) = h(i) + i$. Then h_1 is a monotonically increasing function on the integers. If $h_1^{-1}(N)$ denote the largest positive integer n such that $h(n) \leq m$, then clearly, $h_1^{-1}(N) \to \infty$ as $N \to \infty$. We have thus shown:

Lemma 38 *There exists a function $M(N) : \mathbb{N} \to \mathbb{N}$ such that $M(N) \to \infty$ as $N \to \infty$ for which the following holds:*
For any geodesic $\lambda \subset \widetilde{S} \times \{0\} \subset \widetilde{B_0}$, a fixed reference point $p \in \widetilde{S} \times \{0\} \subset \widetilde{B_0}$ and any x on a B_λ-admissible path,

$$d(\lambda, p) \geq N \Rightarrow d(x, p) \geq M(N).$$

As pointed out before, the discussion and Lemmas of the previous two subsections go through just as well in the context of *weak amalgamation geometry* manifolds. We make this explicit in the case of Lemma 38 above.

Corollary 39 *Let M be a* model manifold of **weak amalgamation geometry**. *Then there exists a function $M(N) : \mathbb{N} \to \mathbb{N}$ such that $M(N) \to \infty$ as $N \to \infty$ for which the following holds:*
Given any geodesic $\lambda \subset \widetilde{S} \times \{0\}$, let B_λ be as before. For $\lambda \subset \widetilde{S} \times \{0\} \subset \widetilde{B_0}$, a fixed reference point $p \in \widetilde{S} \times \{0\} \subset \widetilde{B_0}$ and any x on a B_λ-admissible path,

$$d(\lambda, p) \geq N \Rightarrow d(x, p) \geq M(N).$$

5.4 Joining the Dots

Recall that **admissible paths** in a model manifold of bounded geometry consist of:

1) Horizontal segments along some $\widetilde{S} \times \{i\}$ for $i = \{0, 1, 2, 3\}$ (thin blocks) or $i = \{0, 1\}$ (thick blocks).
2) Vertical segments $x \times [0, 1]$ or $x \times [2, 3]$ for amalgamated blocks.
3) Vertical segments of length $\leq l_i$ joining $x \times \{1\}$ to $x \times \{2\}$ for amalgamated blocks.
4) Vertical segments of length 1 joining $x \times \{0\}$ to $x \times \{1\}$ for thick blocks.

Our strategy in this subsection is:

- 1 Start with an electric geodesic β_e in $\widetilde{M_{Gel}}$ joining the end-points of λ.
- 2 Replace it by an *admissible quasigeodesic*, i.e. an admissible path that is a quasigeodesic.
- 3 Project the intersection of the admissible quasigeodesic with the horizontal sheets onto B_λ.
- 4 The result of step 3 above is disconnected. *Join the dots* using B_λ-admissible paths.

The end product is an electric quasigeodesic built up of B_λ admissible paths. Now for the first two steps:

- Since $\widetilde{B}$ (for a thick block B) has thickness 1, any path lying in a thick block can be perturbed to an admissible path lying in $\widetilde{B}$, changing the length by at most a bounded multiplicative factor.
- For B amalgamated, we decompose paths into horizontal paths lying in some $\widetilde{S} \times \{j\}$, for $j = 0, \cdots 3$ and vertical paths of types (2) or (3) above. This can be done without altering electric length within $\widetilde{S} \times [1, 2]$. To see this, project any path $\overline{ab}$ beginning and ending on $\widetilde{S} \times \{1, 2\}$ onto $\widetilde{S} \times \{1\}$ along the fibers. To connect this to the starting and ending points a, b, we have to at most adjoin vertical segments through a, b. Note that this does not increase the electric length of $\overline{ab}$, as the electric length is determined by the number of amalgamation blocks that $\overline{ab}$ traverses.
- For paths lying in $\widetilde{S} \times [0, 1]$ or $\widetilde{S} \times [2, 3]$, we can modify the path into an admissible path, changing lengths by a bounded multiplicative constant. The result is therefore an electric quasigeodesic.
- Without loss of generality, we can assume that the electric quasigeodesic is one without back-tracking (as this can be done without increasing the length of the geodesic - see [Far98] or [Kla99] for instance).
- Abusing notation slightly, assume therefore that β_e is an admissible electric quasigeodesic without backtracking joining the end-points of λ.
  This completes Steps 1 and 2.
- Now act on $\beta_e \cap \widetilde{M_H}$ by Π_λ. From Theorem 31, we conclude, by restricting Π_λ to the horizontal sheets of $\widetilde{M_{Gel}}$ that the image $\Pi_\lambda(\beta_e)$ is a 'dotted electric quasigeodesic' lying entirely on B_λ. This completes step 3.
- Note that since β_e consists of admissible segments, we can arrange so that two nearest points on $\beta_e \cap \widetilde{M_H}$ which are not connected to each other form the end-points of a vertical segment of type (2), (3) or (4). Let $\Pi_\lambda(\beta_e) \cap B_\lambda = \beta_d$, be the dotted quasigeodesic lying on B_λ. We want to join the dots in

β_d converting it into a **connected** electric quasigeodesic built up of B_λ-**admissible paths**.

- For vertical segments of type (4) joining p, q (say), $\Pi_\lambda(p), \Pi_\lambda(q)$ are a bounded hyperbolic distance apart. Hence, by the proof of Lemma 29, we can join $\Pi_\lambda(p), \Pi_\lambda(q)$ by a B_λ-admissible path of length bounded by some C_0 (independent of B, λ).

- For vertical segments of type (2) joining p, q, we note that $\Pi_\lambda(p), \Pi_\lambda(q)$ are a bounded hyperbolic distance apart. Hence, by the proof of Lemma 30, we can join $\Pi_\lambda(p), \Pi_\lambda(q)$ by a B_λ-admissible path of length bounded by some C_1 (independent of B, λ).

- This leaves us to deal with case (3). Such a segment consists of a segment lying within a lift of an amalgamation block. Such a piece has electric length one in the graph model. Its image, too, has electric length one (See for instance, Case (3) of the proof of Lemma 30, where we noted that the projection of any amalgamation component lies within an amalgamation component).

After joining the dots, we can assume further that the quasigeodesic thus obtained does not backtrack (cf [Far98] and [Kla99]).

Putting all this together, we conclude:

Lemma 40 *There exists a function* $M(N) : \mathbb{N} \to \mathbb{N}$ *such that* $M(N) \to \infty$ *as* $N \to \infty$ *for which the following holds:*

For any geodesic $\lambda \subset \widetilde{S} \times \{0\} \subset \widetilde{B_0}$, *and a fixed reference point* $p \in \widetilde{S} \times \{0\} \subset \widetilde{B_0}$, *there exists a connected electric quasigeodesic* β_{adm} *without backtracking, such that*

- β_{adm} *is built up of* B_λ*-admissible paths.*
- β_{adm} *joins the end-points of* λ.
- $d(\lambda, p) \geq N \Rightarrow d(\beta_{adm}, p) \geq M(N)$.

Proof: The first two criteria follow from the discussion preceding this lemma. The last follows from Lemma 38 since the discussion above gives a quasi-geodesic built up out of admissible paths. $\square$

As in the previous subsections, Lemma 40 goes through for **weak amalgamation geometry**. We state this below:

Corollary 41 *Suppose that M is a manifold of* **weak amalgamation geometry**. *There exists a function* $M(N) : \mathbb{N} \to \mathbb{N}$ *such that* $M(N) \to \infty$ *as* $N \to \infty$ *for which the following holds:*

For any geodesic $\lambda \subset \widetilde{S} \times \{0\} \subset \widetilde{B_0}$, *and a fixed reference point* $p \in \widetilde{S} \times \{0\} \subset \widetilde{B_0}$, *there exists a connected electric quasigeodesic* β_{adm} *without backtracking, such that*

- β_{adm} *is built up of* B_λ-*admissible paths.*
- β_{adm} *joins the end-points of* λ.
- $d(\lambda, p) \geq N \Rightarrow d(\beta_{adm}, p) \geq M(N)$.

5.5 Admissible Quasigeodesics and Electro-ambient Quasigeodesics

Definition: We next define (as before) a (k, ϵ) electro-ambient quasigeodesic γ in $\widetilde{M}$ relative to the amalgamation components $\widetilde{K}$ to be a (k, ϵ) quasigeodesic in the graph model of $\widetilde{M}$ such that in an ordering (from the left) of the amalgamation components that γ meets, each $\gamma \cap \widetilde{K}$ is a (k, ϵ) - quasigeodesic in the induced path-metric on $\widetilde{K}$.

This subsection is devoted to extracting an electro-ambient quasigeodesic β_{ea} from a B_λ-admissible quasigeodesic β_{adm}. β_{ea} shall satisfy the property indicated by Lemma 40 above. We shall prove this lemma under the assumption of (strong) amalgamation geometry. However, a weaker assumption (which we shall discuss later, while weakening *amalgamation geometry* to **graph amalgamation geometry**) is enough for the main lemma of this subsection to go through.

Lemma 42 *There exist* κ, ϵ *and a function* $M'(N) : \mathbb{N} \to \mathbb{N}$ *such that* $M'(N) \to \infty$ *as* $N \to \infty$ *for which the following holds:*
For any geodesic $\lambda \subset \widetilde{S} \times \{0\} \subset \widetilde{B_0}$, *and a fixed reference point* $p \in \widetilde{S} \times \{0\} \subset \widetilde{B_0}$, *there exists a* (κ, ϵ) *electro-ambient quasigeodesic* β_{ea} *without backtracking, such that*

- β_{ea} *joins the end-points of* λ.
- $d(\lambda, p) \geq N \Rightarrow d(\beta_{ea}, p) \geq M'(N)$.

Proof: From Lemma 40, we have a B_λ - admissible quasigeodesic β_{adm} and a function $M(N)$ without backtracking satisfying the conclusions of the lemma. Since β_{adm} does not backtrack, we can decompose it as a union of non-overlapping segments $\beta_1, \cdots \beta_k$, such that each β_i is either an admissible (hyperbolic) quasigeodesic lying outside amalgamation components, or a B_λ-admissible quasigeodesic lying entirely within some amalgamation component

$\widetilde{K}_i$. Further, since β_{adm} does not backtrack, we can assume that all K_i's are distinct.

We modify β_{adm} to an electro-ambient quasigeodesic β_{ea} as follows:

1) β_{ea} coincides with β_{adm} outside amalgamation components.
2) There exist κ, ϵ such that if some β_i lies within an amalgamation component $\widetilde{K}_i$ then, by uniform quasiconvexity of the K_i's, it may be replaced by a (κ, ϵ) (hyperbolic) quasigeodesic β_i^{ea} joining the end-points of β_i and lying within $\widetilde{K}_i$.

The resultant path β_{ea} is clearly an electro-ambient quasigeodesic without backtracking. Next, each component β_i^{ea} lies in a C_i neighborhood of β_i, where C_i depends only on the thickness l_i of the amalgamation component K_i.

We let $C(n)$ denote the maximum of the values of C_i for $K_i \subset B_n$. Then, as in the proof of Lemma 38, we have for any $z \in \beta_{ea} \cap B_n$,

$$d(z, p) \geq \max (n, M(N) - C(n)).$$

Again, as in Lemma 38, this gives us a (new) function $M'(N) : \mathbb{N} \to \mathbb{N}$ such that $M'(N) \to \infty$ as $N \to \infty$ for which

$$d(\lambda, p) \geq N \Rightarrow d(\beta_{ea}, p) \geq M'(N).$$

This proves the lemma. $\qquad\square$

Note: We have essentially used the following two properties of amalgamation components in concluding Lemma 42:

(1) any path lying inside an amalgamation component $\widetilde{K}$ may be replaced by a (uniform) hyperbolic quasigeodesic joining its end-points and lying within the same $\widetilde{K}$.
(2) Each electro-ambient quasigeodesic joining the end-points of an admissible quasigeodesic in $\widetilde{K} \subset \widetilde{B}_n$ lies in a (hyperbolic) $C(n)$-neighborhood of the latter.

We shall have occasion to use this when we discuss **graph-quasiconvexity**.

6 Cannon-Thurston Maps for Surfaces Without Punctures

It is now time to introduce hyperbolicity of $\widetilde{M}$, global quasiconvexity of amalgamation components, (and hence) model manifolds of (strong) amalgamation geometry. We shall assume till the end of this section that

1) there exists a hyperbolic manifold M and a homeomorphism from $\widetilde{M}$ to $\widetilde{S} \times \mathbb{R}$. We identify $\widetilde{M}$ with $\widetilde{S} \times \mathbb{R}$ via this homeomorphism.
2) $\widetilde{S} \times \mathbb{R}$ admits a quasi-isometry g to a model manifold of *amalgamated geometry*.
3) g preserves the fibers over $\mathbb{Z} \subset \mathbb{R}$.

We shall henceforth ignore the quasi-isometry g and think of $\widetilde{M}$ itself as the universal cover of a model manifold of *amalgamated geometry*.

6.1 Electric Geometry Revisited

We note the following properties of the pair $(X, \mathcal{H})$ where X is the graph model of $\widetilde{M}$ and $\mathcal{H}$ consists of the amalgamation components. There exist C, D, Δ such that

1) Each amalgamation component is C-quasiconvex.
2) Any two amalgamation components are 1-separated.
3) $\widehat{M_{Gel}} = X_{Gel}$ is Δ-hyperbolic, (where $\widehat{M_{Gel}} = X_{Gel}$ is the electric metric on $\widetilde{M} = X$ obtained by electrocuting all amalgamation components, i.e. all members of $\mathcal{H}$).
4) Given K, ϵ, there exists D_0 such that if γ be a (K, ϵ) hyperbolic quasi-geodesic joining a, b and if β be a (K, ϵ) electro-ambient quasigeodesic joining a, b, then γ lies in a D_0 neighborhood of β.

The first property follows from the definition of a manifold of amalgamation geometry.
The second follows from the construction of the graph model.
The third follows from Lemma 11.
The fourth follows from Lemma 15.

6.2 Proof of Theorem

We shall now assemble the proof of the main theorem.

Theorem 43 *Let M be a 3 manifold homeomorphic to $S \times J$ (for $J = [0, \infty)$ or $(-\infty, \infty)$). Further suppose that M has amalgamated geometry, where $S_0 \subset B_0$ is the lower horizontal surface of the building block B_0. Then the inclusion $i : \widetilde{S} \to \widetilde{M}$ extends continuously to a map $\hat{i} : \widehat{S} \to \widehat{M}$. Hence the limit set of $\widetilde{S}$ is locally connected.*

Proof: Suppose $\lambda \subset \widetilde{S}$ lies outside a large N-ball about p. By Lemma 42 we obtain an electro-ambient quasigeodesic without backtracking β_{ea} lying outside an $M(N)$-ball about p (where $M(N) \to \infty$ as $N \to \infty$).

Suppose that β_{ea} is a (κ, ϵ) electro-ambient quasigeodesic. Note that κ, ϵ depend on 'the Lipschitz constant' of Π_λ and hence only on $\widetilde{S}$ and $\widetilde{M}$.

From Property (4) above, (or Lemma 15) we find that if β^h denote the hyperbolic geodesic in $\widetilde{M}$ joining the end-points of λ, then β^h lies in a (uniform) C' neighborhood of β_{ea}.

Let $M_1(N) = M(N) - C'$. Then $M_1(N) \to \infty$ as $N \to \infty$. Further, the hyperbolic geodesic β^h lies outside an $M_1(N)$-ball around p. Hence, by Lemma 6, the inclusion $i : \widetilde{S} \to \widetilde{M}$ extends continuously to a map $\hat{i} : \widehat{S} \to \widehat{M}$.

Since the continuous image of a compact locally connected set is locally connected (see [HY61]) and the (intrinsic) boundary of $\widetilde{S}$ is a circle, we conclude that the limit set of $\widetilde{S}$ is locally connected.

This proves the theorem. □

7 Weakening the Hypothesis I: Graph Quasiconvexity and Graph Amalgamation Geometry

We now proceed to weaken the hypothesis of amalgamation geometry in the hope of capturing all Kleinian surface groups. Recall that in the definition of amalgamation geometry, two criteria were used - *local and global quasiconvexity of amalgamation components*. We shall retain local quasiconvexity, and replace global quasiconvexity by a **weaker condition** which we shall term **graph quasiconvexity**. The rationale behind this terminology shall be made clear later. We first modify the definition of amalgamation geometry as follows, retaining only local quasiconvexity. We first recall the definition of *weak amalgamation geometry*.

A manifold M homeomorphic to $S \times J$, where $J = [0, \infty)$ or $J = (-\infty, \infty)$, is said to be a model of **weak amalgamation geometry** if

1) there is a fiber preserving homeomorphism from M to $\widetilde{S} \times J$ that lifts to a quasi-isometry of universal covers.
2) there exists a sequence I_i of intervals (with disjoint interiors) and blocks B_i where the metric on $S \times I_i$ is the same as that on some building block B_i. Each block is either thick or has amalgamation geometry.
3) $\bigcup_i I_i = J$.

4) There exists $C > 0$ such that for all amalgamated blocks B_i and geometric cores $K \subset B_i$, all amalgamation components of $\widetilde{K}$ are C-quasiconvex in $\widetilde{B}_i$.

Definition 44 An amalgamation component $K \subset B_n$ is said to be (m, κ) **graph - quasiconvex** if there exists a κ-quasiconvex (in the hyperbolic metric) subset $CH(K)$ containing K such that

1) $CH(K) \subset N_m^G(K)$ where $N_m^G(K)$ denotes the m neighborhood of K in the graph model of M.
2) For each K there exists C_K such that K is C_K-quasiconvex in $CH(K)$.

Since the quasiconvex sets (thought of as convex hulls of K) lie within a bounded distance from K in the *graph model* we have used the term *graph-quasiconvex*.

Definition 45 A manifold M_{ga} of weak amalgamation geometry is said to be a model of **graph amalgamation geometry** if there exist m, κ such that each amalgamation geometry component is (m, κ) -graph - quasiconvex.

A manifold N is said to have **graph amalgamation geometry** if there is a level-preserving homeomorphism from N to a model manifold of *graph amalgamation geometry* that lifts to a quasi-isometry at the level of universal covers.

As before, we shall deal only with closed surfaces.

Now, let us indicate the modifications necessary to carry out the proof of the Cannon-Thurston Property for manifolds of graph amalgamation geometry (suppressing the quasi-isometry to a model manifold). As in Theorem 43, the proof consists of two steps:

1) Constructing a quasiconvex set B_λ in an auxiliary electric space (the **graph model**), and from this an admissible electric quasigeodesic β.
2) Recovering from β and its intersection pattern, information about the hyperbolic geodesic joining its end-points.

The first step is the same as that for models of *amalgamation geometry* as it goes through for *weak amalgamation geometry*. Then from Corollary 32 we have:

Step 1A: Given $\lambda \subset \widetilde{S} \times \{0\}$, construct B_λ, Π_λ as before. There exists $C > 0$ such that the the retraction $\Pi_\lambda : \widetilde{M_{ga}} \to B_\lambda$ satisfies:
$d_{Gel}(\Pi_\lambda(x), \Pi_\lambda(y)) \le C d_{Gel}(x, y) + C$, where d_{Gel} denotes the metric in the graph model.

Again, from Corollary 41 we have:

Step 1B: There exists a function $M(N) : \mathbb{N} \to \mathbb{N}$ such that $M(N) \to \infty$ as $N \to \infty$ for which the following holds:

For any geodesic $\lambda \subset \widetilde{S} \times \{0\} \subset \widetilde{B_0}$, and a fixed reference point $p \in \widetilde{S} \times \{0\} \subset \widetilde{B_0}$, there exists a connected B_λ-admissible quasigeodesic β_{adm} without backtracking, such that

- β_{adm} is built up of B_λ-admissible paths.
- β_{adm} joins the end-points of λ.
- $d(\lambda, p) \geq N \Rightarrow d(\beta_{adm}, p) \geq M(N)$. ($d$ is the ordinary, non-electric metric.)

Summary of Step 2:

Now we come to the second step: **recovering a hyperbolic geodesic from an electric geodesic**.

This step can be further subdivided into two parts. Let $\widetilde{M_1}$ denote $\widetilde{M_{ga}}$ with the graph metric obtained by electrocuting amalgamation components. Next, let $\widetilde{M_2}$ denote $\widetilde{M_{ga}}$ with an electric metric obtained by electrocuting the family of sets $CH(\widetilde{K})$ (for amalgamation components K) appearing in the definition of **graph amalgamation geometry**. We show that the spaces $\widetilde{M_1}$ and $\widetilde{M_2}$ are quasi-isometric. In fact we show that the identity map on the underlying subset $\widetilde{M_{ga}}$ is a quasi-isometry. This step requires only the first condition in the definition of *graph quasiconvexity*. The second stage extracts information about an electro-ambient quasi-geodesic in $\widetilde{M_2}$ from an admissible path in $\widetilde{M_1}$. It is at this second stage that we require the second condition: (not necessarily uniform) quasi-convexity of amalgamation components.

We now furnish the details.

Step 2A:

Lemma 46 *The identity map on the underlying set $\widetilde{M_{ga}}$ from M_1 to M_2 induces a quasi-isometry of universal covers $\widetilde{M_1}$ and $\widetilde{M_2}$.*

Proof: Let d_1, d_2 denote the electric metrics on $\widetilde{M_1}$ and $\widetilde{M_2}$. Since $K \subset CH(K)$ for every amalgamation component, we have right off

$$d_1(x, y) \leq d_2(x, y) \text{ for all } x, y \in \widetilde{M}.$$

To prove a reverse inequality with appropriate constants, it is enough to show that each set $CH(K)$ (of diameter one in M_2) has uniformly bounded diameter in M_1. To see this, note that by definition of graph-quasiconvexity, there exists n such that for all K and each point a in $CH(K)$, there exists a point $b \in K$ with $d_1(x, y) \leq n$. Hence by the triangle inequality,

$$d_2(x, y) \leq 2n + 1 \text{ for all } x, y \in \widetilde{CH(K)}.$$

Therefore,

$$d_2(x, y) \leq (2n + 1)d_1(x, y) \text{ for all } x, y \in \widetilde{M}.$$

This proves the lemma.   □

Step 2B:

Now let β_{adm} denote an admissible B_λ quasigeodesic in $\widetilde{M}_1$, which does not backtrack relative to the amalgamation components. By Lemma 46 above, β_{adm} is a quasigeodesic in $\widetilde{M}_2$. As in Lemma 42, using the Note following it, we conclude:

There exists a κ, ϵ-electro-ambient quasigeodesic β_{ea} in $\widetilde{M}_2$ (as opposed to $\widetilde{M}_1$, which is what we needed in the *amalgamation geometry* case). (See Lemma 42.) Note that in $\widetilde{M}_2$, we electrocute the lifts of the sets $CH(K)$ rather than $\widetilde{K}$'s.

We thus obtain, as in Lemma 42 a function $M'(N) : \mathbb{N} \to \mathbb{N}$ such that $M'(N) \to \infty$ as $N \to \infty$ for which the following holds:

For any geodesic $\lambda \subset \widetilde{S} \times \{0\} \subset \widetilde{B}_0$, and a fixed reference point $p \in \widetilde{S} \times \{0\} \subset \widetilde{B}_0$, there exists a (κ, ϵ) electro-ambient quasigeodesic β_{ea} without backtracking, such that

- β_{ea} joins the end-points of λ.
- $d(\lambda, p) \geq N \Rightarrow d(\beta_{ea}, p) \geq M'(N)$.

Finally, as in the proof of Theorem 43, we use Lemma 15 to conclude that the hyperbolic geodesic in $\widetilde{M}$ joining the end-points of λ lies in a uniform hyperbolic neighborhood of β_{ea}. This gives us Theorem 43 with **graph amalgamation geometry** replacing *amalgamation geometry*.

Theorem 47 *Let M_{ga} be a 3 manifold homeomorphic to $S \times J$ (for $J = [0, \infty)$ or $(-\infty, \infty)$). Further suppose that M_{ga} has graph amalgamation geometry, where $S_0 \subset B_0$ is the lower horizontal surface of the building block B_0. Then the inclusion $i : \widetilde{S} \to \widetilde{M}$ extends continuously to a map $\hat{i} : \widetilde{S} \to \widehat{M}$. Hence the limit set of $\widetilde{S}$ is locally connected.*

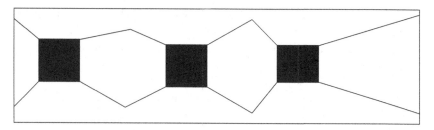

Figure 4 *Building Block for Generalized Amalgamation Geometry*

8 Weakening the Hypothesis II: Split Geometry

In this section, we shall weaken the hypothesis of *graph amalgamation geometry* further to include the possibility of Margulis tubes cutting across the blocks B_i. But before we do this, let us indicate a straightforward generalization of *amalgamation geometry* or *graph amalgamation geometry*.

8.1 More Margulis Tubes in a Block

A straightforward generalization of Theorem 43 (or Theorem 47) is to the case where more than one Margulis tube is allowed per block B, and each of these tubes splits the block B locally. On the surface S, this corresponds to a number of disjoint (uniformly) bounded length curves. As before we require that each amalgamation component be uniformly quasiconvex (or graph quasiconvex) in $\widetilde{M_{ga}}$ for the proof of Theorem 43 (or Theorem 47) to go through. See the figure below for a schematic rendering of the model block of amalgamation geometry.

8.2 Motivation for Split Geometry

So far, we have assumed that the boundaries of *amalgamated geometry blocks* or *graph amalgamated geometry blocks* are all of bounded geometry. This assumption needs to be relaxed to accommodate general surface Kleinian groups. Before we define the objects of interest, we shall first informally analyze what went into the construction of the hyperbolic ladder B_λ. We require:

1) Horizontal surfaces S_i, all abstractly homeomorphic to each other.
2) A block decomposition $M = \cup B_i$, where $B_{i-1} \cap B_i = S_i$.
3) Given a geodesic $\lambda_i \subset \widetilde{S}_i$, we require a (uniformly) large-scale retract π_i of $\widetilde{S}_i$ onto λ_i and a prescription to construct $\lambda_{i+1} \subset \widetilde{S}_{i+1}$. Thus, starting

with $\lambda_0 \subset \widetilde{S}_0$, we first construct π_0 and then inductively construct the pairs (λ_i, π_i).

4) Each block B_i has an auxiliary metric or pseudometric which induces the given path metrics on S_{i-1}, S_i.

We want to relax the assumption that S_i's have bounded geometry, while retaining the essential properties of bounded geometry. As elsewhere in this paper we invoke the following (uncomfortably dictatorial) policy that we have adopted:

Policy: *Electrocute anything that gives trouble.*

What this policy means is that whenever some construction possibly gives rise to non-uniformity of some parameter(s), locate the source of non-uniformity and electrocute it. Then, at the end of the game, re-instate the original geometry by using comparison properties between ordinary hyperbolic geometry and electric geometry.

Thus, each S_i is now allowed to have a pseudometric where a finite number of disjoint, bounded length (uniformly, independent of i) collection of simple closed geodesics are electrocuted. Then, instead of geodesics $\lambda_i \subset \widetilde{S}_i$, we shall require the λ_i to be only electro-ambient geodesics. This will allow us to go ahead with the construction of B_λ.

One further comment as to how this solves the problem. Let us fix a small (less than Margulis constant) ϵ_0. Given any hyperbolic surface S^h, we can simply electrocute *thin parts*, i.e. tubular neighborhoods of short (less than ϵ_0) geodesics with boundaries of length ϵ_0. Alternately, we can first cut out the interiors of these thin parts. Next, corresponding to each *Margulis annulus* that has been cut out, glue the corresponding boundary components of length ϵ_0 together, and then electrocute the resulting closed curves.

This construction is adapted to the construction of *split level surfaces* in Minsky [Min10], and Brock-Canary-Minsky [BCM12].

8.3 Definitions

Topologically, a **split subsurface** S^s of a surface S is a (possibly disconnected, proper) subsurface with boundary such that $S - S^s$ consists of a non-empty family of non-homotopic annuli, which in turn are not homotopic into the boundary of S^s.

Geometrically, we assume that S is given some finite volume hyperbolic structure. A split subsurface S^s of S has bounded geometry, i.e.

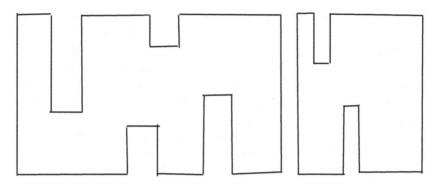

Figure 5 *Split Block with hanging tubes*

1) each boundary component of S^s is of length ϵ_0, and is in fact a component of the boundary of $N_k(\gamma)$, where γ is a hyperbolic geodesic on S, and $N_k(\gamma)$ denotes its k-neighborhood.
2) For any closed geodesic β on S, either $\beta \subset S - S^s$, or, the length of any component of $\beta \cap (S - S^s)$ is greater than ϵ_0.

Topologically, a **split block** $B^s \subset B = S \times I$ is a topological product $S^s \times I$ for some *connected* S^s. However, its upper and lower boundaries need not be $S^s \times 1$ and $S^s \times 0$. We only require that the upper and lower boundaries be split subsurfaces of S^s. This is to allow for Margulis tubes starting (or ending) within the split block. Such tubes would split one of the horizontal boundaries but not both. We shall call such tubes **hanging tubes**. See figure below:

Geometrically, we require that the metric on a split block induces a path metric on its *upper and lower horizontal* boundary components, which are subsurfaces of $S^s \times \partial I$, such that each horizontal boundary component is a (geometric) split surface. Further, the metric on B^s induces on each *vertical* boundary component of a Margulis tube $\partial S^s \times I$ the product metric. Each boundary component for Margulis tubes that 'travel all the way from the lower to the upper boundary' is an annulus of height equal to length of I. We demand further that *hanging tubes* have *length uniformly bounded below by $\eta_0 > 0$*. Further, each such annulus has cross section a round circle of length ϵ_0. This leaves us to decide the metric on lower and upper boundaries of hanging tubes. Such boundaries are declared to have a metric equal to that on $S^1 \times [-\eta, \eta]$, where S^1 is a round circle of length ϵ_0 and η is a sufficiently small number.

Note: In the above definition, we *do not* require that the upper (or lower) horizontal boundary of a split block B^s be connected for a connected B^s. This happens due to the presence of *hanging tubes*.

We further require that the distance between horizontal boundary components is at least 1, i.e. for a component R of S^s $d(R \times 0, R \times 1) \geq 1$. We define the **thickness** of a split block to be the supremum of the lengths of $x \times I$ for $x \in S^s$ and demand that it be finite (which holds under all reasonable conditions, e.g. a smooth metric; however, since we shall have occasion to deal with possibly discontinuous pseudometrics, we make this explicit). We shall denote the thickness of a split block B^s by l_B.

Each component of a split block shall be called a **split component**. We further require that the 'vertical boundaries' (corresponding to Euclidean annuli) of split components be uniformly (independent of choice of a block and a split component) quasiconvex in the corresponding split component.

Note that the boundary of each split block has an intrinsic metric that is flat and corresponds to a Euclidean torus.

A lift of a split block to the universal cover of the block $B = S \times I$ shall be termed a **split component** of $\tilde{B}$.

Remark The notion of *split components* we deal with here is closely related to the notion of **bands** described by Bowditch in [Bow05a], [Bow05b] and also to the notion of **scaffolds** introduced by Brock, Canary and Minsky in [BCM12].

We define a **welded split block** to be a split block with identifications as follows: Components of $\partial S^s \times 0$ are glued together if and only if they correspond to the same geodesic in $S - S^s$. The same is done for components of $\partial S^s \times 1$. A simple closed curve that results from such an identification shall be called a **weld curve**. For hanging tubes, we also weld the boundary circles of their *lower or upper boundaries* by simply collapsing $S^1 \times [-\eta, \eta]$ to $S^1 \times \{0\}$.

This may be done topologically or geometrically while retaining Dehn twist information about the curves. To record information about the Dehn twists, we have to define (topologically) a map that takes the lower boundary of a welded split block to the upper boundary. We define a map that takes $x \times 0$ to $x \times 1$ for every point in S^s. This clearly induces a map from the lower boundary of a welded split block to its upper boundary. However, this is not enough to give a well-defined map on paths. To do this, we have to record *twist information* about *weld curves*. The way to do this is to define a map on transversals to weld curves. The map is defined on transversals by recording the number of times a transversal to a weld curve $\gamma \times 0$ twists around $\gamma \times 1$ on the upper boundary of the welded split block. (A related context in which such transversal information is important is that of markings described in Minsky [Min10].)

Let the metric product $S^1 \times [0, 1]$ be called the **standard annulus** if each horizontal S^1 has length ϵ_0. For hanging tubes the standard annulus will be taken to be $S^1 \times [0, 1/2]$.

Next, we require another pseudometric on B which we shall term the **tube-electrocuted metric**. We first define a map from each boundary annulus $S^1 \times I$ (or $S^1 \times [0, 1/2]$ for hanging annuli) to the corresponding standard annulus that is affine on the second factor and an isometry on the first. Now glue the mapping cylinder of this map to the boundary component. The resulting 'split block' has a number of standard annuli as its boundary components. Call the split block B^s with the above mapping cylinders attached, the *stabilized split block* B^{st}.

Glue boundary components of B^{st} corresponding to the same geodesic together to get the **tube electrocuted metric** on B as follows. Suppose that two boundary components of B^{st} correspond to the same geodesic γ. In this case, these boundary components are both of the form $S^1 \times I$ or $S^1 \times [0, \frac{1}{2}]$ where there is a projection onto the horizontal S^1 factor corresponding to γ. Let $S_l^1 \times J$ and $S_r^1 \times J$ denote these two boundary components (where J denotes I or $[0, \frac{1}{2}]$). Then each $S^1 \times \{x\}$ has length ϵ_0. Glue $S_l^1 \times J$ to $S_r^1 \times J$ by the natural 'identity map'. Finally, on each resulting $S^1 \times \{x\}$ put the zero metric. Thus the annulus $S^1 \times J$ obtained via this identification has the zero metric in the *horizontal direction* $S^1 \times \{x\}$ and the Euclidean metric in the *vertical direction* J. The resulting block will be called the **tube-electrocuted block** B_{tel} and the pseudometric on it will be denoted as d_{tel}. Note that B_{tel} is homeomorphic to $S \times I$. The operation of obtaining a *tube electrocuted block and metric* (B_{tel}, d_{tel}) from a split block B^s shall be called *tube electrocution*.

Next, fix a hyperbolic structure on a Riemann surface S and construct the metric product $S \times \mathbb{R}$. Fix further a positive real number l_0.

Definition 48 An annulus A will be said to be **vertical** if it is of the form $\sigma \times J$ for σ a geodesic of length less than l_0 on S and $J = [a, b]$ a closed sub-interval of $\mathbb{R}$. J will be called the **vertical interval** for the vertical annulus A.

A disjoint collection of annuli is said to be a **vertical system** of annuli if each annulus in the collection is vertical.

The above definition is based on a definition due to Bowditch, see [Bow05a],[Bow05b].

Suppose now that $S \times \mathbb{R}$ is equipped with a vertical system $\mathcal{A}$ of annuli. We shall call $z \in \mathbb{R}$

1) a **beginning level** if z is the lower bound of a vertical interval for some annulus $A \in \mathcal{A}$.
2) an **ending level** if z is the lower bound of a vertical interval for some annulus $A \in \mathcal{A}$.

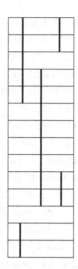

Figure 6 *Vertical Annulus Structure*

3) an **intermediate level** if z is an interior point of a vertical interval for some annulus $A \in \mathcal{A}$.

In the figure below (where for convenience, all appropriate levels are marked with integers), 2, 5, 11 and 14 are *beginning levels*, 4, 7, 13 and 16 are *ending levels*, 3, 6, 9, 12 and 15 are *intermediate levels*. We shall also allow Dehn twists to occur while going along the annulus.

A slight modification of the vertical annulus structure will sometimes be useful.

Replacing each geodesic γ on S by a neighborhood $N_\epsilon(\gamma)$ for sufficiently small ϵ, we obtain a **vertical Margulis tube** structure after taking products with vertical intervals. The family of Margulis tubes shall be denoted by $\mathcal{T}$ and the union of their interiors as $Int\mathcal{T}$. The union of $Int\mathcal{T}$ and its horizontal boundaries (corresponding to neighborhoods of geodesics $\gamma \subset S$) shall be denoted as $Int^+\mathcal{T}$.

Thick Block

Fix constants D, ϵ and let $\mu = [p, q]$ be an ϵ-thick Teichmuller geodesic of length less than D. μ is ϵ-thick means that for any $x \in \mu$ and any closed geodesic η in the hyperbolic surface S_x over x, the length of η is greater than ϵ. Now let B denote the universal curve over μ reparametrized such that the length of μ is covered in unit time. Thus $B = S \times [0, 1]$ topologically.

B is given the path metric and is called a **thick building block**.

Note that after acting by an element of the mapping class group, we might as well assume that μ lies in some given compact region of Teichmuller space. This is because the marking on $S \times \{0\}$ is not important, but rather its position relative to $S \times \{1\}$ Further, since we shall be constructing models only up to quasi-isometry, we might as well assume that $S \times \{0\}$ and $S \times \{1\}$ *lie in the orbit* under the mapping class group of some fixed base surface. Hence μ can be further simplified to be a Teichmuller geodesic joining a pair (p, q) amongst a finite set of points in the orbit of a fixed hyperbolic surface S.

Weak Split Geometry

A manifold M_{ws} homeomorphic to $S \times \mathbb{R}$ equipped with a vertical Margulis tube structure is said to be a model of **weak split geometry**, if it is equipped with a new metric satisfying the following conditions:

1) $S \times [m, m + 1] \cap Int\mathcal{T} = \emptyset$ (for $m \in \mathbb{Z} \subset \mathbb{R}$) implies that $S \times [m, m + 1]$ is a thick block.
2) $S \times [m, m + 1] \cap Int\mathcal{T} \neq \emptyset$ (for $m \in \mathbb{Z} \subset \mathbb{R}$) implies that $S \times [m, m + 1] - Int^+\mathcal{T}$ is (geometrically) a split block.
3) There exists a uniform upper bound on the lengths of vertical intervals for vertical Margulis tubes.
4) The metric on each component Margulis tube T of $\mathcal{T}$ is hyperbolic.

Note 1: Dehn twist information can still be implicitly recorded in a model of *weak split geometry* by the Dehn filling information corresponding to tubes T.

Note 2: The metric on a model of *weak split geometry* is possibly discontinuous along the boundary tori of Margulis tubes. If necessary, one could smooth this out. But we would like to carry on with the above metric.

Removing the interiors of Margulis tubes and tube electrocuting each block, we obtain a new pseudo-metric on M_{ws} called the **tube electrocuted metric** d_{tel} on M_{ws}. The pseudometric d_{tel} may also be lifted to $\widetilde{M_{ws}}$.

The induced pseudometric on $\widetilde{S_i}$'s shall be referred to as **split electric metrics**. The notions of *electro-ambient metrics, geodesics and quasigeodesics* go through in this context.

Next, we shall describe a **graph metric** on $\widetilde{M_{ws}}$ which is almost (but not quite) the metric on the nerve of the covering of $\widetilde{M_{ws}}$ by split components (where each edge is assigned length 1). This is not strictly true as thick blocks are retained with their usual geometry in the graph metric. However the analogy with the nerve is exact if all blocks have *weak split geometry*.

For each split component $\widetilde{K}$ assign a single vertex v_K and construct a cone of height $1/2$ with base $\widetilde{K}$ and vertex v_K. The metric on the resulting space (coned-off or electric space in the sense of Farb [Far98]) shall be called the **graph metric** on $\widetilde{M_{ws}}$.

The union of a split component of $\widetilde{B}$ and the lifts of Margulis tubes (to $\widetilde{M_{ws}}$) that intersect its boundary shall be called a **split amalgamation component** in $\widetilde{M_{ws}}$.

Definition 49 A split amalgamation component K is said to be $(\mathbf{m}. \kappa)$–**graph quasiconvex** if there exists a κ-quasiconvex (in the hyperbolic metric) subset $CH(K)$ containing K such that

(1) $CH(K) \subset N_m^G(K)$ where $N_m^G(K)$ denotes the m neighborhood of K in the graph metric on M.
(2) For each K there exists C_K such that K is C_K-quasiconvex in $CH(K)$.

A model manifold M_{ws} of weak split geometry is said to be a model of **split geometry** if there exist m, κ such that each split amalgamation component is (m, κ) - graph quasiconvex.

8.4 The Cannon-Thurston Property for Manifolds of Split Geometry

We shall first extract information about geodesics in the *tube electrocuted* model. As with Theorem 43 and Theorem 47, the proof splits into two parts:

Step 1: Construction of B_λ and its quasiconvexity in an auxiliary graph metric. The end-product of this step is an electro-ambient quasigeodesic in the graph model.

Step 2: Extraction of information about a hyperbolic geodesic and its intersection pattern with blocks from the electro-ambient quasigeodesic constructed in Step 1 above.

Details of Step 1:

Step 1A: Construction of B_λ

It is at this stage that the construction differs somewhat from the construction of B_λ for manifolds of *graph amalgamated geometry*.

We start with the (tube-electrocuted) metric d_{tel} on the model manifold of *split geometry*. Then there exists a sequence of split surfaces S_i exiting the end(s).

Recall that in the construction of B_λ (for all preceding cases) we are *not* interested in the metric on each $\widetilde{S}_i$ per se, but in geodesics on $\widetilde{S}_i$.

The metric d_{tel} on the model manifold induces the **split electric metric** on each S_i obtained by electrocuting the **weld curves**. The natural geodesics to consider on $\widetilde{S}_i$ are therefore the *electro-ambient* quasigeodesics where the electrocuted subsets correspond to geodesics representing the weld curves.

Thus we start off with a hyperbolic geodesic λ in $\widetilde{S}_0$ joining a, b say. We let λ_0 denote the electro-ambient quasigeodesic joining a, b in the split electric metric on $\widetilde{S}_0$. Now construct B_λ inductively as follows:

- Each split block B_i and hence $\widetilde{B}_i$ comes equipped with a (topological) product structure. Thus there is a canonical map $\Phi_i : \widetilde{S}_i \to \widetilde{S}_{i+1}$ which maps each (x, i) to a point $(x, i+1)$ by lifting the map from S_i to S_{i+1} ($i \geq 0$ corresponding to the product structure).

- Next, if λ_i is an electro-ambient quasi-geodesic in the split electric metric on $\widetilde{S}_i$ joining (a, i) and (b, i) we let λ_{i+1} denote the electro-ambient quasi-geodesic in the split-electric metric on $\widetilde{S}_{i+1}$ joining $(a, i+1)$ and $(b, i+1)$. This gives us a prescription for constructing λ_{i+1} from λ_i for $i \geq 0$. Similarly, for $i \leq 0$ (in the totally degenerate case) we can construct λ_{i-1} from λ_i. Then as before, define

$$B_\lambda = \bigcup_i \lambda_i.$$

- Again, $\pi_i : \widetilde{S}_i \to \lambda_i$ is defined as the retraction that minimizes the ordered pair of distances in the split electric metric and the hyperbolic metric (without electrocuting weld curves). Π_λ is obtained in the graph metric by defining it on the horizontal sheets $\widetilde{S}_i$ as

$$\Pi_\lambda(x) = \pi_i(x) \quad \text{for } x \in \widetilde{S}_i.$$

- Then as before we conclude that in the graph model for $\widetilde{M_{ws}}$, with the metric d_{Gel}, Π_λ does not stretch distances much, i.e. there exists a uniform $C \geq 0$ such that

$$d_{Gel}(\Pi_\lambda(x), \Pi_\lambda(y)) \leq Cd_{Gel}(x, y) + C.$$

Step 1B: Construction of admissible quasigeodesic

The above construction of Π_λ may be used to construct a B_λ- admissible quasigeodesic β_{adm} in the tube-electrocuted model. As before we have:

There exists a function $M(N) : \mathbb{N} \to \mathbb{N}$ such that $M(N) \to \infty$ as $N \to \infty$ for which the following holds:

For any geodesic $\lambda \subset \widetilde{S} \times \{0\} \subset \widetilde{B}_0$, and a fixed reference point $p \in \widetilde{S} \times \{0\} \subset \widetilde{B}_0$, there exists a connected B_λ-admissible quasigeodesic β_{adm} without backtracking, such that

- β_{adm} is built up of B_λ-admissible paths.
- β_{adm} joins the end-points of λ.
- If $d(\lambda, p) \geq N$ then for any $x \in \beta_{adm} - Int\mathcal{T}$, $d(x, p) \geq M(N)$. (d is the ordinary, hyperbolic, or non-electric metric.)

Step 2: Recovering a quasigeodesic in the tube electrocuted model from an admissible quasigeodesic

We now follow the proof of Theorem 47.

Step 2A:

As in Step 2A in the proof of Theorem 47 we construct a second auxiliary space M_2 by electrocuting the elements $CH(K)$ for split components K. The spaces $\widetilde{M}_1$ and $\widetilde{M}_2$ are quasi-isometric by uniform *graph quasiconvexity* of split components. In fact the identity map on the underlying subset $\widetilde{M}_{ws}$ is a quasi-isometry as in Lemma 46.

Step 2B:

Next, as in Step 2B in the proof of Theorem 47, we extract information about an electro-ambient quasi-geodesic in $\widetilde{M}_2$ from an admissible path in $\widetilde{M}_1$. It is at this second stage that we require the condition that split components are (not necessarily uniformly) quasi-convex in the hyperbolic metric, and hence in the tube electrocuted metric d_{tel}.

We may assume that β_{adm} does not backtrack relative to the split components. From Step 2A above, β_{adm} is a quasigeodesic in $\widetilde{M}_2$. Then we conclude:

There exists a κ, ϵ-electro-ambient quasigeodesic β_{tea} in $\widetilde{M}_2$ (Note that in $\widetilde{M}_2$, we electrocute the lifts of the sets $CH(K)$ rather than $\widetilde{K}$'s).

We finally obtain a function $M'(N) : \mathbb{N} \to \mathbb{N}$ such that $M'(N) \to \infty$ as $N \to \infty$ for which the following holds:
For any geodesic $\lambda \subset \widetilde{S} \times \{0\} \subset \widetilde{B}_0$, and a fixed reference point $p \in \widetilde{S} \times \{0\} \subset \widetilde{B}_0$, there exists a (κ, ϵ) electro-ambient quasigeodesic β_{tea} (in the **tube electrocuted metric**) without backtracking, such that

- β_{tea} joins the end-points of λ.
- If λ lies outside a large ball about a fixed reference point $p \in \widetilde{S}_0$, then each point of $\beta_{tea} \cap (\widetilde{M} - Int\mathcal{T})$ also lies outside a large ball about p.

Step 3: Recovering a hyperbolic geodesic from the tube electrocuted quasigeodesic β_{tea}

This is a new step that comes from the extra phenomenon of tube electrocution which makes the metric d_{tel} an 'intermediate' metric between the hyperbolic metric d and the graph metric d_{Gel}.

Observe that lifts of Margulis tubes to $(\widetilde{M_{ws}}, d_{Gel})$ have uniformly bounded diameter in the metric d_{Gel} and consequently in the metric d_{tel} by uniform boundedness of vertical intervals of vertical Margulis tubes. Hence the *tube electrocuted metric* d_{tel} on $\widetilde{M_{ws}}$ is quasi-isometric to the metric d_{fe} where lifts of Margulis tubes are electrocuted (i.e. fully electrocuted rather than just tube electrocuted, and hence each tube has diameter 1). Let $\widetilde{M}_{fe}$ denote $\widetilde{M_{ws}}$ equipped with this new metric. Then geodesics without backtracking in the tube electrocuted metric become (uniform) quasi-geodesics without backtracking in $\widetilde{M}_{fe}$.

Note: It is at this (rather late) stage that we need to assume that $\widetilde{M_{ws}}$ is a hyperbolic metric space.

Let γ^h denote a hyperbolic geodesic joining the end-points of β_{tea} and hence λ. By Lemma 23, γ^h and β_{tea} track each other off Margulis tubes. Hence $\gamma^h \cap (\widetilde{M_{ws}} - Int\mathcal{T})$ lies outside a large ball about p. In particular, this is true for entry and exit points of γ^h with respect to Margulis tubes. This implies (See for instance Lemma 7.3 of [Mj11]) that the parts of λ^h lying within Margulis tubes also lie outside large balls about p. As before, by Lemma 6 we infer the Cannon-Thurston property for manifolds of *split geometry*.

Theorem 50 *Let M_{ws} be a 3 manifold homeomorphic to $S \times J$ (for $J = [0, \infty)$ or $(-\infty, \infty)$). Further suppose that M_{ws} has* split geometry, *where $S_0 \subset B_0$ is the lower horizontal surface of the building block B_0. Then the inclusion $i : \widetilde{S} \to \widetilde{M_{ws}}$ extends continuously to a map $\hat{i} : \widehat{S} \to \widehat{M_{ws}}$. Hence the limit set of $\widetilde{S}$ is locally connected.*

Mahan Mj
RKM Vivekananda University, Belur Math, WB-711 202, INDIA
Email: mahan.mj@gmail.com; mahan@rkmvu.ac.in

References

[Abi76] W. Abikoff. Two theorems on totally degenerate Kleinian groups. *American Journal of Mathematics, Vol. 98, No. 1 (Spring, 1976)*, pages 109–118, 1976.

[BCM12] J. F. Brock, R. D. Canary, and Y. N. Minsky. The Classification of Kleinian surface groups II: The Ending Lamination Conjecture. *Ann. of Math. 176 (1)*, pages 1–149, 2012.

[Bes04] M. Bestvina. Geometric group theory problem list. *M. Bestvina's home page: http:math.utah.edu/ bestvina*, 2004.

[Bow02] B. H. Bowditch. Stacks of hyperbolic spaces and ends of 3 manifolds. *preprint, Southampton*, 2002.

[Bow05a] B. H. Bowditch. End invariants of hyperbolic manifolds. *preprint, Southampton*, 2005.

[Bow05b] B. H. Bowditch. Model geometries for hyperbolic manifolds. *preprint, Southampton*, 2005.

[Bow07] B. H. Bowditch. The Cannon-Thurston map for punctured-surface groups. *Mathematische Zeitschrift, vol. 255, no. 1*, pages 35–76, 2007.

[CDA90] M. Coornaert, T. Delzant, and A.Papadopoulos. Geometrie et theorie des groupes. *Lecture Notes in Math.,vol.1441,Springer Verlag*, 1990.

[CT85] J. Cannon and W. P. Thurston. Group Invariant Peano Curves. *preprint, Princeton*, 1985.

[CT07] J. Cannon and W. P. Thurston. Group Invariant Peano Curves. *Geometry and Topology vol 11*, pages 1315–1356, 2007.

[DM10] S. Das and M. Mj. Semiconjugacies Between Relatively Hyperbolic Boundaries. *arXiv:1007.2547*, 2010.

[Far98] B. Farb. Relatively hyperbolic groups. *Geom. Funct. Anal. 8*, pages 810–840, 1998.

[Flo80] W. J. Floyd. Group Completions and Limit Sets of Kleinian Groups. *Invent. Math. vol.57*, pages 205–218, 1980.

[GdlH90] E. Ghys and P. de la Harpe(eds.). Sur les groupes hyperboliques d'apres Mikhael Gromov. *Progress in Math. vol 83, Birkhauser, Boston Ma.*, 1990.

[Gro85] M. Gromov. Hyperbolic Groups. *in Essays in Group Theory, ed. Gersten, MSRI Publ.,vol.8, Springer Verlag*, pages 75–263, 1985.

[Gro93] M. Gromov. Asymptotic Invariants of Infinite Groups. *in Geometric Group Theory,vol.2; Lond. Math. Soc. Lecture Notes 182, Cambridge University Press*, 1993.

[HY61] J. G. Hocking and G. S. Young. Topology. *Addison Wesley*, 1961.

[Kla99] E. Klarreich. Semiconjugacies between Kleinian group actions on the Riemann sphere. *Amer. J. Math 121*, pages 1031–1078, 1999.

[McM01] C. T. McMullen. Local connectivity, Kleinian groups and geodesics on the blow-up of the torus. *Invent. math.*, 97:95–127, 2001.

[Min94] Y. N. Minsky. On Rigidity, Limit Sets, and End Invariants of Hyperbolic 3-Manifolds. *Jour. Amer. Math. Soc. 7*, pages 539–588, 1994.

[Min99] Y. N. Minsky. The Classification of Punctured Torus Groups. *Ann. of Math. 149*, pages 559–626, 1999.

[Min10] Y. N. Minsky. The Classification of Kleinian surface groups I: Models and bounds. *Ann. of Math. 171 (1)*, pages 1–107, 2010.

[Mit97] M. Mitra. Maps on boundaries of hyperbolic metric spaces. *PhD Thesis, U.C. Berkeley*, 1997.

[Mit98a] M. Mitra. Cannon-Thurston Maps for Hyperbolic Group Extensions. *Topology 37*, pages 527–538, 1998.

[Mit98b] M. Mitra. Cannon-Thurston Maps for Trees of Hyperbolic Metric Spaces. *Jour. Diff. Geom.48*, pages 135–164, 1998.

[Mj05] M. Mj. Cannon-Thurston Maps for Surface Groups I: Amalgamation Geometry and Split Geometry. *preprint, arXiv:math.GT/0512539 v5*, 2005.

[Mj09] M. Mj. Cannon-Thurston Maps for Pared Manifolds of Bounded Geometry. *Geom. Topol. 13*, pages 189–245, 2009.

[Mj10a] M. Mj. Cannon-Thurston Maps and Bounded Geometry. *in Teichmuller Theory and Moduli Problems, Proceedings of Workshop at HRI, Allahabad, Ramanujan Mathematical Society Lecture Notes Series Number 10, arXiv:math.GT/0603729*, pages 489–511, 2010.

[Mj10b] M. Mj. Cannon-Thurston Maps for Kleinian Groups. *preprint, arXiv:math 1002.0996*, 2010.

[Mj11] M. Mj. Cannon-Thurston Maps, i-bounded Geometry and a Theorem of McMullen. *Actes du séminaire Théorie spectrale et géométrie 28, Année 2009-10, arXiv:math.GT/0511104*, pages 63–108, 2011.

[Mj14a] M. Mj. Cannon-Thurston Maps for Surface Groups. *Ann. of Math. 179 (1)*, pages 1–80, 2014.

[Mj14b] M. Mj. Ending Laminations and Cannon-Thurston Maps, with an appendix by S. Das and M. Mj. *to appear in Geom. Funct. Anal.*, 2014.

[MP11] M. Mj and A. Pal. Relative Hyperbolicity, Trees of Spaces and Cannon-Thurston Maps. *Geom. Dedicata 151, arXiv:0708.3578*, pages 59–78, 2011.

8

Counting visible circles on the sphere and Kleinian groups

HEE OH[1] AND NIMISH SHAH[2]

Abstract

For a circle packing $\mathcal{P}$ on the sphere invariant under a nonelementary Kleinian group satisfying certain finiteness conditions, we describe the asymptotic distribution of circles in $\mathcal{P}$ of spherical curvature at most T as T tends to infinity.

1 Introduction

In the unit sphere $\mathbb{S}^2 = \{x^2 + y^2 + z^2 = 1\}$ with the Riemannian metric induced from $\mathbb{R}^3$, the distance (or *the spherical distance*) between two points is simply the angle between the rays connecting them to the origin o.

Let $\mathcal{P}$ be a collection of circles on the sphere $\mathbb{S}^2$, also called a *circle packing* on $\mathbb{S}^2$. The *visual size* of a circle C in $\mathbb{S}^2$ can be measured by its spherical radius $0 < \theta(C) \le \pi/2$, that is, the half of the visual angle of C from the origin $o = (0, 0, 0)$. We label the circles by their spherical curvatures given by

$$\mathrm{Curv}_{\mathbb{S}^2}(C) := \cot\theta(C).$$

We suppose that $\mathcal{P}$ is *locally finite* in the sense that for any $T > 0$,

$$\#\{C \in \mathcal{P} : \mathrm{Curv}_{\mathbb{S}^2}(C) < T\} < \infty.$$

In the beautiful book *Indra's pearls*, Mumford, Series and Wright ask the following question: ([13, Section 5.4, pg.155])

How many visible circles are there?

To address this question, for any subset $E \subset \mathbb{S}^2$ and $T > 0$, we define

$$N_T(\mathcal{P}, E) := \#\{C \in \mathcal{P} : C \cap E \ne \emptyset, \ \mathrm{Curv}_{\mathbb{S}^2}(C) < T\}.$$

[1] Oh was partially supported by NSF grant 0629322.
[2] Shah was partially supported by NSF grant 1001654.

272

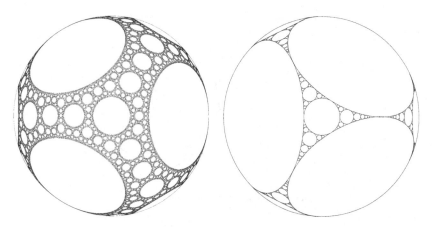

Figure 1 Sierpinski curve and Apollonian gasket (by C. McMullen)

The main goal of this article is to obtain an asymptotic formula for $N_T(\mathcal{P}, E)$ as $T \to \infty$ when $\mathcal{P}$ is invariant under a Kleinian group satisfying certain finiteness assumptions. Our formula involves notions from hyperbolic geometry. Consider the Poincare ball model $\mathbb{B} = \{x_1^2 + x_2^2 + x_3^2 < 1\}$ of the hyperbolic 3-space with the metric given by $\frac{2 \cdot \sqrt{dx_1^2 + dx_2^2 + dx_3^2}}{1 - (x_1^2 + x_2^2 + x_3^2)}$. The geometric boundary of $\mathbb{B}$ naturally identifies with $\mathbb{S}^2$.

In this article, let G denote the group of orientation preserving isometries of $\mathbb{B}$ and $\Gamma < G$ a non-elementary (= non virtually-abelian) Kleinian group. We denote by $\Lambda(\Gamma) \subset \mathbb{S}^2$ the limit set of Γ, and by $\delta = \delta_\Gamma$ the critical exponent of Γ. Let $\{\nu_x : x \in \mathbb{B}\}$ be a Patterson-Sullivan density, i.e., a Γ-invariant conformal density of dimension δ on $\Lambda(\Gamma)$. We denote by m^{BMS} the Bowen-Margulis-Sullivan measure on the unit tangent bundle $\mathrm{T}^1(\Gamma \backslash \mathbb{B})$ associated to $\{\nu_x\}$, see Section 2.2.

For a vector $u \in \mathrm{T}^1(\mathbb{B})$, denote by $u^+ \in \mathbb{S}^2$ the forward end point of the geodesic determined by u, and by $\pi(u) \in \mathbb{B}$ the base point of u. For $x_1, x_2 \in \mathbb{B}$ and $\xi \in \mathbb{S}^2$, $\beta_\xi(x_1, x_2)$ denotes the signed distance between horospheres based at ξ and passing through x_1 and x_2.

Definition 1 (Skinning size of $\mathcal{P}$) For a circle packing $\mathcal{P}$ on $\mathbb{S}^2$ invariant under Γ, we define $0 \le \mathrm{sk}(\mathcal{P}) \le \infty$ by

$$\mathrm{sk}(\mathcal{P}) := \sum_{i \in I} \int_{s \in \mathrm{Stab}_\Gamma(C_i^\dagger) \backslash C_i^\dagger} e^{\delta \beta_{s^+}(x, \pi(s))} d\nu_x(s^+)$$

where $x \in \mathbb{B}$, $\{C_i : i \in I\}$ is a set of representatives of Γ-orbits in $\mathcal{P}$ and $C_i^\dagger \subset \mathrm{T}^1(\mathbb{B})$ is the set of unit normal vectors to the convex hull of C_i.

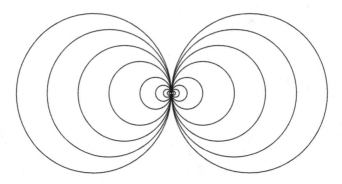

Figure 2 Infinite bouquet of tangent circles

By the conformal property of $\{v_x\}$, the definition of sk($\mathcal{P}$) is independent of the choice of $x \in \mathbb{B}$ and the choice of representatives $\{C_i\}$.

Theorem 1 *Let $\mathcal{P}$ be a locally finite Γ-invariant circle packing on the sphere $\mathbb{S}^2$ with finitely many Γ-orbits. Suppose that $|m^{\text{BMS}}| < \infty$ and sk($\mathcal{P}$) $< \infty$. Then for any Borel subset $E \subset \mathbb{S}^2$ with $v_o(\partial E) = 0$,*

$$\lim_{T \to \infty} \frac{N_T(\mathcal{P}, E)}{T^\delta} = \frac{2^\delta \cdot \text{sk}(\mathcal{P})}{\delta \cdot |m^{\text{BMS}}|} \cdot v_o(E).$$

where $o = (0, 0, 0)$. If $\mathcal{P}$ is infinite, sk($\mathcal{P}$) > 0.

Remark (1) If Γ is *geometrically finite*, that is, if Γ admits a finite sided fundamental domain in $\mathbb{B}$, then $|m^{\text{BMS}}| < \infty$ [21].

By [16, Lemma 1.13 and Theorem 1.14(2)], if $\delta > 1$ then sk($\mathcal{P}$) $< \infty$; and $\delta \leq 1$, then we have sk($\mathcal{P}$) $< \infty$ if and only if $\mathcal{P}$ does not contains an *infinite bouquet of tangent circles glued at a parabolic fixed point* of Γ (see Fig. 2, or [15, Definition 1.3]). We note that by [15, Proposition 3.4] the nonexistence of such infinite bouquets corresponds to Γ-parabolic-coranks of $\hat{C}$ for each $C \in \mathcal{P}$ being equal to zero, and in this case [16, Theorem 1.14(2)] implies that sk($\mathcal{P}$) $< \infty$.

(2) Under the assumption of $|m^{\text{BMS}}| < \infty$, v_o is atom-free by [18, Sec.1.5], and hence the above theorem works for any Borel subset E whose boundary intersects $\Lambda(\Gamma)$ in countably many points. If Γ is Zariski dense in G, then any proper real subvariety of $\mathbb{S}^2$ has zero v_o-measure [6, Cor. 1.4] and hence Theorem 1 applies to any Borel subset E of $\mathbb{S}^2$ whose boundary is contained in a countable union of real algebraic curves.

We combine the above remarks in the following:

Corollary 2 *Let* Γ *be a geometrically finite Zariski dense discrete subgroup of* G. *Let* $\mathcal{P}$ *be a locally finite* Γ-*invariant circle packing on* $\mathbb{S}^2$ *which is a union of finitely many* Γ-*orbits. Let* E *be a Borel subset of* $\mathbb{S}^2$ *such that* $\partial E \cap \Lambda(\Gamma)$ *is contained in a union of countably many proper real algebraic sub-varieties of* $\mathbb{S}^2$. *Then*

$$\lim_{T \to \infty} \frac{N_T(\mathcal{P}, E)}{T^\delta} = \frac{2^\delta \operatorname{sk}(\mathcal{P})}{\delta |m^{\mathrm{BMS}}|} \cdot v_o(E).$$

1.1 Examples

(1) If X is a finite volume hyperbolic 3 manifold with totally geodesic boundary, its fundamental group $\Gamma := \pi_1(X)$ is geometrically finite and X is homeomorphic to $\Gamma \backslash (B \cup \Omega(\Gamma))$ where $\Omega(\Gamma)$ is the domain of discontinuity for Γ [8]. The universal cover $\tilde{X}$ developed in $\mathbb{B}$ has geodesic boundary components which are Euclidean hemispheres normal to $\mathbb{S}^2$. Then $\Omega(\Gamma)$ is the union of a countably many disjoint open disks corresponding to the geodesic boundary components of $\tilde{X}$. The Ahlfors finiteness theorem [1] implies that the circle packing $\mathcal{P}$ on $\mathbb{S}^2$ consisting of the geodesic boundary components of $\tilde{X}$ is locally finite and has finitely many Γ-orbits. Moreover, $\operatorname{sk}(\mathcal{P}) < \infty$ as $\mathcal{P}$ contains no infinite bouquet of tangent circles.

In the case when $\pi_1(X)$ is convex co-compact, then no disks in $\Omega(\Gamma)$ are tangent to each other and $\Lambda(\Gamma)$ is known to be homeomorphic to a Sierpinski curve [4] (see Fig. 1).

(2) Starting with four mutually tangent circles on the sphere $\mathbb{S}^2$, one can inscribe into each of the curvilinear triangle a unique circle by an old theorem of Apollonius of Perga (c. BC 200). Continuing to inscribe the circles this way, one obtains an Apollonian circle packing on $\mathbb{S}^2$ (see Fig. 1). Apollonian circle packings are examples of circle packing obtained in the way described in (1) (cf. [5] and [10].).

(3) Take $k \geq 1$ pairs of mutually disjoint closed disks $\{(D_i, D_i') : 1 \leq i \leq k\}$ in $\mathbb{S}^2$. For each $1 \leq i \leq k$, choose $\gamma_i \in G$ which maps the interior of D_i to the exterior of D_i' and vice versa. The group $\Gamma := \langle \gamma_i : 1 \leq i \leq k \rangle$ is called a Schottky group of genus k (cf. [11, Sec. 2.7]). Let $\mathcal{P} := \cup_{1 \leq i \leq k} \Gamma(C_i) \cup \Gamma(C_i')$, where C_i and C_i' are the boundaries of D_i and D_i', respectively. Then $\mathcal{P}$ is locally finite, as the Γ-orbit of the disks nest down onto the limit set $\Lambda(\Gamma)$, which is totally disconnected. Such a collection $\mathcal{P}$ is called a *Schottky dance* (see Fig. 3 or [13, Fig. 4.11]).

The common exterior of hemispheres above the initial disks D_i and D_i' is a fundamental domain for Γ in $\mathbb{B}$ and hence Γ is geometrically finite.

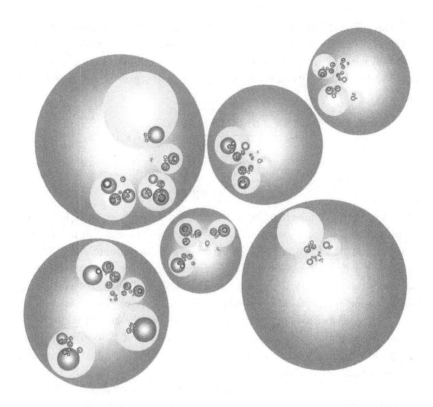

Figure 3 Schottky dance (from Indra's Pearls, by D.Mumford, C. Series and
D. Wright, copyright Cambridge University Press 2002)

Since $\mathcal{P}$ has no infinite bouquet of tangent circles, Theorem 1 applies by
Remark 1 (1).

1.2 Counting in terms of the visual distance

Let $\hat{C} \subset \mathbb{B}$ denote the convex hull of C. Then (cf. [22, P.24]) since $o = (0, 0, 0)$,
we have

$$\sin \theta(C) = 1/\cosh d(\hat{C}, o) \quad \text{and} \quad \mathrm{Curv}_S(C) = \sinh d(\hat{C}, o). \quad (1)$$

Since $2\sinh(T) \sim e^T$ as $T \to \infty$, Theorem 1 follows from the following
result for $x = o$.

Theorem 3 *Keeping the same assumption as in Theorem 1, we have, for any*
$x \in \mathbb{B}$,

$$\lim_{T \to \infty} \frac{\#\{C \in \mathcal{P} : C \cap E \neq \emptyset, \, d(\hat{C}, x) < T\}}{e^{\delta \cdot T}} = \frac{\mathrm{sk}(\mathcal{P})}{\delta \cdot |m^{\mathrm{BMS}}|} \cdot \nu_x(E). \quad (2)$$

Using this result, in the last section we also obtain the counting estimate in terms of Euclidean curvatures of circles in $\mathcal{P}$ (see Theorem 8), and thus provide a shorter proof of [15, Theorem 1.4].

Acknowledgments

We are very grateful to Curt McMullen for generously sharing his intuition and ideas. The applicability of our results in [16] to the question addressed in this paper came up in the conversation of the first named author with him. We also thank Yves Benoist, Jeff Brock, and Richard Schwartz for useful conversations. Our sincere thanks are due to the referees of this article for careful reading and useful suggestions.

2 Equidistribution of normal translates of a hyperbolic surface

In this section, we set up notations as well as recall a result from [16] on limiting distribution of normal geodesic evolution of a hyperbolic surface.

2.1 Group theoretic notations for hyperbolic space and hyperbolic surfaces

Fix a circle $C_0 \subset \mathbb{S}^2$. Denote by $\hat{C}_0 \subset \mathbb{B}$ the convex hull of C_0, that is, $\hat{C}_0$ is the smallest convex set in $\mathbb{B}$ containing all geodesics whose end points are in C_0. Then $\hat{C}_0$ is a two dimensional hyperbolic disc isometrically imbedded in $\mathbb{B}$. Let $C_0^\dagger \subset \mathrm{T}^1(\mathbb{B})$ denote the set of unit normal vectors to $\hat{C}_0$. Let $p_0 \in \hat{C}_0$ and $X_0 \in C_0^\dagger$ based at p_0.

Let K be the stabilizer subgroup of p_0 in G and $M \subset K$ be the stabilizer of X_0 in G. Then under the maps $gK \mapsto gp_0$ and $gM \mapsto gX_0$ we identify G/K with $\mathbb{B}$ and G/M with $\mathrm{T}^1(\mathbb{B})$, respectively. Let

$$H = \{h \in G : hC_0 = C_0\} = \{h \in G : h\hat{C}_0 = \hat{C}_0\}.$$

Then $H \cdot p_0 = \hat{C}_0$ and $H \cdot X_0 = C_0^\dagger$. Here H has two connected components, one of which is the group of orientation preserving hyperbolic isometries of $\hat{C}_0$. Also $M \subset H$.

Proposition 4 ([15, Lemma 3.2]) *Let Γ be a discrete subgroup of G.*

(1) *If $\Gamma(C_0) = \{\gamma C_0 : \gamma \in \Gamma\}$ is infinite, then $[\Gamma : H \cap \Gamma] = \infty$.*

(2) *$\Gamma(C_0)$ is a locally finite packing*
   $\Leftrightarrow$ the natural projection map $(\Gamma \cap H)\backslash\hat{C}_0 \to \Gamma\backslash\mathbb{B}$ is proper
   $\Leftrightarrow$ the natural inclusion $(\Gamma \cap H)\backslash H \to \Gamma\backslash G$ is proper.

2.2 Patterson-Sullivan conformal density and BMS and BR measures

For $u \in T^1(\mathbb{B})$, we define $u^+ \in \mathbb{S}^2 = \partial\mathbb{B}$ (resp. $u^- \in \mathbb{S}^2$) the forward (resp. backward) endpoint of the geodesic determined by u and $\pi(u) \in \mathbb{B}$ the basepoint.

Let Γ be a non-elementary Klenian group. Let $\{\nu_x : x \in \mathbb{B}\}$ be a Γ-invariant conformal density on $\mathbb{S}^2$ of dimension $\delta = \delta_\Gamma$; that is, each ν_x is a finite measure on $\mathbb{B}$ and

$$\gamma_* \nu_x = \nu_{\gamma x} \text{ for all } \gamma \in \Gamma \text{ and} \tag{3}$$

$$\frac{d\nu_y}{d\nu_x}(\xi) = e^{-\delta\beta_\xi(y,x)} \text{ for all } \xi \in \mathbb{S}^2, \tag{4}$$

where $\gamma_* \nu_x(R) = \nu_x(\gamma^{-1}R)$ and the *Busemann function* $\beta_\xi(y,x)$ is given by

$$\beta_\xi(y,x) = \lim_{t\to\infty} d(y, \xi_t) - d(x, \xi_t), \tag{5}$$

for any geodesic ray $\{\xi_t\}$ such that $\lim_{t\to\infty} \xi_t = \xi$.

Let $\{m_x : x \in \mathbb{B}\}$ be the G-invariant (Lebesgue) probability conformal density of dimension 2 on $\mathbb{S}^2$.

We define the Bowen-Margulis-Sullivan measure m^{BMS} ([2], [12], [21]) and the Burger-Roblin measure m^{BR} ([3], [18]) associated to $\{\nu_x\}$ and $\{m_x\}$ to be the measures on $T^1(\Gamma\backslash\mathbb{B})$ induced by the following Γ-invariant measures on $T^1(\mathbb{B})$ respectively: for $x \in \mathbb{B}$,

$$d\tilde{m}^{\text{BMS}}(u) = e^{\delta\beta_{u^+}(x,\pi(u))} e^{\delta\beta_{u^-}(x,\pi(u))} d\nu_x(u^+)d\nu_x(u^-)dt;$$

$$d\tilde{m}^{\text{BR}}(u) = e^{2\beta_{u^+}(x,\pi(u))} e^{\delta\beta_{u^-}(x,\pi(u))} dm_x(u^+)d\nu_x(u^-)dt.$$

By the conformal properties (4) of $\{\nu_x\}$ and $\{m_x\}$, these definitions are independent of the choice of $x \in \mathbb{B}$. Moreover both of these measures are invariant under the left action of Γ on $T^1(\mathbb{B})$. Let m^{BMS} and m^{BR} denote the corresponding measures on $\Gamma\backslash T^1(\mathbb{B}) = T^1(\Gamma\backslash\mathbb{B})$.

2.3 Comparison of visual density of circles corresponding to different base points

Let $\mathcal{P}$ and Γ be as in the statement of Theorem 3. For any $y \in \mathbb{B}$ and any $E \subset \mathbb{S}$, set

$$\mathcal{N}_T^y(\mathcal{P}, E) = \#\{C \in \mathcal{P} : C \cap E \neq \emptyset, \ d(y, \hat{C}) < T\}. \tag{6}$$

Proposition 5 *Choose $x \in \mathbb{B}$. Suppose that there exists a constant $D \geq 0$ such that for any $F \subset \mathbb{S}^2$ with $v_o(\partial F) = 0$, we have*

$$\lim_{T \to \infty} \frac{\mathcal{N}_T^x(\mathcal{P}, F)}{e^{\delta T}} = D \cdot v_x(F). \tag{7}$$

Then for any $y \in \mathbb{B}$ and any $E \subset \mathbb{S}^2$ with $v_o(\partial E) = 0$, we get

$$\lim_{T \to \infty} \frac{\mathcal{N}_T^y(\mathcal{P}, E)}{e^{\delta T}} = D \cdot v_y(E).$$

Remark (1) By the conformality property (4), v_o and v_y are absolutely continuous with respect to each other, so $v_o(\partial F) = 0$ if and only if $v_y(\partial F) = 0$.

(2) Due to Proposition 5, in order to prove Theorem 3, it is enough to derive its conclusion (2) for a particular choice $x = p_0$.

Proof Let $\epsilon > 0$ be given. For any $\xi \in \mathbb{S}^2$ there exists an open disc D in $\mathbb{S}^2$ centered at ξ such that

$$|\beta_{\xi_1}(x, y) - \beta_{\xi_2}(x, y)| \leq \epsilon, \text{ for all } \xi_1, \xi_2 \in D, \tag{8}$$

and $v_o(\partial D) = 0$; such a disc D exists because we can let the radius of D tend to 0 over an uncountable set of radii, the boundaries of concentric discs of distinct radii are disjoint, and v_o is finite. By (4) and (8), for any $\xi \in D$ we have

$$v_y(F)e^{-\delta\epsilon} \leq v_x(F)e^{\delta\beta_\xi(x,y)} \leq v_y(F)e^{\delta\epsilon}, \text{ for any measurable } F \subset D. \tag{9}$$

We cover $\mathbb{S}^2$ by finitely many such discs D_i, $1 \leq i \leq k$, and let

$$E_i = E \cap D_i \setminus (\cup_{j<i} D_j), \text{ for } 1 \leq i \leq k.$$

For subsets A and B of $\mathbb{S}^2$, $\partial(A \cap B) \subset \partial(A) \cup \partial(B)$. Therefore

$$\partial(E_i) \subset \partial(E) \cup \cup_{j=1}^{k} \partial(D_j).$$

Therefore $v_o(\partial(E_i)) = 0$, and hence $v_y(\partial E_i) = 0$ by Remark 2.3(1), and in particular $v_y(\text{int}(E_i)) = v_y(E_i)$.

For each i, choose a closed set $F_i = (E_i)_\eta^- := \{\xi \in E_i : d(\xi, \partial(E_i)) \geq \eta\}$ such that $v_y(F_i) \geq e^{-\epsilon} v_y(E_i)$ and $v_o(\partial F_i) = 0$. Such F_i exists because $(E_i)_\eta^- \uparrow \text{int}(E)$ and hence $v_y((E_i)_\eta^-) \to v_y(\text{int}(E_i)) = v_y(E_i)$ as $\eta \to 0$, and $\partial((E_i)_\eta^-)$ are disjoint for distinct η's, so we can find arbitrarily small $\eta > 0$ such that $v_o(\partial((E_i)_\eta^-)) = 0$.

By (5), there exists $T_\epsilon > 0$ such that if $C \in \mathcal{P}$ with $d(x, \hat{C}) \geq T_\epsilon$ and $C \cap E_i \neq \emptyset$ for some i, then for any $\xi \in E_i$,

$$|d(y, \hat{C}) - (d(x, \hat{C}) + \beta_\xi(x, y))| \leq \epsilon, \tag{10}$$

and further if

$$C \cap F_i \neq \emptyset \text{ then } C \cap F_j = \emptyset, \text{ for all } j \neq i, \tag{11}$$

because for sufficiently large $T_\epsilon > 0$, $d(x, \hat{C}) > T_\epsilon$ implies that the spherical diameter of C is less than the minimum of spherical distances between distinct nonempty F_i and F_j.

Then by (10) and (11), for any $\xi \in E_i$,

$$\mathcal{N}_T^y(\mathcal{P}, E_i) - \mathcal{N}_{T_\epsilon}^y(\mathcal{P}, E_i) \leq \mathcal{N}_{T+\beta_\xi(x,y)+\epsilon}^x(\mathcal{P}, E_i) \text{ and} \tag{12}$$

$$\mathcal{N}_T^y(\mathcal{P}, F_i) \geq N_{T+\beta_\xi(x,y)-\epsilon}^x(\mathcal{P}, F_i) - N_{T_\epsilon+d(x,y)+1}^x(\mathcal{P}, F_i). \tag{13}$$

By local finiteness of $\mathcal{P}$, $\mathcal{N}_{T_\epsilon}^y(\mathcal{P}, E_i) < \infty$. Hence by (12), (7) and (9),

$$\limsup_{T\to\infty} \frac{\mathcal{N}_T^y(\mathcal{P}, E_i)}{e^{\delta T}} \leq D \cdot \nu_x(E_i) e^{\delta(\beta_\xi(x,y)+\epsilon)} \leq D \cdot \nu_y(E_i) e^{\delta(2\epsilon)},$$

and by (13), (7) and (9),

$$\liminf_{T\to\infty} \frac{\mathcal{N}_T^y(\mathcal{P}, F_i)}{e^{\delta T}} \geq D \cdot \nu_x(F_i) e^{\delta(\beta_\xi(x,y)-\epsilon)} \geq D \cdot \nu_y(E_i) e^{-(1+2\delta)\epsilon}.$$

By summing over $1 \leq i \leq k$, we get

$$\limsup_{T\to\infty} \frac{\mathcal{N}_T^y(\mathcal{P}, E)}{e^{\delta T}} - \liminf_{T\to\infty} \frac{\mathcal{N}_T^y(\mathcal{P}, E)}{e^{\delta T}} \leq D \cdot \nu_y(E)(e^{2\delta\epsilon} - e^{-(1+2\delta)\epsilon}),$$

and the conclusion follows by taking the limit as $\epsilon \to 0$. $\square$

2.4 Equidistribution for orthogonal translates of a hyperbolic surface

We consider the following two measures on $H/M \cong H \cdot X_0 = C_0^\dagger$: Choose any $x \in \mathbb{B}$, and define

$$d\mu_{C_0^\dagger}^{\text{Leb}}(s) = e^{2\beta_{s+}(x,\pi(s))} dm_x(s^+) \text{ and } d\mu_{C_0^\dagger}^{\text{PS}}(s) := e^{\delta\beta_{s+}(x,\pi(s))} d\nu_x(s^+). \tag{14}$$

We note that the map $s \mapsto s^+$ from $C_0^\dagger \to \mathbb{S}^2 \setminus C_0$ is a diffeomorphism [16, Lemma 2.1]. These definitions are in fact independent of the choice of x. The measures $\mu_{C_0^\dagger}^{\text{Leb}}$ and $\mu_{C_0^\dagger}^{\text{PS}}$ are invariant under the action of H and $H \cap \Gamma$, respectively. We will denote the corresponding measures on the quotient space $(H \cap \Gamma)\backslash C_0^\dagger$ by μ^{Leb} and μ^{PS}, respectively.

Let $\mathcal{G}^t$ denote the geodesic flow on $\mathrm{T}^1(\mathbb{B})$. Then for any $v \in C_0^\dagger$, $t \mapsto \mathcal{G}^t(v)$ is the geodesic orthogonal to C_0 with tangent v. And the image of $\mathcal{G}^t(C_0^\dagger)$ in $\mathbb{B}$ is a union of two connected codimension one submanifolds in $\mathbb{B}$ consisting of points at distance t from C_0 on each side of C_0. In the next result we describe the limiting distribution of the geodesic evolution of $C_0^\dagger$ modulo Γ in $\Gamma \backslash T^1(\mathbb{B}) = T^1(\Gamma \backslash \mathbb{B})$.

Let $A = \{a_t : t \in \mathbb{R}\}$ be the one-parameter subgroup of G such that $a_t X_0 = \mathcal{G}^t(X_0)$ for all $t \in \mathbb{R}$. Then M is the centralizer of A in K. Now if we write $v \in C_0^\dagger = H \cdot X_0 \cong H/M$ as $v = sX_0 = sM = [s]$ for $s \in H$, then $\mathcal{G}^t(v) = sa_t X_0 = [sa_t]$.

Theorem 6 ([16, Theorem 1.2]) *Suppose that the natural projection map* $(\Gamma \cap H) \backslash \hat{C}_0 \to \Gamma \backslash \mathbb{B}$ *is proper. If* $|m^{\text{BMS}}| < \infty$ *and* $|\mu^{\text{PS}}| := \mu^{\text{PS}}(H \cap \Gamma \backslash C_0^\dagger) < \infty$, *then for any* $\psi \in C_c(\Gamma \backslash G/M)$, *we have*

$$\lim_{t \to \infty} e^{(2-\delta)t} \int_{[s] \in (\Gamma \cap H) \backslash C_0^\dagger} \psi([sa_t]) d\mu^{\text{Leb}}([s]) = \frac{|\mu^{\text{PS}}|}{|m^{\text{BMS}}|} m^{\text{BR}}(\psi) \quad as \ t \to \infty.$$

Moreover if $[\Gamma : H \cap \Gamma] = \infty$ *then* $|\mu^{\text{PS}}| > 0$.

Note that due to Proposition 4, the properness condition in Theorem 6 is satisfied if ΓC_0 is a locally finite circle packing.

2.5 Haar measure on G in terms of $\mu^{\text{Leb}}_{C_0^\dagger}$

Let $A^+ = \{a_t : t \geq 0\}$. We have the following generalized Cartan decomposition ([19, Prop. 7.1.3]): $G = HA^+K$, in the sense that every element of $g \in G$ can be written as $g = ha_tk$, where $t \geq 0$, $h \in H$ and $k \in K$. Also if $ha_tk = h'a_{t'}k'$, with $t, t' > 0$, then $t = t'$, $h = h'm$, and $k = m^{-1}k'$ for some $m \in M$.

Let dm correspond the Haar probability measure on M. Writing $h = sm \in C_0^\dagger \times M$, let $dh = d\mu^{\text{Leb}}_{C_0^\dagger}(s)dm$ and let $dk = dm_{p_0}(k.X_0^-)dm$, then dh and dk correspond to Haar measures on H and K, respectively. Then the following defines a Haar measure on G [19, Prop. 8.1.1]: for any $\psi \in C_c(G)$,

$$\int_G \psi(g)dg = \int_{HA^+K} \psi(ha_tk) \, 4 \sinh t \cdot \cosh t \, dhdtdk. \tag{15}$$

We denote by $d\lambda$ the unique right G-invariant measure on $H \backslash G$ which is compatible with the choice of dg and dh: for $\psi \in C_c(G)$ and $\bar{\psi} \in C_c(H \backslash G)$ given by $\bar{\psi}[g] := \int_{h \in H} \psi(hg)dh$,

$$\int_G \psi \, dg = \int_{[g] \in H \backslash G} \overline{\psi}([g])d\lambda[g].$$

Hence $d\lambda([a_tk]) = (4 \sinh t \cdot \cosh t) \, dtdk$.

2.6 An asymptotic property relating m^{BR} to ν_{p_0}

Fixing a left-invariant metric on G, we denote by U_ϵ an ϵ-ball around e, and for $S \subset G$, we set $S_\epsilon = S \cap U_\epsilon$. For each small $\epsilon > 0$, we choose a non-negative function $\psi_\epsilon \in C_c(G)$ supported inside U_ϵ and $\int_G \psi_\epsilon dg = 1$ and define $\Psi_\epsilon \in C_c(\Gamma \backslash G)$ by

$$\Psi_\epsilon(g) = \sum_{\gamma \in \Gamma} \psi_\epsilon(\gamma g). \tag{16}$$

For a Borel subset $E \subset \mathbb{S}^2$, let

$$E_{X_0} := \{k \in K : kX_0^- \in E\} \subset K, \tag{17}$$

and define functions ψ_ϵ^E on G/M and Ψ_ϵ^E on $\Gamma \backslash G/M$ by

$$\psi_\epsilon^E(g) = \int_{k \in (E_{X_0})^{-1}} \psi_\epsilon(gk)dk \text{ and } \Psi_\epsilon^E(g) = \int_{k \in (E_{X_0})^{-1}} \Psi_\epsilon(gk)dk. \tag{18}$$

Proposition 7 *If* $\nu_{p_0}(\partial E) = 0$, *then*

$$\lim_{\epsilon \to 0} m^{\mathrm{BR}}(\Psi_\epsilon^E) = \nu_{p_0}(E).$$

Proof We have $m^{\mathrm{BR}}(\Psi_\epsilon^E) = \tilde{m}^{\mathrm{BR}}(\psi_\epsilon^E)$. Let $\Omega = (E_{X_0})^{-1}$. Then

$$\nu_{p_0}(\partial(\Omega^{-1}X_0^-)) = \nu_{p_0}(\partial E) = 0.$$

Set $f = \chi_K$, the characteristic function of K. Then for any $g \in G$,

$$\psi_\epsilon^E(g) = \int_{k \in \Omega} f(k)\psi_\epsilon(gk)dk =: f *_\Omega \psi_\epsilon(g),$$

as per the notation of [16, eq.(7.4)]. By [16, Prop. 7.5],

$$\lim_{\epsilon \to 0} \tilde{m}^{\mathrm{BR}}(f *_\Omega \psi_\epsilon) = \int_{k \in \Omega^{-1}} f(k^{-1})d\nu_{p_0}(kX_0^-) = \nu_{p_0}(E),$$

here we note that the choice of the Haar measure dg considered in (15) is same as the one considered for [16, Prop. 7.5] due to [16, Section 8]. $\qquad\square$

3 Proof of Theorem 3

It is enough to prove the theorem under the assumption that $\mathcal{P} = \Gamma C_0$.

Let $t_0 = d(o, p_0)$ and $\epsilon_0 > 0$ be such that $\sin(\epsilon_0) = 1/\cosh(t_0)$ (see (1)). Given $0 < \epsilon < \epsilon_0$, let $\mathcal{P}_\epsilon = \{C \in \mathcal{P} : \theta(C) \le \epsilon/2\}$, and let $T_\epsilon > 0$ be such that $\sin(\epsilon/2) = 1/\cosh(T_\epsilon)$. Note that $T_\epsilon > t_0$.

Given $E \subset \mathbb{S}^2$, define

$$E_\epsilon^+ = \{x \in \mathbb{S}^2 : \text{dist}(x, \overline{E}) < \epsilon\} \text{ and } E_\epsilon^- = \mathbb{S}^2 \setminus (\mathbb{S}^2 \setminus E)_\epsilon^+.$$

Then E_ϵ^+ is open and E_ϵ^- is compact, and as $\epsilon \to 0$,

$$E_\epsilon^+ \downarrow \overline{E} \text{ and } E_\epsilon^- \uparrow \text{int}(E). \tag{19}$$

For $T > 0$, define (see (17))

$$B_T(E) = H A_T^+(E_{X_0})^{-1}, \text{ where } A_T^+ = \{a_t : 0 \le t < T\}.$$

Let $c_\epsilon^+ = \#(\mathcal{P} \setminus \mathcal{P}_\epsilon) < \infty$ and $c_\epsilon^- = \#([e]\Gamma \cap [e]A_{t_0+T_\epsilon}^+ K) < \infty$, where $[e]$ represents the coset of identity in $(\Gamma \cap H)\backslash G$, and note that $[e]\Gamma$ is discrete and $A_{t_0+T_\epsilon}^+ K$ is relatively compact.

Lemma 2 (Basic counting) *Given $T > T_\epsilon$,*

$$\#([e](\Gamma \cap B_T(E_\epsilon^-))) - c_\epsilon^- \le \mathcal{N}_T^{p_0}(\mathcal{P}, E) \le \#([e](\Gamma \cap B_T(E_\epsilon^+))) + c_\epsilon^+. \tag{20}$$

Proof Let $C \in \mathcal{P}$. Assume that $\theta(C) < \epsilon$, or equivalently $d(\hat{C}, o) > T_\epsilon > t_0$. Let $\text{int}(C)$ denote the smaller of the two open discs in $\mathbb{S}^2$ bounded by C. Then the following statements are equivalent:

(1) $d(\hat{C}, p_0) = t$;
(2) the distance between the orthogonal projection x of p_0 onto $\hat{C}$ is t;
(3) there exists $\xi \in \text{int}(C)$ such that the directed geodesic from p_0 to ξ intersects $\hat{C}$ perpendicularly at a distance t from p_0;
(4) there exists $k \in K$ such that $\xi = kX_0^- \in \text{int}(C)$ and $ka_{-t}X_0 \in C^\dagger$;
(5) there exists $\gamma \in \Gamma$ such that $\gamma C_0 = C \in \mathcal{P}_\epsilon$ and there exists $k \in K$ such that $kX_0^- \in \text{int}(\gamma C_0)$, and $ka_{-t}X_0 \in \gamma C_0^\dagger = \gamma H X_0$;
(6) $t > T_\epsilon$, $k \in (\text{int}(\gamma C_0))_{X_0}$, and $\gamma \in \Gamma \cap ka_{-t}H$.

Therefore if $C = \gamma C_0 \in \mathcal{P}_\epsilon$ for some $\gamma \in \Gamma, d(\hat{C}, p_0) < T$ and $C \cap E \ne \emptyset$ then $\text{int}(C) \subset E_\epsilon^+$, and

$$\gamma \in \Gamma \cap (E_\epsilon^+)_{X_0}(A_T^+)^{-1}H = \Gamma \cap B_T(E_\epsilon^+)^{-1}.$$

Also $\gamma C_0 = \gamma' C_0$ for some $\gamma' \in \Gamma$ if and only if $\gamma^{-1}\gamma' \in H \cap \Gamma$. Therefore

$$\#\{\gamma C_0 \in \mathcal{P}_\epsilon : \gamma \in \Gamma, d(\gamma \hat{C}_0, p_0) < T, \gamma C_0 \cap E \ne \emptyset\}$$
$$\le \#\{(\Gamma \cap B_T(E_\epsilon^+)^{-1})/(H \cap \Gamma)\} = \#\{(\Gamma \cap H)\backslash(\Gamma \cap B_T(E_\epsilon^+))\}.$$

This gives the second inequality in (20).

Conversely, if $\gamma \in \Gamma \cap (E_\epsilon^-)_{X_0}(A_T^+ \setminus A_{t_0+T_\epsilon}^+)^{-1}H$ then $\text{int}(\gamma C_0) \cap E_\epsilon^- \ne \emptyset$, $d(\gamma \hat{C}_0, p_0) < T$ and $d(\gamma \hat{C}_0, o) > T_\epsilon$. Therefore $\theta(\gamma C_0) < \epsilon/2$, and hence $\gamma C_0 \subset E$. Therefore $\gamma C_0 \in \mathcal{P}$. This leads to the first inequality in (20).   □

Strong wavefront lemma

By [7, Thm. 1.6], there exists $\epsilon_1 > 0$, such that given $0 < \epsilon < \epsilon_1$, there exists $T'_\epsilon > 0$ such that

$$Ha_t kU_\epsilon \subset Ha_t A_{2\epsilon} kK_{2\epsilon}, \text{ for all } t \geq T'_\epsilon, \text{ and } k \in K. \tag{21}$$

There exists $0 < \alpha < 1$ (depending only on $d(o, p_0)$) such that for any $\epsilon > 0$ and $k \in K_{2\alpha\epsilon}$, we have $\operatorname{dist}(k\xi, \xi) < \epsilon$ for all $\xi \in \mathbb{S}^2$. Therefore if $T > T'_{\alpha\epsilon}$ then by (21)

$$B_T(E_\epsilon^+) \subset B_{T+2\epsilon}(E_{2\epsilon}^+)U_{\alpha\epsilon} \text{ and } B_{T-2\epsilon}(E_{2\epsilon}^-)U_{\alpha\epsilon} \subset B_T(E_\epsilon^-).$$

Fix $0 < \epsilon < \min(\epsilon_0, \epsilon_1)$. Define the counting functions $F_T^\pm$ on $\Gamma\backslash G$ by

$$F_T^\pm(g) := \sum_{[\gamma] \in (\Gamma \cap H)\backslash \Gamma} \chi_{B_{T\pm 2\epsilon}(E_{2\epsilon}^\pm)}(\gamma g).$$

Then for any $g \in U_{\alpha\epsilon}$ and $T > T'_{\alpha\epsilon}$,

$$F_T^-(g) - d_1 \leq \#[e](\Gamma \cap B_T(E_\epsilon^-)), \quad F_T^+(g) \geq \#[e](\Gamma \cap B_T(E_\epsilon^+)) - d_1,$$

where $d_1 = \#(\Gamma \cap A_{T'_{\alpha\epsilon}}^+ KU_{\epsilon_1}) < \infty$. Put $m^\pm = d_1 + c_\epsilon^\pm$. By Lemma 2, for all $T > \max(T_\epsilon, T'_{\alpha\epsilon})$ we have

$$F_T^-(g) - m^- \leq \mathcal{N}_T^{p_0}(\mathcal{P}, E) \leq F_T^+(g) + m^+.$$

Integrating against Ψ_ϵ (see (16)), we obtain

$$\langle F_T^-, \Psi_\epsilon \rangle - m^- \leq \mathcal{N}_T^{p_0}(\mathcal{P}, E) \leq \langle F_T^+, \Psi_\epsilon \rangle + m^+,$$

where the inner product is taken with respect dg.

Setting $\Xi_t = 4\sinh t \cdot \cosh t$, we have

$$\langle F_T^\pm, \Psi_\epsilon \rangle = \int_{g \in \Gamma \cap H\backslash G} \chi_{B_{T\pm 2\epsilon}}(g)\Psi_\epsilon(g)\, dg$$

$$= \int_{k \in ((E_{2\epsilon}^\pm)x_0)^{-1}} \int_0^{T\pm 2\epsilon} \int_{s \in \Gamma \cap H\backslash C_0^\dagger} \left(\int_{m \in M} \Psi_\epsilon(sa_t mk)\, dm \right) \Xi_t\, d\mu^{\mathrm{Leb}}(s) dt dk$$

$$= \int_{k \in ((E_{2\epsilon}^\pm)x_0)^{-1}} \left(\int_0^{T\pm 2\epsilon} \Xi_t \int_{s \in \Gamma \cap H\backslash C_0^\dagger} \Psi_k^\epsilon(sa_t)\, d\mu^{\mathrm{Leb}}(s) dt \right) dk,$$

where $\Psi_{g_1}^\epsilon \in C_c(\Gamma\backslash G)^M$ is given by $\Psi_{g_1}^\epsilon(g) = \int_{m \in M} \Psi_\epsilon(gmg_1)\, dm$.

Hence by Theorem 6, and using $\Xi_t \sim e^{2t}$, and $\delta > 0$, we deduce that

$$\lim_{T \to \infty} e^{-\delta(T \pm 2\epsilon)} \langle F_{T \pm 2\epsilon}^{\pm}, \Psi_\epsilon \rangle = \frac{|\mu_{C_0^\dagger}^{\mathrm{PS}}|}{\delta \cdot |m^{\mathrm{BMS}}|} \int_{k \in ((E_{2\epsilon}^{\pm}) x_0)^{-1}} m^{\mathrm{BR}}(\Psi_k^\epsilon) \, dk$$

$$= \frac{|\mu_{C_0^\dagger}^{\mathrm{PS}}|}{\delta \cdot |m^{\mathrm{BMS}}|} m^{\mathrm{BR}}(\Psi_\epsilon^{E_{2\epsilon}^{\pm}}), \text{ by (18)}.$$

Hence

$$\frac{|\mu_{C_0^\dagger}^{\mathrm{PS}}|}{\delta \cdot |m^{\mathrm{BMS}}|} m^{\mathrm{BR}}(\Psi_\epsilon^{E_{2\epsilon}^{-}}) e^{\delta(-2\epsilon)} \le \liminf_{T \to \infty} \frac{N_T^{p_0}(\mathcal{P}, E)}{e^{\delta T}} \tag{22}$$

$$\frac{|\mu_{C_0^\dagger}^{\mathrm{PS}}|}{\delta \cdot |m^{\mathrm{BMS}}|} m^{\mathrm{BR}}(\Psi_\epsilon^{E_{2\epsilon}^{+}}) e^{\delta(2\epsilon)} \ge \limsup_{T \to \infty} \frac{N_T^{p_0}(\mathcal{P}, E)}{e^{\delta T}}. \tag{23}$$

Fix any $\eta > 0$. By Proposition 7,

$$\limsup_{\epsilon \to 0} m^{\mathrm{BR}}(\Psi_\epsilon^{E_{2\epsilon}^{+}}) \le \lim_{\epsilon \to 0} m^{\mathrm{BR}}(\Psi_\epsilon^{E_\eta^{+}}) = \nu_{p_0}(E_\eta^{+})$$

$$\liminf_{\epsilon \to 0} m^{\mathrm{BR}}(\Psi_\epsilon^{E_{2\epsilon}^{-}}) \ge \lim_{\epsilon \to 0} m^{\mathrm{BR}}(\Psi_\epsilon^{E_\eta^{-}}) = \nu_{p_0}(E_\eta^{-}).$$

By (19)

$$\lim_{\eta \to 0} \nu_{p_0}(E_\eta^{+} \setminus E_\eta^{-}) = \nu_{p_0}(\partial E).$$

Therefore if we assume that $\nu_{p_0}(\partial E) = 0$, then

$$\lim_{T \to \infty} \frac{N_T^{p_0}(\mathcal{P}, E)}{e^{\delta T}} = \frac{|\mu^{\mathrm{PS}}|}{\delta \cdot |m^{\mathrm{BMS}}|} \nu_{p_0}(E). \qquad \square$$

4 Counting with respect to Euclidean curvature

Let $o = (0, 1) \in \mathbb{R}^2 \times \mathbb{R}_{>0} \cong \mathbb{H}^2$. Let $\xi \in \mathbb{R}^2$ and consider unit speed hyperbolic geodesic $[o, \xi) := \{\xi_t : t \ge 0\}$ joining o to ξ. For any $t > 0$, let $C_\xi(t)$ be the circle in $\mathbb{R}^2$ such that the hyperbolic geodesic joining $\xi(t)$ with any point of $C_\xi(t)$ is perpendicular to the geodesic $[o, \xi)$. Then $d(\hat{C}_\xi(t), o) = t$. Let $\mathrm{Curv}(C_\xi(t))$ denote the Euclidean curvature of $C_\xi(t)$. Then

$$\lim_{t \to \infty} \mathrm{Curv}(C_\xi(t))/e^t = 1/(1 + |\xi|^2),$$

and the converge is uniform for ξ in a compact set. Now using the arguments as in the Proof of Proposition 5, it is straightforward to deduce the following result from Theorem 3.

Theorem 8 ([15, Theorem 1.4]) *Let $\mathcal{P}$ be a locally finite packing of circles in $\mathbb{R}^2$ invariant under a non-elementary Klenian group Γ with finitely many Γ-orbits. Suppose that $|m^{\text{BMS}}| < \infty$ and $\text{sk}(\mathcal{P}) < \infty$. Then for any bounded set $E \subset \mathbb{R}^2$ with $v_o(\partial E) = 0$, we have*

$$\lim_{T \to \infty} \frac{\#\{C \in \mathcal{P} : \text{Curv}(C) < T, \ C \cap E \neq \emptyset\}}{T^\delta}$$
$$= \frac{\text{sk}(\mathcal{P})}{\delta \cdot |m^{\text{BMS}}|} \int_{\xi \in E} (1 + |\xi|^2)^\delta \, dv_o(\xi).$$

Hee Oh
Mathematics Department, 10 Hillhouse Ave, P.O. Box 208283
Yale University, New Haven, CT 06520 U.S.A., and
Korea Institute for Advanced Study (KIAS), Seoul, KOREA
E-mail: hee.oh@yale.edu

Nimish Shah
Department of Mathematics, The Ohio State University,
Columbus, OH 43210 U.S.A.
E-mail: shah@math.osu.edu

References

[1] Lars V. Ahlfors. Finitely generated Kleinian groups. *Amer. J. Math.*, 86:413–429, 1964.

[2] Rufus Bowen. Periodic points and measures for Axiom A diffeomorphisms. *Trans. Amer. Math. Soc.*, 154:377–397, 1971.

[3] Marc Burger. Horocycle flow on geometrically finite surfaces. *Duke Math. J.*, 61(3):779–803, 1990.

[4] Schieffelin Claytor. Topological immersion of Peanian continua in a spherical surface. *Ann. of Math. (2)*, 35(4):809–835, 1934.

[5] Nicholas Eriksson and Jeffrey C. Lagarias. Apollonian circle packings: number theory. II. Spherical and hyperbolic packings. *Ramanujan J.*, 14(3):437–469, 2007.

[6] Livio Flaminio and Ralph J. Spatzier. Geometrically finite groups, Patterson-Sullivan measures and Ratner's rigidity theorem. *Invent. Math.*, 99(3):601–626, 1990.

[7] Alex Gorodnik, Hee Oh, and Nimish Shah. Strong wavefront lemma and counting lattice points in sectors. *Israel J. Math*, 176:419–444, 2010.

[8] Sadayoshi Kojima. Polyhedral decomposition of hyperbolic 3-manifolds with totally geodesic boundary. In *Aspects of low-dimensional manifolds*, volume 20 of *Adv. Stud. Pure Math.*, pages 93–112. Kinokuniya, Tokyo, 1992.

[9] Alex Kontorovich and Hee Oh. Apollonian circle packings and closed horospheres on hyperbolic 3-manifolds (with an appendix by Oh and Shah). *Journal of AMS*, 24:603–648, 2011.

[10] Jeffrey C. Lagarias, Colin L. Mallows, and Allan R. Wilks. Beyond the Descartes circle theorem. *Amer. Math. Monthly*, 109(4):338–361, 2002.

[11] Albert Marden. *Outer circles*. Cambridge University Press, Cambridge, 2007. An introduction to hyperbolic 3-manifolds.

[12] Gregory Margulis. On some aspects of the theory of Anosov systems. *Springer Monographs in Mathematics*. Springer-Verlag, Berlin, 2004.

[13] David Mumford, Caroline Series, and David Wright. Indra's pearls. *Cambridge University Press*, New York, 2002.

[14] Hee Oh. Dynamics on Geometrically finite hyperbolic manifolds with applications to Apollonian circle packings and beyond. *Proc. of I.C.M. (Hyperabad, 2010)* Vol III, 1308–1331.

[15] Hee Oh and Nimish Shah. The asymptotic distribution of circles in the orbits of Kleinian groups. *Inventiones Math.* 187:1–35, 2012.

[16] Hee Oh and Nimish Shah. Equidistribution and counting for orbits of geometrically finite hyperbolic groups. *J. Amer. Math. Soc.*, 26:511–562, 2013.

[17] S.J. Patterson. The limit set of a Fuchsian group. *Acta Mathematica*, 136:241–273, 1976.

[18] Thomas Roblin. Ergodicité et équidistribution en courbure négative. *Mém. Soc. Math. Fr. (N.S.)*, 95:vi+96, 2003.

[19] Henrik Schlichtkrull. *Hyperfunctions and harmonic analysis on symmetric spaces*, volume 49 of *Progress in Mathematics*. Birkhäuser Boston Inc., Boston, MA, 1984.

[20] Dennis Sullivan. The density at infinity of a discrete group of hyperbolic motions. *Inst. Hautes Études Sci. Publ. Math.*, 50:171–202, 1979.

[21] Dennis Sullivan. Entropy, Hausdorff measures old and new, and limit sets of geometrically finite Kleinian groups. *Acta Math.*, 153(3-4):259–277, 1984.

[22] William Thurston. *The Geometry and Topology of Three-Manifolds*. available at www.msri.org/publications/books. Electronic version-March 2002.

9

Counting arcs in negative curvature

JOUNI PARKKONEN AND FRÉDÉRIC PAULIN

Abstract

Let M be a complete Riemannian manifold with negative curvature, and let C_-, C_+ be two properly immersed closed convex subsets of M. We survey the asymptotic behaviour of the number of common perpendiculars of length at most s from C_- to C_+, giving error terms and counting with weights, starting from the work of Huber, Herrmann, Margulis and ending with the works of the authors. We describe the relationship with counting problems in circle packings of Kontorovich, Oh, Shah. We survey the tools used to obtain the precise asymptotics (Bowen-Margulis and Gibbs measures, skinning measures). We describe several arithmetic applications, in particular the ones by the authors on the asymptotics of the number of representations of integers by binary quadratic, Hermitian or Hamiltonian forms.[1]

1 Introduction

Let M be a complete connected Riemannian manifold with negative sectional curvature. Let C_- and C_+ be two properly immersed closed convex subsets of M (see the end of Section 3.1 for a precise definition). For instance, C_- and C_+ could be points, totally geodesic immersed submanifolds, Margulis cusp neighbourhoods, or images in M of convex hulls in a universal Riemannian cover of M of limit sets of subgroups of the fundamental group of M. A *common perpendicular* from C_- to C_+ is a locally geodesic path $c : [a, b] \to M$

---

[1] **Keywords:** counting, geodesic arc, common perpendicular, convexity, equidistribution, mixing, rate of mixing, decay of correlation, negative curvature, convex hypersurfaces, skinning measure, Bowen-Margulis measure, Gibbs measure.
AMS codes: 37D40, 37A25, 53C22, 20H10, 20G20, 11R52, 11N45, 11E39, 30F40.

289

such that $\dot{c}(a)$ is an outer unit normal vector to C_- and $-\dot{c}(b)$ is an outer unit normal vector to C_+ (see Section 3.1 and Section 3.2 for precise definitions when the boundary of $C_\pm$ is not smooth). The (appropriately indexed) set of lengths of these common perpendiculars is called the *(marked) ortholength spectrum* of (C_-, C_+), and variations on it have been introduced in particular cases of constant curvatures by Basmajian, Bridgeman, Bridgeman-Kahn, Martin-McKee-Wambach, Meyerhoff, Mirzakhani (see Section 3.2 for precisions).

The aim of this survey is to present several results on the asymptotic behaviour as s tends to $+\infty$ of the number $\mathcal{N}(s)$ of common perpendiculars (counted with multiplicities, see Section 3.2) between C_- and C_+ with length at most s.

We describe the first results on this problem by Huber [Hub], Herrmann [Herr] and Margulis [Mar1], see also the surveys [Bab2, Sha] and their references. We explain in Section 3 how several works of Duke-Rudnick-Sarnak on counting integral points on hyperboloids and of Kontorovich, Oh and Shah on counting problems in circle and sphere packings are related to counting problems of common perpendiculars. We particularly emphasize the arithmetic applications developped in [PaP4, PaP3, PaP5] to the counting of the representations of integers by quadratic, Hermitian or quaternionic binary forms, see Section 5.

A few results are new, in particular the deduction of the equidistribution of the outer unit normal bundles of equidistant hypersurfaces of totally geodesic submanifolds in a hyperbolic manifold from Eskin-McMullen's [EM] work (see Section 4), and the computations of the constant relating the Bowen-Margulis measure and the Liouville measure in constant curvature and finite volume, as well as the one relating the skinning measure and the Riemannian measure on the unit normal bundle of a totally geodesic submanifold in constant curvature and finite volume (see Section 7).

We survey the main tools used for the counting results: the geometry of negatively curved manifolds in Section 2 on one end, and on the other hand, in Section 6, the various measures, as the Patterson-Sullivan densities, the Bowen-Margulis measures, the skinning measures, that are needed to explicit the multiplicative constant in front of the exponential term in the asymptotics of the number $\mathcal{N}(s)$ of common perpendiculars. Skinning measures have first been introduced by Oh-Shah in constant curvature for immersed balls, horoballs and totally geodesic submanifolds, and are developed in general in [PaP6].

We give in Section 8 a sketch of the proof of the main counting result of [PaP7], which seems to contain as particular cases all previous results on the asymptotics of $\mathcal{N}(s)$, and to give many new ones. We conclude the paper by studying the error term to this asymptotic equivalent under hypotheses of exponential mixing of the geodesic flow (see Section 9), and by giving counting results when weights have been added to the common perpendiculars, by means of a potential function, the main tools being then the Gibbs measures (see Section 10).

To keep this paper to a reasonable length, we have chosen not to develop the related counting problems of closed geodesics with lengths at most s, which have been studied extensively (see for instance [Bowe, PaPo, Par, Rob2, PauPS], and the surveys [Bab2, Sha] and their references).

Acknowledgment. The authors thank the FIM of the ETH in Zürich, where most of this paper has been written, in particular for the support of the first author during the year 2011-2012. We thank Hee Oh for the instigation to write Section 4.

2 Geometry and dynamics in negative curvature

In this section, we survey briefly the required background on the geometry and dynamics of negatively curved Riemannian manifolds, considered as locally CAT($-\kappa$) spaces (using for instance [BriH] as a general reference), with a particular emphasis on the metric aspects and the regularity properties.

For every $n \geq 2$, we denote by $\mathbb{H}^n_\mathbb{R}$ the real hyperbolic space of dimension n and constant sectional curvature -1.

For every $\epsilon > 0$, we denote by $\mathcal{N}_\epsilon A$ the closed ϵ-neighbourhood of a subset A of a metric space, and by convention $\mathcal{N}_0 A = \overline{A}$. Recall that a map $f : X \to Y$ between two metric spaces is called α-*Hölder*-continuous, where $\alpha \in]0, 1]$, if there exists $c, c' > 0$ such that for every x, y in X with $d(x, y) \leq c'$, we have $d(f(x), f(y)) \leq c\, d(x, y)^\alpha$, and *Hölder*-continuous if there exists $\alpha \in]0, 1]$ such that f is α-Hölder-continuous.

2.1 Geometry of the unit tangent bundle

We denote by $\pi : TN \to N$ the tangent bundle of any (smooth) Riemannian manifold N, and again by $\pi : T^1N \to N$ its unit tangent bundle. Recall that the Levi-Civita connexion ∇ of N gives a decomposition $TTN = V \oplus H$ of the vector bundle $TTN \to TN$ into the direct sum of two smooth vector subbundles $V \to TN$ and $H \to TN$, called vertical and horizontal, such that if $\pi_H : TTN \to V$ is the linear projection of TTN onto V parallelly to H, if H_v and V_v are the fibers of H and V above $v \in TN$, then

- we have $V_v = \operatorname{Ker} T_v \pi = T_v(T_{\pi(v)}N) = T_{\pi(v)}N$;
- the restriction $T\pi_{|H_v} : H_v \to T_{\pi(v)}N$ of the tangent map of π to H_v is a linear isomorphism;
- for every smooth vector field X on N, we have $\nabla_v X = \pi_H \circ TX(v)$.

The manifold TN has a unique Riemannian metric, called *Sasaki's metric*, such that for every $v \in TM$, the map $T\pi_{|H_v} : H_v \to T_{\pi(v)}N$ is isometric, the restriction to V_v of Sasaki's scalar product is the Riemannian scalar product on $T_{\pi(v)}N$, and the decomposition $T_v TN = V_v \oplus H_v$ is orthogonal. We endow the smooth submanifold $T^1 N$ of TN with the induced Riemannian metric, also called *Sasaki's metric*. The fiber $T_x^1 N$ of every $x \in N$ is then isometric to the standard unit sphere $\mathbb{S}^{n-1}$ of the standard Euclidean space $\mathbb{R}^n$, if n is the dimension of N.

The Riemannian measure $d\operatorname{Vol}_{T^1 N}$ of $T^1 N$, called *Liouville's measure*, disintegrates under the fibration $\pi : T^1 N \to N$ over the Riemannian measure $d\operatorname{Vol}_N$ of N, as

$$d\operatorname{Vol}_{T^1 N} = \int_{x \in N} d\operatorname{Vol}_{T_x^1 N} \, d\operatorname{Vol}_N(x),$$

where $d\operatorname{Vol}_{T_x^1 N}$ is the spherical measure on the fiber $T_x^1 N$ of π above $x \in N$. In particular,

$$\operatorname{Vol}(T^1 N) = \operatorname{Vol}(\mathbb{S}^{n-1}) \operatorname{Vol}(N).$$

2.2 Hölder structure on the boundary at infinity

Let $\widetilde{M}$ be a complete simply connected Riemannian manifold with dimension at least 2 and sectional curvature at most -1, and let $x_0 \in \widetilde{M}$. To shorten the exposition, we assume in this survey that $\widetilde{M}$ has pinched negative sectional curvature $-b^2 \leq K \leq -1$ (where $b \in [1, +\infty[$), which is used in particular when working with Gibbs measures in Section 10, see [PaP6] for the extensions. The error term estimates of Section 9 require another geometric assumption, that the sectional curvature of M has bounded derivatives.

We denote by $\partial_\infty \widetilde{M}$ the boundary at infinity of $\widetilde{M}$, with its usual Hölder structure when the sectional curvature of M has bounded derivatives and its usual conformal structure, which we describe below. Recall that a *Hölder structure* on a topological manifold X is a maximal atlas of charts (U, φ), where $\varphi : U \to \varphi(U)$ is a homeomorphism between an open subset U of X and an open subset of a fixed smooth Riemannian manifold, such that the transition maps are α-Hölder homeomorphisms for some $\alpha > 0$.

Two geodesic rays $\rho, \rho' : [0, +\infty[\to \widetilde{M}$ are *asymptotic* if their images are at finite Hausdorff distance, or equivalently if there exist $c > 0$ and $t_0 \in \mathbb{R}$ such that the inequality $d(\rho(t), \rho'(t + t_0)) \le c\,e^{-t}$ holds for all $t \in [\max\{0, -t_0\}, +\infty[$. The *boundary at infinity* $\partial_\infty \widetilde{M}$ of $\widetilde{M}$ is the quotient topological space of the space of geodesic rays, endowed with the compact-open topology, by the equivalence relation "to be asymptotic to". The asymptotic class of a geodesic ray is called its *point at infinity*. For all $x \in M$ and $\xi \in \partial_\infty \widetilde{M}$, there exists a unique geodesic ray with origin x and point at infinity ξ, whose image we denote by $[x, \xi[$. Given two distinct points at infinity $\xi, \eta \in \partial_\infty \widetilde{M}$, there exists a unique (up to translation on the source) geodesic line $\rho : \mathbb{R} \to \widetilde{M}$ such that the points at infinity of the geodesic rays $t \mapsto \rho(-t)$ and $t \mapsto \rho(t)$, $t \in [0, +\infty[$, are ξ and η, respectively. We denote the image of such a geodesic line by $]\xi, \eta[$.

For every $x \in \widetilde{M}$, the map $\theta_x : T^1_x \widetilde{M} \to \partial_\infty \widetilde{M}$, which sends $v \in T^1_x \widetilde{M}$ to the point at infinity of the geodesic ray with initial tangent vector v, is a homeomorphism. We define the angle $\angle_x(y, z)$ of two geodesic segments or rays with the same origin x and endpoints $y, z \in (\widetilde{M} - \{x\}) \cup \partial_\infty \widetilde{M}$ as the angle of their tangent vectors at x. The disjoint union $\widetilde{M} \cup \partial_\infty \widetilde{M}$ has a unique compact metrisable topology, inducing the original topologies on $\widetilde{M}$ and on $\partial_\infty \widetilde{M}$, such that a sequence of points $(y_n)_{n \in \mathbb{N}}$ in $\widetilde{M}$ converges to a point $\xi \in \partial_\infty \widetilde{M}$ if and only if $\lim_{n \to +\infty} d(y_n, x_0) = +\infty$ and $\lim_{n \to +\infty} \angle_{x_0}(y_n, \xi) = 0$. An isometry γ of $\widetilde{M}$ extends uniquely to a homeomorphism of $\widetilde{M} \cup \partial_\infty \widetilde{M}$, and we will also denote by γ its extension to the boundary at infinity.

When the sectional curvature of M, besides being pinched negative, has bounded derivatives, it is known since Anosov (see also [Brin], [PauPS, Section 7.1]) that the maps $\theta_x^{-1} \circ \theta_{x'} : T^1_{x'} M \to T^1_x M$, for all $x, x' \in \widetilde{M}$, are α-Hölder homeomorphisms for some $\alpha > 0$. Hence there is then a unique Hölder structure on the topological manifold $\partial_\infty \widetilde{M}$ such that θ_x is a Hölder homeomorphism, for every $x \in \widetilde{M}$. Furthermore, the isometries of $\widetilde{M}$ are then α-Hölder homeomorphisms of $\partial_\infty \widetilde{M}$ for some $\alpha > 0$.

2.3 Conformal structure on the boundary at infinity

Let us now define the natural conformal structure on $\partial_\infty \widetilde{M}$. Recall that two distances d and δ on a set Z are called *conformally equivalent* if they induce the same topology and if for every $z_0 \in Z$, the limit $\lim_{x \to z_0, x \ne z_0} \frac{d(x, z_0)}{\delta(x, z_0)}$ exists and is strictly positive. The relation "to be conformally equivalent to" is an equivalence relation on the set of distances on Z, and a *conformal structure* on Z is an equivalence class thereof.

Let Z and Z' be two sets endowed with a conformal structure, and let d and d' be distances on Z and Z' representing them. A bijection $\gamma : Z \to Z'$ is *conformal* if the distances d and $\gamma^* d' : (x, y) \mapsto d'(\gamma x, \gamma y)$ are conformally equivalent. This does not depend on the choice of representatives d and d' of the conformal structures of Z and Z'.

The *Busemann cocycle* of $\widetilde{M}$ is the continuous map $\beta : \widetilde{M} \times \widetilde{M} \times \partial_\infty \widetilde{M} \to \mathbb{R}$ defined by

$$(x, y, \xi) \mapsto \beta_\xi(x, y) = \lim_{t \to +\infty} d(\rho_t, x) - d(\rho_t, y),$$

where $\rho : t \mapsto \rho_t$ is any geodesic ray with point at infinity ξ. The above limit exists and is independent of ρ. The Busemann cocycle is Hölder-continuous when the sectional curvature of M has bounded derivatives, and satisfies the following equivariance and cocycle properties:

$$\beta_{\gamma\xi}(\gamma x, \gamma y) = \beta_\xi(x, y) \quad \text{and} \quad \beta_\xi(x, y) + \beta_\xi(y, z) = \beta_\xi(x, z), \qquad (1)$$

for all $\xi \in \partial_\infty \widetilde{M}$, all $x, y, z \in \widetilde{M}$ and every isometry γ of $\widetilde{M}$. In particular, $\beta_\xi(y, x) = -\beta_\xi(x, y)$. By the triangular inequality, we have

$$|\beta_\xi(x, y)| \leq d(x, y). \qquad (2)$$

If y is a point in the (image of the) geodesic ray from x to ξ, then $\beta_\xi(x, y) = d(x, y)$. For every $y \in \widetilde{M}$ and $\xi \in \partial_\infty \widetilde{M}$, the map $x \mapsto \beta_\xi(x, y)$ is smooth and 1-Lipschitz.

For every $\xi \in \partial_\infty \widetilde{M}$, the *horospheres centered at* ξ are the level sets of the map $y \mapsto \beta_\xi(y, x_0)$ from $\widetilde{M}$ to $\mathbb{R}$, and the (closed) *horoballs centered at* ξ are its sublevel sets. Horoballs are closed (strictly) convex subsets of $\widetilde{M}$. A horosphere centered at ξ is a smooth hypersurface of $\widetilde{M}$, orthogonal to the geodesic lines having ξ as a point at infinity. In the upper halfspace model of the real hyperbolic space $\mathbb{H}^n_\mathbb{R}$, the horospheres are the horizontal affine hyperplanes therein or the Euclidean spheres therein tangent to the horizontal coordinate hyperplane, with the point of tangency removed. In the ball model of the real hyperbolic space $\mathbb{H}^n_\mathbb{R}$, the horospheres (respectively horoballs) are the Euclidean spheres (respectively balls) tangent to the unit sphere and contained in the unit ball, with the point of tangency removed.

The horoballs are limits of big balls (and their centres the limits of the centres thereof, explaining the terminology). More precisely, if ρ is a geodesic ray in $\widetilde{M}$ with point at infinity ξ, if $B(t)$ is the ball of centre $\rho(t)$ and radius t, then the map $t \mapsto B(t)$ converges to the horoball HB of centre ξ containing $\rho(0)$ in its boundary, for Chabauty's topology on closed subsets of $\widetilde{M}$ (that is, for the Hausdorff convergence on compact subsets: for every compact subset K

of $\widetilde{M}$, as $t \to +\infty$, the closed subset $(B(t) \cap K) \cup {}^c\overline{K}$ converges to the closed subset $(HB \cap K) \cup {}^c\overline{K}$ for the Hausdorff distance).

For every $x \in \widetilde{M}$, for all distinct $\xi, \eta \in \partial_\infty \widetilde{M}$, the *visual distance* between ξ and η seen from x is

$$d_x(\xi, \eta) = \lim_{t \to +\infty} e^{-\frac{1}{2}\left(d(x, \rho_\xi(t)) + d(x, \rho_\eta(t)) - d(\rho_\xi(t), \rho_\eta(t))\right)},$$

for any geodesic rays ρ_ξ and ρ_η with point at infinity ξ and η, respectively. Equivalently, with $t \mapsto \rho_t$ and $t \mapsto \rho'_t$ the geodesic rays with origin x converging to ξ and η, we have

$$d_x(\xi, \eta) = \lim_{t \to +\infty} e^{\frac{1}{2}d(\rho_t, \rho'_t) - t}.$$

Again equivalently, if u is any point on the geodesic line between ξ and η, then

$$d_x(\xi, \eta) = e^{-\frac{1}{2}(\beta_\xi(x, u) + \beta_\eta(x, u))}. \tag{3}$$

Define $d_x(\xi, \eta) = 0$ if $\xi = \eta$.

For every $x \in \widetilde{M}$, the above limits exist and the three formulas coincide. The map $d_x : \partial_\infty \widetilde{M} \times \partial_\infty \widetilde{M} \to [0, +\infty[$ is a distance, inducing the original topology on $\partial_\infty \widetilde{M}$. For all $x, y \in \widetilde{M}$, for all distinct $\xi, \eta \in \partial_\infty \widetilde{M}$, and for every isometry γ of $\widetilde{M}$, we have

$$e^{-d(x, \,]\xi, \, \eta[)} \le d_x(\xi, \eta) \le (1 + \sqrt{2}) e^{-d(x, \,]\xi, \, \eta[)},$$

$$\frac{d_x(\xi, \eta)}{d_y(\xi, \eta)} = e^{-\frac{1}{2}(\beta_\xi(x, y) + \beta_\eta(x, y))}, \tag{4}$$

$$d_{\gamma x}(\gamma \xi, \gamma \eta) = d_x(\xi, \eta). \tag{5}$$

It follows from Equation (4) that the visual distances d_x for $x \in \widetilde{M}$ belong to the same conformal structure on $\partial_\infty \widetilde{M}$. It follows from Equation (4) and Equation (5) that the (boundary extensions of) the isometries of $\widetilde{M}$ are conformal bijections for this conformal structure. Furthermore, these equations and Equation (2) imply that the isometries of $\widetilde{M}$ are bilipschitz homeomorphisms for any visual distance: for all $x \in \widetilde{M}$, for all $\xi, \eta \in \partial_\infty \widetilde{M}$, we have

$$e^{-d(x, \gamma x)} d_x(\xi, \eta) \le d_x(\gamma \xi, \gamma \eta) \le e^{d(x, \gamma x)} d_x(\xi, \eta).$$

2.4 Stable and unstable leaves of the geodesic flow

We now turn to the description of the dynamics of the geodesic flow on $\widetilde{M}$.

The unit tangent bundle $T^1 N$ of a complete Riemannian manifold N can be identified with the set of locally geodesic lines (parametrised by arclength) $\ell : \mathbb{R} \to N$ in N, endowed with the compact-open topology. More precisely, we identify a locally geodesic line ℓ and its (unit) tangent vector $\dot{\ell}(0)$ at time $t = 0$ and, conversely, any $v \in T^1 N$ is the tangent vector at time $t = 0$ of a unique locally geodesic line. We will use this identification without mention in this survey. In particular, the base point projection $\pi : T^1 N \to N$ is given by $\pi(\ell) = \ell(0)$.

The *geodesic flow* on $T^1 N$ is the smooth 1-parameter group $(g^t)_{t \in \mathbb{R}}$, where $g^t \ell(s) = \ell(s + t)$, for all $\ell \in T^1 N$ and $s, t \in \mathbb{R}$. We denote by $\iota : T^1 N \to T^1 N$ the *antipodal (flip) map* $v \mapsto -v$. We have $\iota \circ g^t = g^{-t} \circ \iota$. The isometry group of N acts on the space of locally geodesic lines in N by postcomposition: $(\gamma, \ell) \mapsto \gamma \circ \ell$, and this action commutes with the geodesic flow and the antipodal map.

For every unit tangent vector $v \in T^1 \widetilde{M}$, let $v_- = v(-\infty)$ and $v_+ = v(+\infty)$ be the two endpoints in the sphere at infinity of the geodesic line defined by v. Let $\partial^2_\infty \widetilde{M}$ be the open subset of $\partial_\infty \widetilde{M} \times \partial_\infty \widetilde{M}$ which consists of pairs of distinct points at infinity, with the restriction of the product Hölder structure when the sectional curvature of M has bounded derivatives. *Hopf's parametrisation* (see [Hop]) of $T^1 \widetilde{M}$ is the homeomorphism from $T^1 \widetilde{M}$ to $\partial^2_\infty \widetilde{M} \times \mathbb{R}$ sending $v \in T^1 \widetilde{M}$ to the triple $(v_-, v_+, t) \in \partial^2_\infty \widetilde{M} \times \mathbb{R}$, where t is the signed (algebraic) distance of $\pi(v)$ from the closest point p_{v, x_0} to x_0 on the (oriented) geodesic line defined by v. Hopf's parametrisation is a Hölder homeomorphism when the sectional curvature of M has bounded derivatives. In this survey, we will identify an element of $T^1 \widetilde{M}$ with its image by Hopf's parametrisation. The geodesic flow acts by $g^s(v_-, v_+, t) = (v_-, v_+, t + s)$ and, for every isometry γ of $\widetilde{M}$, the image of γv is $(\gamma v_-, \gamma v_+, t + t_{\gamma, v})$, where $t_{\gamma, v}$ is the signed distance from $\gamma p_{v, x_0}$ to $p_{\gamma v, x_0}$. Furthermore, in these coordinates, the antipodal map ι is $(v_-, v_+, t) \mapsto (v_+, v_-, -t)$.

The *strong stable manifold* of $v \in T^1 \widetilde{M}$ is

$$W^{\mathrm{ss}}(v) = \{v' \in T^1 \widetilde{M} : d(v(t), v'(t)) \to 0 \text{ as } t \to +\infty\},$$

and the *strong unstable manifold* of v is

$$W^{\mathrm{su}}(v) = \{v' \in T^1 \widetilde{M} : d(v(t), v'(t)) \to 0 \text{ as } t \to -\infty\},$$

The projections in $\widetilde{M}$ of the strong unstable and strong stable manifolds of $v \in T^1 \widetilde{M}$, denoted by $H_-(v) = \pi(W^{\mathrm{su}}(v))$ and $H_+(v) = \pi(W^{\mathrm{ss}}(v))$, are called, respectively, the *unstable and stable horospheres* of v, and are the horospheres containing $\pi(v)$ centered at v_- and v_+, respectively. The unstable horosphere of v coincides with the zero set of the map $f_- : x \mapsto \beta_{v_-}(x, \pi(v))$, and, similarly, the stable horosphere of v coincides with the zero set of $f_+ : x \mapsto \beta_{v_+}(x, \pi(v))$. The corresponding sublevel sets $HB_-(v) = f_-^{-1}(] - \infty, 0])$ and $HB_+(v) = f_+^{-1}(] - \infty, 0])$ are called the *unstable and stable horoballs* of v.

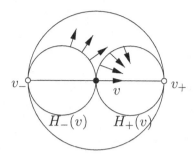

The union for $t \in \mathbb{R}$ of the images under g^t of the strong stable manifold of $v \in T^1 \tilde{M}$ is the *stable manifold* $W^s(v) = \bigcup_{t \in \mathbb{R}} g^t W^{ss}(v)$ of v, which consists of the elements $v' \in T^1 \tilde{M}$ with $v'_+ = v_+$. Similarly, the union of the images under the geodesic flow at all times of the strong unstable manifold of v is the *unstable manifold* $W^u(v)$ of v, which consists of the elements $v' \in T^1 \tilde{M}$ with $v'_- = v_-$.

The subspaces $W^{ss}(v)$ and $W^{su}(v)$ (which are the lifts by the inner and outer unit normal vectors of the unstable and stable horospheres of v, respectively), as well as $W^s(v)$ and $W^u(v)$, are smooth submanifolds of $T^1 \tilde{M}$. The maps from $\mathbb{R} \times W^{ss}(v)$ to $W^s(v)$ defined by $(t, v') \mapsto g^t v'$ and from $\mathbb{R} \times W^{su}(v)$ to $W^u(v)$ defined by $(t, v') \mapsto g^t v'$ are smooth diffeomorphisms. We have $W^{ss}(\iota v) = \iota W^{su}(v)$.

Hamenstädt's distance on stable and unstable leaves

For every $v \in T^1 \tilde{M}$, let $d_{W^{ss}(v)}$ be *Hamenstädt's distance* on the strong stable leaf of v, defined as follows (see [Ham1], [HeP1, Appendix], as well as [HeP3, Section 2.2] for a generalisation when the horosphere $H_+(v)$ is replaced by the boundary of any nonempty closed convex subset): for all $w, w' \in W^{ss}(v)$, we have

$$d_{W^{ss}(v)}(w, w') = \lim_{t \to +\infty} e^{\frac{1}{2} d(w(-t),\, w'(-t)) - t}.$$

This limit exists, and Hamenstädt's distance is a distance inducing the original topology on $W^{ss}(v)$. For all $w, w' \in W^{ss}(v)$ and for every isometry γ of $\tilde{M}$, we have

$$d_{W^{ss}(\gamma v)}(\gamma w, \gamma w') = d_{W^{ss}(v)}(w, w').$$

For all $v \in T^1 \tilde{M}$, $s \in \mathbb{R}$ and $w, w' \in W^{ss}(v)$, we have

$$d_{W^{ss}(g^s v)}(g^s w, g^s w') = e^{-s} d_{W^{ss}(v)}(w, w').$$

For every horosphere H in $\widetilde{M}$ with center H_∞, we also have a distance d_H on the open subset $\partial_\infty \widetilde{M} - \{H_\infty\}$ defined by

$$d_H(\xi, \eta) = \lim_{t \to +\infty} e^t \, d_{\rho_t}(\xi, \eta) = \lim_{t \to +\infty} e^{\frac{1}{2}d(\xi_t, \eta_t)-t},$$

where $t \mapsto \rho_t$ is any geodesic ray with origin a point of H and point at infinity H_∞, and $t \mapsto \xi_t$ and $t \mapsto \eta_t$ are the geodesic lines in $\widetilde{M}$ with origin H_∞, passing at time $t = 0$ through H, and with endpoints ξ and η, respectively. Using the homeomorphism from $W^{ss}(v)$ to $\partial_\infty \widetilde{M} - \{v_+\}$ defined by $w \mapsto w_-$, we have

$$d_{W^{ss}(v)}(w, w') = d_{H_+(v)}(w_-, w'_-).$$

The distance d_H and the restriction of any visual distance to $\partial_\infty \widetilde{M} - \{H_\infty\}$ are conformally equivalent, since for all $x \in H$ and $\xi, \eta \in \partial_\infty \widetilde{M} - \{H_\infty\}$, with the above notation $t \mapsto \xi_t$ and $t \mapsto \eta_t$, we have

$$\frac{d_H(\xi, \eta)}{d_x(\xi, \eta)} = e^{-\frac{1}{2}(\beta_\xi(\xi_0, x)+\beta_\eta(\eta_0, x))}.$$

The Anosov property of the geodesic flow

The strong stable manifolds, stable manifolds, strong unstable manifolds and unstable manifolds are the (smooth) leaves of continuous foliations on $T^1\widetilde{M}$, invariant under the geodesic flow and the isometry group of $\widetilde{M}$, denoted by $\mathcal{W}^{ss}$, $\mathcal{W}^s$, $\mathcal{W}^{su}$ and $\mathcal{W}^u$, respectively. They are Hölder foliations when the sectional curvature of M has bounded derivatives (see for instance [PauPS, Section 7.1]). When $\widetilde{M}$ is a symmetric space (that is, up to homothety, when $\widetilde{M}$ is isometric to the real, complex, quaternionic hyperbolic n-space or to the octonionic hyperbolic plane), then the strong stable, stable, strong unstable and unstable foliations are smooth. But in general, the Hölder regularity cannot be much improved, as we will explain in Section 2.5.

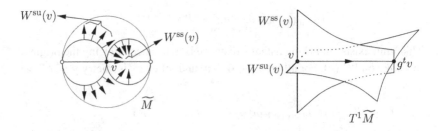

Let $N = T^1\widetilde{M}$. The vector field $Z : N \to TN$ defined by $v \mapsto Z(v) = \frac{d}{dt}\big|_{t=0} g^t v$ is called the *geodesic vector field*. The geodesic flow $(g^t)_{t\in\mathbb{R}}$ on the

Riemannian manifold N is a *contact Anosov flow*. That is, the vector bundle $TN \to N$ is the direct sum of three topological vector subbundles $TN = E_{\mathrm{su}} \oplus E_0 \oplus E_{\mathrm{ss}}$ that are invariant under $(g^t)_{t \in \mathbb{R}}$, where $E_0 \cap T_v N = \mathbb{R}Z(v)$, $E_{\mathrm{su}} \cap T_v N = T_v W^{\mathrm{su}}(v)$, $E_{\mathrm{ss}} \cap T_v N = T_v W^{\mathrm{ss}}(v)$, and there exist two constants $c, \lambda > 0$ such that for every $t > 0$, we have (see the above picture on the right)

$$\|T_v g^t{}_{|E_{\mathrm{ss}}}\| \leq c\, e^{-\lambda t} \text{ and } \|T_v g^{-t}{}_{|E_{\mathrm{su}}}\| \leq c\, e^{-\lambda t}.$$

Furthermore, if α is the differential 1-form on N defined by $\alpha_{|E_{\mathrm{su}} \oplus E_{\mathrm{ss}}} = 0$ and $\alpha(Z) = 1$, called *Liouville's* 1-*form*, then $\alpha \wedge (d\alpha)^{n-1}$, where n is the dimension of M, is a volume form on N, which is invariant under the geodesic flow. Thus, the strong stable leaves are contracted by the geodesic flow, and the strong unstable leaves are dilated. See for instance [KaH] for more information.

2.5 Discrete isometry groups

Let Γ be a discrete group of isometries of $\widetilde{M}$, which is *nonelementary*, that is, it preserves no set of one or two points in $\widetilde{M} \cup \partial_\infty \widetilde{M}$. To shorten the exposition, we will assume in this survey that Γ has no torsion, though this assumption is not necessary (see [PaP4, PaP6, PaP7, BrPP] for the extension), and is useful for some arithmetic applications.

Let us denote the quotient space of $\widetilde{M}$ under Γ by $M = \Gamma \backslash \widetilde{M}$, which is a smooth Riemannian manifold since Γ is torsion free. We also say that the manifold M is *nonelementary* if Γ is nonelementary.

We denote by $\Lambda\Gamma$ the *limit set* of Γ, that is, the set of accumulation points in $\partial_\infty \widetilde{M}$ of any orbit Γx of a point x of $\widetilde{M}$ under Γ. It is the smallest nonempty closed Γ-invariant subset of $\partial_\infty \widetilde{M}$. The *critical exponent* of Γ is

$$\delta_\Gamma = \lim_{n \to +\infty} \frac{1}{n} \ln \mathrm{Card}\{\gamma \in \Gamma : d(x_0, \gamma x_0) \leq n\}.$$

The above limit exists (see [Rob1]), and the critical exponent is positive (since Γ is nonelementary), finite (since M has a finite lower bound on its sectional curvatures, see for instance [Bowd]), and independent of the base point x_0.

Since Γ acts without fixed points on $\widetilde{M}$, we have an identification $\Gamma \backslash T^1 \widetilde{M} = T^1 M$, and we again denote by $\mathcal{W}^{\mathrm{ss}}$, $\mathcal{W}^{\mathrm{s}}$, $\mathcal{W}^{\mathrm{su}}$ and $\mathcal{W}^{\mathrm{u}}$ the continuous foliations of $T^1 M$ induced by the corresponding ones in $T^1 \widetilde{M}$ (which are Hölder foliations when the sectional curvature of M has bounded derivatives). We use the notation $(g^t)_{t \in \mathbb{R}}$ also for the geodesic flow on $T^1 M$. We again denote by $\iota : T^1 M \to T^1 M$ the antipodal (flip) map $v \mapsto -v$, which also anti-commutes with the geodesic flow.

Let us conclude this section by explaining some rigidity results on the regularity of the foliations $\mathcal{W}^{ss}$, $\mathcal{W}^s$, $\mathcal{W}^{su}$ and $\mathcal{W}^u$. Anosov has proved that if M is compact, then the vector subbundles E_{su} and E_{ss} are Hölder-continuous (see for instance [PauPS, Th. 7.3] when M is only assumed to have pinched negative sectional curvature with bounded derivatives). If M is a compact surface, Hurder and Katok [HuK, Theo. 3.1, Coro. 3.5] have proved that these subbundles are $C^{1,\alpha}$ for every $\alpha \in {]}0, 1{[}$ (see also [HiP]), and that if they are $C^{1,1}$, then they are C^∞. Ghys [Ghy, p. 267] has proved that if M is a compact surface, and if the stable foliation of $T^1 M$ is C^2, then the geodesic flow is C^∞-conjugate to the geodesic flow of a hyperbolic surface. In higher dimension, we have the following result.

Theorem 1 (Benoist-Foulon-Labourie [BFL]) *Let M be a compact negatively curved Riemannian manifold. If the stable foliation of $T^1 M$ is smooth, then the geodesic flow of M is C^∞-conjugate to the geodesic flow of a Riemannian symmetric manifold with negative curvature.*   □

3 Common perpendiculars of convex sets

Let M be a complete nonelementary connected Riemannian manifold of dimension at least 2, with pinched negative sectional curvature $-b^2 \leq K \leq -1$. Let $\widetilde{M} \to M$ be a universal Riemannian cover of M, so that $\widetilde{M}$ is complete simply connected with the same curvature bounds, and let Γ be its covering group, so that Γ is a discrete, torsionfree, nonelementary group of isometries of $\widetilde{M}$.

3.1 Convex subsets

Let $\widetilde{C}$ be a nonempty *proper* (that is, different from $\widetilde{M}$) closed convex subset of $\widetilde{M}$. Recall that a subset A of $\widetilde{M}$ is said to be *convex* if (the image of) any geodesic segment with endpoints in A is contained in A. We denote by $\partial\widetilde{C}$ the boundary of $\widetilde{C}$ in $\widetilde{M}$ and by $\partial_\infty\widetilde{C}$ its set of points at infinity (the set of endpoints of geodesic rays contained in $\widetilde{C}$). We say that the Γ-orbit of $\widetilde{C}$ is *locally finite* if, with $\Gamma_{\widetilde{C}}$ the stabiliser of $\widetilde{C}$ in Γ, for every compact subset K of $\widetilde{M}$, the number of right cosets $[\gamma] \in \Gamma/\Gamma_{\widetilde{C}}$ such that $\gamma\widetilde{C}$ meets K is finite.

Natural examples of convex subsets of $\widetilde{M}$ include the points, the balls, the horoballs, the totally geodesic subspaces of $\widetilde{M}$ and the convex hulls in $\widetilde{M}$ of the limit sets of nonelementary subgroups of Γ. Recall that the *convex hull* of a subset A of $\partial_\infty\widetilde{M}$ with at least two points is the smallest closed convex subset of $\widetilde{M}$ that contains A in its set of points at infinity.

Let $P_{\widetilde{C}} : \widetilde{M} \cup (\partial_\infty \widetilde{M} - \partial_\infty \widetilde{C}) \to \widetilde{C}$ be the closest point map: if $\xi \in \partial_\infty \widetilde{M} - \partial_\infty \widetilde{C}$, then $P_{\widetilde{C}}(\xi)$ is the unique point in $\widetilde{C}$ that minimises the map $x \mapsto \beta_\xi(x, x_0)$ from $\widetilde{C}$ to $\mathbb{R}$. For every isometry γ of $\widetilde{M}$, we have $P_{\gamma \widetilde{C}} \circ \gamma = \gamma \circ P_{\widetilde{C}}$. The closest point map is continuous in the topology of $\widetilde{M} \cup \partial_\infty \widetilde{M}$.

The *outer unit normal bundle* $\partial_+^1 \widetilde{C}$ of $\widetilde{C}$ is the subspace of $T^1 \widetilde{M}$ consisting of the geodesic lines $v \colon \mathbb{R} \to \widetilde{M}$ with $v(0) \in \partial \widetilde{C}$, $v_+ \notin \partial_\infty \widetilde{C}$ and $P_{\widetilde{C}}(v_+) = v(0)$. Note that $\pi(\partial_+^1 \widetilde{C}) = \partial \widetilde{C}$, and that for all isometries γ of $\widetilde{M}$, we have $\partial_+^1(\gamma \widetilde{C}) = \gamma \, \partial_+^1 \widetilde{C}$. In particular, $\partial_+^1 \widetilde{C}$ is invariant under the isometries of $\widetilde{M}$ that preserve $\widetilde{C}$. When $\widetilde{C} = HB_-(v)$ is the unstable horoball of $v \in T^1 \widetilde{M}$, then $\partial_+^1 \widetilde{C}$ is the strong unstable manifold $W^{\mathrm{su}}(v)$ of v, and similarly, $W^{\mathrm{ss}}(v) = \iota \, \partial_+^1 HB_+(v)$.

The restriction of $P_{\widetilde{C}}$ to $\partial_\infty \widetilde{M} - \partial_\infty \widetilde{C}$ (which is not necessarily injective) has a natural lift to a homeomorphism

$$P_{\widetilde{C}}^+ : \partial_\infty \widetilde{M} - \partial_\infty \widetilde{C} \to \partial_+^1 \widetilde{C}$$

such that $\pi \circ P_{\widetilde{C}}^+ = P_{\widetilde{C}}$. The inverse of $P_{\widetilde{C}}^+$ is the *positive endpoint map* $v \mapsto v_+$ from $\partial_+^1 \widetilde{C}$ to $\partial_\infty \widetilde{M} - \partial_\infty \widetilde{C}$. In particular, $\partial_+^1 \widetilde{C}$ is a topological submanifold of $T^1 \widetilde{M}$, and a Hölder submanifold when the sectional curvature of M has bounded derivatives. For every $s \geq 0$, the geodesic flow induces a homeomorphism $g^s : \partial_+^1 \widetilde{C} \to \partial_+^1 \mathcal{N}_s \widetilde{C}$, which is a Hölder homeomorphism when the sectional curvature of M has bounded derivatives. For every isometry γ of $\widetilde{M}$, we have $P_{\gamma \widetilde{C}}^+ \circ \gamma = \gamma \circ P_{\widetilde{C}}^+$. When $\widetilde{C}$ has nonempty interior and $\mathrm{C}^{1,1}$ boundary, then $\partial_+^1 \widetilde{C}$ is the Lipschitz submanifold of $T^1 \widetilde{M}$ consisting of the outer unit normal vectors to $\partial \widetilde{C}$, and the map $P_{\widetilde{C}}$ itself is a homeomorphism (between $\partial_\infty \widetilde{M} - \partial_\infty \widetilde{C}$ and $\partial \widetilde{C}$). This holds when $\widetilde{C}$ is the closed η-neighbourhood of any nonempty convex subset of $\widetilde{M}$ with $\eta > 0$ (see [Fed, Theo. 4.8(9)], [Wal, p. 272]).

In this survey, we define a *properly immersed closed convex subset C of M* as the data of a nonempty proper closed convex subset $\widetilde{C}$ of $\widetilde{M}$, with locally finite Γ-orbit, and of the locally isometric proper immersion $C = \Gamma_{\widetilde{C}} \backslash \widetilde{C} \to M$ induced by the inclusion of $\widetilde{C}$ in $\widetilde{M}$ and the Riemannian covering map $\widetilde{M} \to M$. (To simplify the exposition, we do not allow in this survey the replacement of $\Gamma_{\widetilde{C}}$ by one of its finite index subgroups as it is done in [PaP7, Section 3.2], and [BrPP] even though this is sometimes useful.) By abuse, when no confusion is possible, we will again denote by C the image of this immersion. We define $\partial_+^1 C = \Gamma_{\widetilde{C}} \backslash \partial_+^1 \widetilde{C}$, which comes with a proper immersion $\partial_+^1 C \to T^1 M$ induced by the inclusion of $\partial_+^1 \widetilde{C}$ in $T^1 \widetilde{M}$ and the covering map $T^1 \widetilde{M} \to T^1 M$. By abuse also, we will again denote by $\partial_+^1 C$ the image of this immersion.

3.2 The general counting problem

Let C_+, C_- be two properly immersed closed convex subsets of M. A locally geodesic path $c : [0, T] \to M$ is a *common perpendicular* from C_- to C_+ if $\dot{c}(0) \in \partial^1_+ C_-$ and $\dot{c}(T) \in \iota\, \partial^1_+ C_+$. For every $s \geq 0$, we denote by $\mathrm{Perp}_{C_-, C_+}(s)$ the set of common perpendiculars from C_- to C_+ of length at most s. Each common perpendicular c from C_- to C_+ has a *multiplicity* $m(c)$, defined as follows. If C_- and C_+ are the images in M of two nonempty proper closed convex subsets $\widetilde{C}_-$ and $\widetilde{C}_+$ of $\widetilde{M}$ with locally finite Γ-orbits, respectively, then $m(c)$ is the number of (left) orbits under Γ of pairs $([\alpha], [\beta])$ in $\Gamma / \Gamma_{\widetilde{C}_-} \times \Gamma / \Gamma_{\widetilde{C}_+}$ such that the closed convex subsets $\alpha \widetilde{C}_-$ and $\beta \widetilde{C}_+$ have a (unique) common perpendicular whose image by $\widetilde{M} \to M$ is c. (Multiplicities are also useful when Γ is allowed to have torsion, or when the stabilizers $\Gamma_{\widetilde{C}_\pm}$ are replaced by finite index subgroups, or when $C_\pm$ is replaced by finite families of such convex subsets, see [PaP7, Section 3.2] and [BrPP] for a general version.)

In particular, any locally geodesic path is a common perpendicular of its endpoints (with multiplicity 1), since the outer unit normal bundle of a point is equal to its unit tangent sphere. If C_- and C_+ have nonempty interior and $C^{1,1}$ smooth boundary (in the appropriate sense for immersed subsets), the above definition of common perpendicular agrees with the usual definition: a common perpendicular exits C_- perpendicularly to the boundary of C_- at its initial point and it enters C_+ perpendicularly to the boundary of C_+ at its terminal point.

We study in this survey the asymptotic behaviour, as $s \to +\infty$, of the number

$$\mathcal{N}(s) = \mathcal{N}_{C_-, C_+}(s) = \sum_{c \in \mathrm{Perp}_{C_-, C_+}(s)} m(c)$$

of common perpendiculars, counted with multiplicities, from C_- to C_+, of length at most s. We refer to [PaP7, Section 3.3] for more general counting functions.

Problems of this kind have been studied in various forms in the literature since the 1950's and in a number of recent works, sometimes in a different guise, as demonstrated in the examples below. These examples indicate that the general form of the counting results is $\mathcal{N}(s) \sim \kappa\, e^{\delta s}$, where $\delta = \delta_\Gamma$ is the critical exponent of Γ and $\kappa > 0$ is a constant. Landau's notation $f(s) \sim g(s)$ (as $s \to +\infty$) means as usual that $g(s) \neq 0$ for s big enough, and that the ratio $\frac{f(s)}{g(s)}$ converges to 1 as $s \to +\infty$.

Observing that for $t \geq 2\epsilon$, we have

$$\mathcal{N}_{\mathcal{N}_\epsilon(C_-), \mathcal{N}_\epsilon(C_+)}(t - 2\epsilon) \leq \mathcal{N}_{C_-, C_+}(t)$$
$$\leq \mathcal{N}_{\mathcal{N}_\epsilon(C_-), \mathcal{N}_\epsilon(C_+)}(t - 2\epsilon) + \mathcal{N}_{C_-, C_+}(2\epsilon),$$

we could replace the convex sets C_- and C_+ by their ϵ-neighbourhoods for some fixed (small) positive ϵ, and then assume that C_- and C_+ have $C^{1,1}$ boundaries and use the more conventional definition of common perpendicular. However, it is more natural to work directly with the given convex sets instead of, for example, replacing points by small balls.

Let $\mathrm{Perp}'(C_-, C_+)$ be the set $\mathrm{Perp}(C_-, C_+) = \bigcup_{s \geq 0} \mathrm{Perp}_{C_-, C_+}(s)$ where each element c has been replaced by $m(c)$ copies of it. The family $(\ell(\alpha))_{\alpha \in \mathrm{Perp}'(C_-,C_+)}$ is called the *marked ortholength spectrum* of (C_-, C_+). The set of lengths (with multiplicities) of elements of $\mathrm{Perp}(C_-, C_+)$ is called the *ortholength spectrum* of (C_-, C_+). This second set has been introduced by Basmajian [Bas] (under the name "full orthogonal spectrum") when M has constant curvature, and C_- and C_+ are disjoint or equal, embedded, totally geodesic hypersurfaces or embedded horospherical cusp neighbourhoods or embedded balls. Using the complex lengths of the common perpendiculars between all closed geodesics available in hyperbolic 3-manifolds, and additional combinatorial data, Meyerhoff [Mey] caracterizes the isometry classes of closed hyperbolic 3-manifolds. See also [Brid, BriK, Cal] for nice identities relating the volume of M and the ortholength spectrum, when M is a compact hyperbolic manifold with totally geodesic boundary and $C_- = C_+ = \partial M$. The aim of this paper is hence to survey the asymptotic properties of the (marked) ortholength spectra.

3.3 Counting orbit points in a ball

If $C_- = \{\bar{p}\}$ and $C_+ = \{\bar{q}\}$ are singletons in M, then

$$\mathcal{N}(s) = \mathrm{Card}(B(p, s) \cap \Gamma q),$$

for any lifts p and q of $\bar{p}$ and $\bar{q}$ in $\widetilde{M}$. When $\widetilde{M} = \mathbb{H}^2_\mathbb{R}$ and M is compact and orientable, Huber [Hub, Satz 3] proved that

$$\mathcal{N}(s) \sim \frac{1}{4(g-1)} e^s,$$

where g is the genus of M. His proof makes use of the Dirichlet series $\sum_{\gamma \in \Gamma} \cosh^{-s} d(p, \gamma q)$ and the Tauberian theorem of Wiener-Ikehara [Kor, Section 4,5].

Margulis [Mar1, Theo. 2] (see also [Pol]) generalised Huber's result for all compact connected negatively curved manifolds of arbitrary dimension $n \geq 2$. Note that $\delta = \delta_\Gamma$ is also the topological entropy of the geodesic flow of M, since M is compact. He showed that

$$\mathcal{N}(s) \sim c(p, q) e^{\delta s}$$

for some constant $c(p, q)$ which depends continuously on p and q. He proved that if $\widetilde{M} = \mathbb{H}_{\mathbb{R}}^n$, then

$$\mathcal{N}(s) \sim \frac{\mathrm{Vol}(\mathbb{S}^{n-1})}{2^{n-1}(n-1)\,\mathrm{Vol}(M)}\, e^{(n-1)s}. \tag{6}$$

This agrees with Huber's result in dimension $n = 2$ because the area of a compact genus g surface is $4\pi(g - 1)$. Margulis's proof of the above result established the approach

$$\text{mixing} \to \text{equidistribution} \to \text{counting}$$

that has been used in most of the subsequent results. Roblin [Rob2, p. 56] generalised Margulis's result when Γ is nonelementary, the sectional curvature of M is at most -1, and the Bowen-Margulis measure of $T^1 M$ is finite, and he has an expression for the constant $c(p, q)$ in terms of the Patterson density and the Bowen-Margulis measure, see Section 8 for more details. Lax and Phillips [LaP] obtained an expression with error bounds for the asymptotic behaviour of $\mathcal{N}(s)$ in terms of the eigenvalues of the Laplacian on $\Gamma \backslash \mathbb{H}_{\mathbb{R}}^n$ when Γ is *geometrically finite* (that is, in constant curvature, when Γ is nonelementary and has a fundamental polyhedron with finitely many sides).

3.4 Counting common perpendiculars from a point to a totally geodesic submanifold

Herrmann [Herr, Theo. I] proved for $\widetilde{M} = \mathbb{H}_{\mathbb{R}}^n$, M compact, $C_- = \{p\}$ a singleton, C_+ a compact totally geodesic submanifold of dimension k, an asymptotic estimate

$$\mathcal{N}(s) \sim \frac{2}{n-1}\,\frac{\pi^{(n-k)/2}}{\Gamma\!\left(\frac{n-k}{2}\right)}\,\frac{\mathrm{Vol}(C_+)}{\mathrm{Vol}(M)}\,\frac{e^{(n-1)s}}{2^{n-1}} \tag{7}$$

$$= \frac{\mathrm{Vol}(\mathbb{S}^{n-k-1})\,\mathrm{Vol}(C_+)}{2^{n-1}\,\mathrm{Vol}(M)}\,\frac{e^{(n-1)s}}{n-1}, \tag{8}$$

as $s \to +\infty$. Furthermore, he showed that the endpoints of the common perpendiculars on the totally geodesic submanifold C_+ are evenly distributed in terms of the Riemannian measure of C_+. More precisely, if Ω_+ is a measurable subset of C_+ with boundary of (Lebesgue) measure 0, and if $\mathcal{N}_{p, \Omega_+}(s)$ is the number of those common perpendiculars of $\{p\}$ and C_+ whose terminal endpoints are contained in Ω_+, then, as $s \to +\infty$,

$$\mathcal{N}_{p, \Omega_+}(s) \sim \frac{\mathrm{Vol}(\mathbb{S}^{n-k-1})\,\mathrm{Vol}(\Omega_+)}{2^{n-1}\,\mathrm{Vol}(M)}\,\frac{e^{(n-1)s}}{n-1}. \tag{9}$$

The method of proof was a generalisation of that used by Huber. The asymptotic (7) was also treated in [EM] as an illustration of their equidistribution result, see Theorem 3.

Oh and Shah [OS3] generalised Herrmann's result in dimension 3 for $\widetilde{M} = \mathbb{H}^3_{\mathbb{R}}$ and Γ geometrically finite (in which case $\delta = \delta_\Gamma$ is the Hausdorff dimension of the limit set of Γ, see for instance [Bou]). They showed that, as $s \to +\infty$,

$$\mathcal{N}(s) \sim c(\{p\}, C_+) e^{\delta s}$$

with a constant $c(\{p\}, C_+)$ generalising that of Roblin's result described above. Again, we postpone the description of the constant $c(\{p\}, C_+)$ until Section 8. This result is used in [OS3] to study Γ-invariant families $(P_i)_{i \in I}$ of possibly intersecting circles in $\mathbb{S}^2$, called "circle packings", that consist of a finite number of Γ-orbits such that the family of totally geodesic planes $(P_i^*)_{i \in I}$ in the ball model of $\mathbb{H}^3_{\mathbb{R}}$ with $\partial_\infty P_i^* = P_i$ is locally finite. They consider the counting function

$$N(T) = \mathrm{Card}\{i \in I \; : \; \mathrm{curv}_{\mathbb{S}}(P_i^*) < T\},$$

where the *spherical curvature* $\mathrm{curv}_{\mathbb{S}} P_i^*$ is the cotangent of the angle, at the origin 0 of the ball model of $\mathbb{H}^3_{\mathbb{R}}$, between the common perpendicular between $\{0\}$ and P_i^*, and any geodesic ray from 0 which is tangent to P_i^* at infinity. Elementary hyperbolic geometry (the angle of parallelism formula, see for instance [Bea, p. 147]) implies that

$$\mathrm{curv}_{\mathbb{S}} P_i^* = \sinh d(0, P_i^*),$$

and thus, the above asymptotic estimate of $\mathcal{N}(s)$ is equivalent to $N(T) \sim c(\{p\}, C_+)(2T)^\delta$ as $T \to +\infty$.

3.5 Counting common perpendiculars between horoballs

When $C_- = C_+ = \mathcal{H}$ is a *Margulis cusp neighbourhood* (that is, the image by $\widetilde{M} \to M$ of a Γ-orbit of horoballs, centered at fixed points of parabolic elements of Γ, with pairwise disjoint interiors), then results of [BHP, HeP2, Cos, Rob2] show that if Γ is geometrically finite (see [Bowd] for a definition in variable curvature), then

$$\mathcal{N}(s) \sim c(\mathcal{H}) \, e^{\delta s},$$

as $t \to +\infty$, for some $c(\mathcal{H}) > 0$. Cosentino obtained explicit expressions for the constant $c(\mathcal{H})$ in special arithmetic cases: $\Gamma = \mathrm{PSL}_2(\mathbb{Z})$ acts on the upper halfplane model of $\widetilde{M} = \mathbb{H}^2_{\mathbb{R}}$ and the quotient space $\mathrm{PSL}_2(\mathbb{Z}) \backslash \mathbb{H}^2_{\mathbb{R}}$ has a unique cusp that corresponds to the orbit of ∞. The orbit of the subset $\widetilde{\mathcal{H}}$ of $\mathbb{H}^2_{\mathbb{R}}$

consisting of points with vertical coordinates at least 1 maps under the quotient map to the maximal Margulis cusp neighbourhood of $\mathrm{PSL}_2(\mathbb{Z})\backslash\mathbb{H}_{\mathbb{R}}^2$. The Γ-orbit of $\widetilde{\mathcal{H}}$ consists of $\widetilde{\mathcal{H}}$ and of the horoballs centered at all rational points $\frac{p}{q}$ with $\gcd(p,q)=1$ and with Euclidean diameter q^{-2}. The number of such horoballs of diameter q^{-2} modulo the stabiliser of ∞ (consisting of translations by the integers) is $\phi(q)$, where ϕ is Euler's totient function. A classical result of Mertens on the average order of ϕ (see for example [HaW, Theo. 330]) implies that

$$\mathcal{N}(s) = \frac{3}{\pi^2}\, e^s + O(se^{s/2}),$$

as $s \to +\infty$. Similarly, if $\mathcal{O}_K$ is the ring of integers in $K = \mathbb{Q}(\sqrt{d})$ with d a negative squarefree integer and if D_K is the discriminant of K, then $\Gamma = \mathrm{PSL}(\mathcal{O}_K)$ acts on $\mathbb{H}_{\mathbb{R}}^3$ by homographies as a cofinite volume discrete group of isometries. Let $\mathcal{H}$ also be the image in $\Gamma\backslash\mathbb{H}_{\mathbb{R}}^3$ of the set of points in $\mathbb{H}_{\mathbb{R}}^3$ with vertical coordinates at least 1. A generalisation of the above argument gives, when $D_K \neq -3, -4$,

$$\mathcal{N}(s) = \frac{\pi}{\sqrt{|D_K|}\,\zeta_K(2)}\, e^{2s} + O(e^{3t/2}), \tag{10}$$

as $s \to +\infty$, where ζ_K is Dedekind's zeta function of K. See [Gro, Satz 2] and Section 6.1, Section 6.2 in [Cos] for a proof, and [PaP8] for further generalisations.

3.6 Counting common perpendiculars of horoballs and totally geodesic submanifolds

When $\widetilde{M} = \mathbb{H}_{\mathbb{R}}^n$, M has finite volume, C_- is a Margulis cusp neighbourhood of M and C_+ is a finite volume totally geodesic immersed submanifold of M of dimension k with $1 \le k < n$, we proved in [PaP4] (see Theorem 1.1 and Lemma 3.1) the following result, announced in [PaP2].

Theorem 2 (Parkkonen-Paulin [PaP4]) *As $s \to +\infty$,*

$$\mathcal{N}(s) \sim \frac{\mathrm{Vol}(\mathbb{S}^{n-k-1})\,\mathrm{Vol}(C_-)\,\mathrm{Vol}(C_+)}{\mathrm{Vol}(\mathbb{S}^{n-1})\,\mathrm{Vol}(M)}\, e^{(n-1)s}$$

$$= \frac{\mathrm{Vol}(\mathbb{S}^{n-k-1})\,\mathrm{Vol}(\partial C_-)\,\mathrm{Vol}(C_+)}{\mathrm{Vol}(\mathbb{S}^{n-1})\,\mathrm{Vol}(M)}\, \frac{e^{(n-1)s}}{n-1}. \qquad \square \tag{11}$$

Oh and Shah [OS1] studied a counting problem analogous to the one described in Section 3.4 for a bounded (to simplify in this survey) family $\mathcal{P}$ of circles in $\mathbb{R}^2$ that consists of one (to simplify in this survey) orbit under a nonelementary subgroup Γ of $PSL_2(\mathbb{C})$ such that the family $\mathcal{P}^*$ of totally geodesic hyperplanes in the upper halfspace model of $\mathbb{H}^3_{\mathbb{R}}$ whose boundaries are the circles of $\mathcal{P}$ is locally finite. For any circle $P \in \mathcal{P}$, let $\operatorname{curv}_S(P)$ be the reciprocal of the radius of P, that is, the curvature of the circle P. For any $T > 0$, let

$$N(T) = \operatorname{Card}\{P \in \mathcal{P} : \operatorname{curv}_S(P) < T\}.$$

Oh and Shah showed [OS3, Theo. 1.2] that if $\delta > 1$ and Γ is geometrically finite (see loc. cit. for more general assumptions)

$$N(T) \sim c(\mathcal{P}) T^\delta \tag{12}$$

for a constant $c(\mathcal{P}) > 0$, as $T \to \infty$. Furthermore, they proved that the endpoints of the common perpendiculars are evenly distributed on $\partial \widetilde{C}_-$ in the same sense as in Equation (9) in terms of a natural measure, which is the skinning measure pushed to the boundary, see Section 8.

Let $\widetilde{C}_-$ be the horoball that consists of the points in the upper halfspace model of $\mathbb{H}^3_{\mathbb{R}}$ whose vertical coordinates are at least 1. Now,

$$d(\widetilde{C}_-, \widetilde{C}_+) = \ln \operatorname{curv}_S(\widetilde{C}_+)$$

for any hyperbolic plane $\widetilde{C}_+$ in $\mathcal{P}^*$. Hence, the result (12) on circle packings has an interpretation as a counting result for the common perpendiculars between the image C_- of $\widetilde{C}_-$ and the image C_+ of the hyperplanes of $\mathcal{P}^*$ in M. We then have

$$N(T) = \mathcal{N}_{C_-, C_+}(\ln T).$$

3.7 The density of integer points on homogeneous varieties

Let us denote a generic element of the Euclidean space $\mathbb{R}^{n+1}$ by $x = (x_0, \bar{x})$, where $x_0 \in \mathbb{R}$ and $\bar{x} = (x_1, \ldots, x_n) \in \mathbb{R}^n$, and consider the quadratic form

$$q(x) = -2x_0^2 + \|x\|^2 = -x_0^2 + \|\bar{x}\|^2 = -x_0^2 + x_1^2 + \cdots + x_n^2$$

of signature $(1, n)$. The identity component $G = SO_0(1, n)$ of the special orthogonal group of the form q is a connected semisimple real Lie group with trivial center when $n \geq 2$.

Let $\mathbb{R}^{1,n} = (\mathbb{R}^{n+1}, \langle \cdot, \cdot \rangle)$ be the $(n + 1)$-dimensional Minkowski space with the (indefinite) inner product

$$\langle x, y \rangle = -x_0 y_0 + \sum_{i=1}^{n} x_i y_i$$

where $x = (x_0, x_1, \ldots, x_n)$, $y = (y_0, y_1, \ldots, y_n) \in \mathbb{R}^{n+1}$. The hyperboloid model of the n-dimensional real hyperbolic space $\mathbb{H}^n_{\mathbb{R}}$ is the upper half $\{x \in \mathbb{R}^{1,n} : q(x) = -1, \ x_0 > 0\}$ of the hyperboloid with equation $q = -1$, endowed with the Riemannian metric of constant sectional curvature -1 induced by the (positive definite) restriction of the indefinite inner product $\langle \cdot, \cdot \rangle$ to the tangent space of the hyperboloid. The hyperbolic distance $d(x, y)$ of two points $x, y \in \mathbb{H}^n_{\mathbb{R}}$ has a simple expression in terms of the indefinite inner product: $\cosh d(x, y) = -\langle x, y \rangle$. The restriction to $\mathbb{H}^n_{\mathbb{R}}$ of the (left) linear action of G on $\mathbb{R}^{1,n}$ is the group of orientation-preserving isometries of $\mathbb{H}^n_{\mathbb{R}}$.

Oh and Shah proved the following counting result (and even a more general one, see [OS2, Th. 1.2]) for linear orbits of (nonelementary torsion free discrete) geometrically finite subgroups Γ of G on the level sets of the form q, generalising a special case of a result of Duke, Rudnick and Sarnak [DRS]: For any $m \in \mathbb{R}$, let $V_m = \{x \in \mathbb{R}^{n+1} : q(x) = m\}$. Let $w \in V_m - \{0\}$ be a vector such that the linear orbit Γw is discrete. If $\delta > 1$, then by [OS2, Coro. 1.6]

$$\mathrm{Card}\{y \in \Gamma w : \|y\| < T\} \sim c(m)T^{\delta}, \tag{13}$$

with a constant $c(m) > 0$ similar to those in the previous cases.

The above counting result is equivalent to three results on counting common perpendiculars, depending on the sign of m, as we will now explain. For convenience, we will restrict to the three essential cases $m \in \{-1, 0, 1\}$. Consider first the case $q(w) = -1$. Now, the orbit of w is contained in $\mathbb{H}^n_{\mathbb{R}}$. For any $y \in \mathbb{H}^n_{\mathbb{R}}$, we have $\langle y, y \rangle = -y_0^2 + \|\bar{y}\|^2 = -1$. Thus, $\|y\|^2 = 2y_0^2 + 1$, and we have $\cosh d(y, (1, 0)) = -\langle (1, 0), y \rangle = y_0 \sim \frac{1}{\sqrt{2}} \|y\|$ as $\|y\| \to +\infty$. Therefore, the asymptotic (13) gives an asymptotic count of orbit points as in the results of Margulis and Roblin, see Section 3.3.

If $q(w) = 0$, then w lies in the light cone of q and it defines a horosphere

$$H_w = \{y \in \mathbb{H}^n_{\mathbb{R}} : \langle y, w \rangle = -1\}.$$

Using a rotation with fixed point at $(0, 1)$, we can assume that $w = (w_0, w_0, 0) \in \mathbb{R} \times \mathbb{R} \times \mathbb{R}^{n-1}$, with $w_0 \neq 0$. Now

$$\begin{aligned}
H_w &= \left\{ y \in \mathbb{H}^n_{\mathbb{R}} : y_1 = y_0 - \frac{1}{w_0} \right\} \\
&= \left\{ y \in \mathbb{R}^{1,n} : y_0 = \frac{1}{2}\left(w_0 + \frac{1}{w_0} \right), \ y_1 = \frac{1}{2}\left(w_0 - \frac{1}{w_0} \right) \right\}.
\end{aligned}$$

By the symmetry of the situation, it is clear that

$$d((1,0), H_w) = d\left((1,0), \left(\frac{1}{2}\left(w_0 + \frac{1}{w_0}\right), \frac{1}{2}\left(w_0 - \frac{1}{w_0}\right), 0\right)\right)$$

$$= \operatorname{arcosh}\left(\frac{1}{2}\left(w_0 + \frac{1}{w_0}\right)\right) = \ln w_0 = \ln \|w\| - \ln 2,$$

so the asymptotic for the norms of points in the orbit of a point in the light cone is equivalent to an asymptotic of the distance of an orbit of horoballs from a point. The same counting problem is also considered by Kontorovich and Oh in [KoO], and earlier in [Kon] in the two-dimensional case.

In the third case, when $q(w) = 1$, the vector w defines a totally geodesic hyperplane $w^\perp = \{y \in \mathbb{H}^n_\mathbb{R} : \langle y, w \rangle = 0\}$ in $\mathbb{H}^n_\mathbb{R}$. As in the two cases above, one can check that this asymptotic is equivalent to the asymptotic count of geodesic arcs starting at a fixed point and ending perpendicularly at an orbit of totally geodesic hyperplanes.

4 Using Eskin-McMullen's equidistribution theorem

In order to prove the kind of asymptotic results described in Section 3, following Margulis [Mar1, Mar2], one usually proves first an appropriate equidistribution result using mixing, and this result is then used to study the common perpendiculars.

Eskin and McMullen [EM, Th. 1.2] proved a very general equidistribution theorem for Lie groups orbits using mixing properties and a technical "wave front lemma" in affine symmetric spaces.

Theorem 3 (Eskin-McMullen) *Let G be a connected semisimple real Lie group with finite center. Let $\sigma : G \to G$ be an involutive Lie group automorphism, and H be its fixed subgroup. Let Γ be a lattice in G and let m be the unique G-invariant probability measure on $\Gamma \backslash G$. Assume that the projection of Γ to G/G' is dense for all noncompact connected normal Lie subgroups G' of G, and that $\Gamma \cap H$ is a lattice in H. Let $Y = (\Gamma \cap H) \backslash H$ and let μ_g be the image by the right multiplication by g of the unique H-invariant probability measure on Y. Then, for every $f : \Gamma \backslash G \to \mathbb{R}$ which is continuous with compact support,*

$$\int_{Yg} f(h) \, d\mu_g(h) \to \int_{\Gamma \backslash G} f(x) \, dm(x),$$

as g goes to infinity in $H \backslash G$. □

This result is used in [EM] to prove a result of Duke, Rudnick and Sarnak [DRS] on counting integral points on homogeneous varieties, see Section 3.7.

In [PaP4], we proved the following equidistribution result using mixing and hyperbolic geometry, as a tool to prove the asymptotic estimate (11). A modification of the proof in [PaP4] enabled us to prove the general equidistribution result in variable curvature in [PaP6], whose tools are also used for the counting results of [PaP7] and [BrPP] that we describe in Section 8 and 10. Here, at the instigation of Hee Oh, we present a different proof using Theorem 3, which also serves as an illustration of the use of Theorem 3.

Theorem 4 (Parkkonen-Paulin [PaP4]) *Let M be a complete connected hyperbolic manifold with finite volume. Let C be a nonempty proper totally geodesic immersed submanifold of M with finite volume. The induced Riemannian measure on $g^t \partial_+^1 C$ equidistributes to the Liouville measure as $t \to +\infty$:*

$$\text{Vol}_{g^t \partial_+^1 C} / \| \text{Vol}_{g^t \partial_+^1 C} \| \overset{*}{\rightharpoonup} \text{Vol}_{T^1 M} / \| \text{Vol}_{T^1 M} \|.$$

More general versions of the above result appear in [OS2, Theo. 1.8], [PaP6, Theo. 17] and [BrPP] in the presence of potentials.

We use below the notation introduced in Section 3.7. Before giving the proof of this result, let us review some preliminaries on the action on $T^1 \mathbb{H}_{\mathbb{R}}^n$ of the orientation-preserving isometry group $G = \text{SO}_0(1, n)$ of the hyperboloid model of $\mathbb{H}_{\mathbb{R}}^n$, where $n \geq 2$. Let $(e_0, e_1, \ldots, e_n)$ be the canonical basis of $\mathbb{R}^{1,n}$, and let $w_0 = (1, 0, \ldots, 0) \in \mathbb{H}_{\mathbb{R}}^n$. For any $1 \leq k < n$, we embed $\mathbb{H}_{\mathbb{R}}^k$ isometrically in $\mathbb{H}_{\mathbb{R}}^n$ as the intersection of $\mathbb{H}_{\mathbb{R}}^n$ with the linear subspace given by the equations $x_{k+1} = x_{k+2} = \cdots = x_n = 0$. For any $p \in \mathbb{N}$, let I_p be the $p \times p$ identity matrix. Let H_k be the subgroup of G that consists of the fixed points of the involution $\sigma_k \colon G \to G$ defined by $\sigma_k(g) = J_k g J_k^{-1}$, where $J_k = \begin{pmatrix} I_{k+1} & 0 \\ 0 & -I_{n-k} \end{pmatrix}$. Note that H_k is isomorphic to $(\text{O}(1, k) \times \text{O}(n-k)) \cap G$, hence contains $\text{SO}_0(1, k) \times \text{SO}(n-k)$ with index 2. Let us identify $\text{SO}(n-1)$ with its image in $\text{SO}(n)$, and similarly $\text{SO}(n)$ with its image in G, by the maps $x \mapsto \begin{pmatrix} 1 & 0 \\ 0 & x \end{pmatrix}$. Let λ_G and λ_{H_k} be fixed left Haar measures on G and H_k.

The group G acts transitively on $T^1 \mathbb{H}_{\mathbb{R}}^n$ and its action commutes with the geodesic flow, the stabiliser of $e_1 \in T^1 \mathbb{H}_{\mathbb{R}}^n$ being $\text{SO}(n-1)$. Note that H_k is the subgroup of G which preserves $\mathbb{H}_{\mathbb{R}}^k$. It acts transitively on the unit normal bundle $\partial_+^1 \mathbb{H}_{\mathbb{R}}^k$.

The orbital map $g \mapsto g e_1$ from G to $T^1 \mathbb{H}_{\mathbb{R}}^n$ induces a diffeomorphism $\overline{\varphi} \colon G / \text{SO}(n-1) \to T^1 \mathbb{H}_{\mathbb{R}}^n$ which is equivariant for the left actions of G. The commutativity of the diagram

$$
\begin{array}{ccc}
G/\operatorname{SO}(n-1) & \longrightarrow & G/\operatorname{SO}(n) \\
\downarrow \simeq \overline{\varphi} & & \downarrow \simeq \\
T^1\mathbb{H}^n_{\mathbb{R}} & \longrightarrow & \mathbb{H}^n_{\mathbb{R}}
\end{array}
$$

and the fact that the Riemannian measure of $\mathbb{S}^{n-1} \simeq \operatorname{SO}(n)/\operatorname{SO}(n-1)$ is the unique (up to multiplication by a positive constant) positive Borel measure which is invariant under rotations imply that the image of λ_G by the smooth map $g \mapsto \overline{\varphi}(g \operatorname{SO}(n-1))$ is a multiple of $\operatorname{Vol}_{T^1\mathbb{H}^n_{\mathbb{R}}}$.

Consider the one-parameter subgroup $(a_t)_{t\in\mathbb{R}}$ of G, where

$$
a_t = \begin{pmatrix} \cosh t & \sinh t & 0 \\ \sinh t & \cosh t & 0 \\ 0 & 0 & I_{n-1} \end{pmatrix}.
$$

The action $g \mapsto g a_t$ of a_t by right translations on G commutes with that of $\operatorname{SO}(n-1)$. A calculation in hyperbolic geometry shows that $a_t e_1$ is the image of the unit tangent vector $e_1 \in T^1_{w_0}\mathbb{H}^n_{\mathbb{R}}$ under the geodesic flow g^t in $T^1\mathbb{H}^n_{\mathbb{R}}$. Thus, by equivariance, $\overline{\varphi}(g a_t \operatorname{SO}(n-1)) = g^t \overline{\varphi}(g \operatorname{SO}(n-1))$ for all $g \in G$ and all $t \in \mathbb{R}$. Let us fix a group element $r \in \operatorname{SO}(n)$ which maps e_1 to e_n. As the measures under consideration are induced by differential forms, homogeneity arguments imply that the image measure of λ_{H_k} by the smooth H_k-equivariant map $h \mapsto \overline{\varphi}(h r a_t \operatorname{SO}(n-1))$ is a multiple of $\operatorname{Vol}_{g^t \partial^1_+ \mathbb{H}^k_{\mathbb{R}}}$.

Proof of Theorem 4. By additivity, we can assume that C is connected. Let $\overline{M} \to M$ be the Riemannian orientation cover of M (which is the identity map if M is orientable), and let $\overline{C} \to C$ be the one of C, so that $\overline{C}$ is a connected immersed totally geodesic submanifold of $\overline{M}$. As the image measures by the finite cover $T^1\overline{M} \to T^1M$ of $\operatorname{Vol}_{g^t \partial^1_+ \overline{C}}$ and $\operatorname{Vol}_{T^1\overline{M}}$ are $\operatorname{Vol}_{g^t \partial^1_+ C}$ and $\operatorname{Vol}_{T^1M}$, respectively, we only have to show that the Riemannian measure of $g^t \partial^1_+ \overline{C}$ equidistributes to the Liouville measure of $T^1\overline{M}$ as $t \to +\infty$. We may, therefore, assume that M and C are oriented.

Let us fix a universal Riemannian cover $\mathbb{H}^n_{\mathbb{R}} \to M$. Its covering group Γ is a lattice in G, since M has finite volume. We may assume that the image of $\mathbb{H}^k_{\mathbb{R}}$ under this covering map is C. We define $H = H_k$ as above. Since C has finite volume, $\Gamma \cap H$ is a lattice in H. Since r fixes e_0 and sends e_1 to e_n which is perpendicular to $\mathbb{H}^k_{\mathbb{R}}$, the map $t \mapsto \pi(r a_t e_1)$ from $[0, +\infty[$ to $\mathbb{H}^n_{\mathbb{R}}$ is a geodesic ray starting perpendicularly to $\mathbb{H}^k_{\mathbb{R}}$. Since H is the stabiliser of $\mathbb{H}^k_{\mathbb{R}}$, the map $t \mapsto H r a_t e_1$ from $[0, +\infty[$ to $H\backslash T^1\mathbb{H}^n_{\mathbb{R}}$ tends to infinity. Hence the map $t \mapsto H r a_t$ from $[0, +\infty[$ to $H\backslash G$ tends to infinity.

Since the connected semi-simple real Lie group G has trivial center, and only one noncompact factor, the projection of Γ to G/G' is dense for all noncompact connected normal Lie subgroups G' of G. (In fact, G has no compact factor, and any lattice in G is irreducible, see for instance [Mos]).

We can now use Theorem 3 to conclude that as t tends to $+\infty$, the measure μ_t on $\Gamma\backslash G$ with support $\Gamma H r a_t$ which is defined to be the translate by $r a_t$ of the unique H-invariant probability measure on $(\Gamma \cap H)\backslash H$, equidistributes towards the probability measure m on $\Gamma\backslash G$ induced by λ_G. Let $p : \Gamma\backslash G \to \Gamma\backslash G/\operatorname{SO}(n-1)$ be the canonical projection. The G-equivariant diffeomorphism $\overline{\varphi} : G/\operatorname{SO}(n-1) \to T^1\mathbb{H}_{\mathbb{R}}^n$ induces a diffeomorphism $\varphi : \Gamma\backslash G/\operatorname{SO}(n-1) \to T^1 M$ such that $\varphi(\Gamma g a_t \operatorname{SO}(n-1)) = g^t\varphi(\Gamma g \operatorname{SO}(n-1))$ for all $g \in G$ and $t \in \mathbb{R}$. By the homogeneity argument just before the beginning of the proof and covering arguments, and since direct images of measures preserve the total masses, we have $\varphi_*(p_*\mu_t) = \frac{1}{\operatorname{Vol}(g^t\partial_+^1 C)} \operatorname{Vol}_{g^t\partial_+^1 C}$ and $\varphi_*(p_*m) = \frac{1}{\operatorname{Vol}(T^1 M)} \operatorname{Vol}_{T^1 M}$. As taking direct images of measures by a given continuous map is continuous in the weak-* topology, the measures $\frac{1}{\operatorname{Vol}(g^t\partial_+^1 C)} \operatorname{Vol}_{g^t\partial_+^1 C} = (\varphi \circ p)_*\mu_t$ equidistribute to $(\varphi \circ p)_*m = \frac{1}{\operatorname{Vol}(T^1 M)} \operatorname{Vol}_{T^1 M}$ in $T^1 M$ as t tends to $+\infty$, which is what we wanted to prove. $\square$

5 Arithmetic applications

If the manifold M is arithmetically defined, many counting results for common perpendiculars have an arithmetic interpretation. In this section, we will review some of these arithmetic applications. The arithmetically defined groups in this section will, in general, have torsion. This is not a problem however, as the geometric counting result used in the various cases below is indeed valid in this more general context. In certain cases, the interaction has also worked in the opposite direction, as evidenced by the results of Cosentino in Section 3.5.

5.1 Counting representations of integers by binary quadratic forms

Let $Q(X, Y) = aX^2 + bXY + cY^2$ be an integral binary quadratic form with *discriminant* $\Delta = b^2 - 4ac$. An element $x \in \mathbb{Z}^2$ is a *representation* of an integer n by Q if $Q(x) = n$, and the representation is *primitive* if the components of x are relatively prime. If Q is positive definite (equivalently, if $\Delta < 0$ and $a > 0$), then the number $N(t)$ of representations of integers that are at most t equals the number of lattice points of $\mathbb{Z}^2$ inside the ellipse defined by the equation $Q(x) = t$. The asymptotics of this number (Gauss' circle problem) have been studied extensively, and the best known result with an error bound

$$N(t) = \frac{2\pi}{\sqrt{-\Delta}} t + O(t^{131/416})$$

is due to Huxley [Hux]. Gauss already had a solution with a worse bound on the error term.

The modular group $\mathrm{SL}_2(\mathbb{Z})$ acts on the right by precomposition on the set of binary quadratic forms, preserving the discriminant, and linearly on the left on $\mathbb{Z}^2$. Let us assume that Q is *primitive* (that is, the coefficients a, b and c are relatively prime), *indefinite* (that is, $\Delta > 0$) and not the product of two integral linear forms. The stabiliser of Q in $\mathrm{SL}_2(\mathbb{Z})$, called the *group of automorphs* of Q, is

$\mathrm{SO}(Q, \mathbb{Z}) = \{\gamma \in \mathrm{SL}_2(\mathbb{Z}) : Q \circ \gamma = Q\}$

$$= \left\{ \gamma_{Q,t,u} = \begin{pmatrix} \dfrac{t - bu}{2} & -cu \\ au & \dfrac{t + bu}{2} \end{pmatrix} : t, u \in \mathbb{Z}, \ t^2 - \Delta u^2 = 4 \right\},$$

see for instance [Lan, Theo. 202]. This group is infinite and thus any nonzero integer that is represented by the form Q is represented infinitely many times. Accordingly, in the generalisation of the circle problem for these forms, one counts the number of orbits of lattice points under the linear action of $\mathrm{SO}(Q, \mathbb{Z})$ between the hyperbolas defined by the equations $|Q(x)| = t$. With $\mathcal{P}$ the set of relatively prime elements of $\mathbb{Z}^2$, let

$$\tilde{\Psi}_Q(t) = \mathrm{Card}\big(\mathrm{SO}(Q, \mathbb{Z}) \backslash \{x \in \mathbb{Z}^2 : |Q(x)| \le t\} \big)$$

and

$$\Psi_Q(t) = \mathrm{Card}\big(\mathrm{SO}(Q, \mathbb{Z}) \backslash \{x \in \mathcal{P} : |Q(x)| \le t\} \big)$$

be the counting functions of all the representations and of the primitive representations by Q. The asymptotics of $\tilde{\Psi}_Q(t)$ are also known, see for example [Coh, p. 164] for a proof. It turns out that the asymptotic result on the counting function $\mathcal{N}(s)$ for a horoball and a totally geodesic subspace can be used to give a different proof of this result.

Let us first observe that an asymptotic result for the primitive representations implies one for all representations. Assume that $\Psi_Q(t) = ct + O(t^{1-\epsilon})$ for some $c, \epsilon > 0$. For any $k \in \mathbb{N}$, let $\Psi_{Q,k}(t)$ equal

$$\text{Card}\big(\text{SO}(Q, \mathbb{Z}) \backslash \{ x = (x_1, x_2) \in \mathbb{Z} \; : \; \gcd(x_1, x_2) = k, \; |Q(x)| \leq t \} \big).$$

Now, $\Psi_{Q,k}(t) = \Psi_Q(k^{-2}t)$ and

$$\tilde{\Psi}_Q(t) = \sum_{k=1}^{\infty} \Psi_{Q,k}(t) = \sum_{k=1}^{\infty} \Psi_Q(k^{-2}t)$$

$$= \sum_{k=1}^{\infty} ck^{-2}t + O((k^{-2}t)^{1-\epsilon}) = c\,\zeta(2)t + O(t^{1-\epsilon}).$$

We will now explain how to obtain an asymptotic estimate for $\Psi_Q(t)$ from the solution of the geometric counting problem of Section 3.6. We use the upper halfplane model of $\mathbb{H}^2_{\mathbb{R}}$. The subgroup $\text{PSO}(Q, \mathbb{Z})$ of $\text{PSL}_2(\mathbb{Z})$ is a cyclic group generated by a hyperbolic element. Its index i_Q in the stabiliser Γ_Q of the geodesic line C_Q invariant under $\text{PSO}(Q, \mathbb{Z})$ is either 2 or 1 depending on whether or not the corresponding locally geodesic line on the modular surface $\text{PSL}_2(\mathbb{Z}) \backslash \mathbb{H}^2_{\mathbb{R}}$ passes through the cone point of order 2. Let (t_Q, u_Q) be the fundamental solution of the Pell-Fermat equation $t^2 - \Delta u^2 = 4$, and let $R_Q = \ln \frac{t_Q + u_Q \sqrt{\Delta}}{2}$ be the *regulator* of Q. It is easy to check that the length of the closed geodesic $\Gamma_Q \backslash C_Q$ is $\frac{2R_Q}{i_Q}$.

The stabiliser U of $(1, 0) \in \mathbb{Z}^2$ for the linear action of $\text{SL}_2(\mathbb{Z})$ is the subgroup that consists of integral upper triangular unipotent matrices. Geometrically, the image Γ_∞ of U in $\text{PSL}_2(\mathbb{Z})$ is the stabiliser in $\text{PSL}_2(\mathbb{Z})$ of the horoball $\mathcal{H} = \{ z \in \mathbb{C} \; : \; \text{Im } z \geq 1 \}$ in $\mathbb{H}^2_{\mathbb{R}}$. The horoball $\mathcal{H}$ is *precisely invariant*, that is, each element of $\text{PSL}_2(\mathbb{Z})$ either preserves $\mathcal{H}$ or maps $\mathcal{H}$ to a horoball whose interior is disjoint from $\mathcal{H}$. As $\mathcal{H}$ is the maximal such horoball at ∞, it corresponds to the maximal Margulis cusp neighbourhood of the unique cusp of the quotient space $\text{PSL}_2(\mathbb{Z}) \backslash \mathbb{H}^2_{\mathbb{R}}$.

The length of the common perpendicular of $\mathcal{H}$ and the hyperbolic line C_Q stabilised by $\text{SO}(Q, \mathbb{Z})$ is $\ln \frac{2|a|}{\sqrt{\Delta}}$. For all $\gamma = \pm \begin{pmatrix} A & B \\ C & D \end{pmatrix}$ in $\text{PSL}_2(\mathbb{Z})$, a simple computation (see [PaP4, Lem. 4.2]) shows that the length of the common perpendicular of $\mathcal{H}$ and the image under γ of C_Q is

$$\ln \frac{2}{\sqrt{\Delta}} |Q(D, -C)|.$$

Thus, Corollaire 3.9 of [PaP4], which generalises the result of Equation (11) to the case of groups with torsion, and [PaP8, Section 5] (see Section 9) for the error term, give that there exists $\kappa > 0$ such that

$$\Psi_Q(s) = i_Q \operatorname{Card} \left\{ [\gamma] \in \Gamma_\infty \backslash \Gamma / \Gamma_Q \ : \ d(\mathcal{H}_\infty, \gamma C_Q) \leq \ln \left(\frac{2}{\sqrt{\Delta}} s \right) \right\}$$

$$\sim i_Q \frac{\operatorname{Vol}(\mathbb{S}^0) \operatorname{Vol}(\Gamma_\infty \backslash \mathcal{H}_\infty) \operatorname{Vol}(\Gamma_Q \backslash C_Q)}{\operatorname{Vol}(\mathbb{S}^1) \operatorname{Vol}\left(\Gamma \backslash \mathbb{H}^2_\mathbb{R}\right)} \left(\frac{2}{\sqrt{\Delta}} s \, (1 + \mathrm{O}(s^{-\kappa})) \right)$$

$$= \frac{12 \, R_Q}{\pi^2 \sqrt{\Delta}} s \, (1 + \mathrm{O}(s^{-\kappa})),$$

see [PaP4, Section 4] for more details and more applications, in particular to counting representations satisfying congruence relations, and to [PaP8] for their error terms.

5.2 Counting representations of integers by binary Hermitian forms

A function $f : \mathbb{C}^2 \to \mathbb{R}$ is a *binary Hermitian form* if there are constants $a, c \in \mathbb{R}$ and $b \in \mathbb{C}$, called the *coefficients* of f, such that for all $u, v \in \mathbb{C}$,

$$f(u, v) = a|u|^2 + 2 \operatorname{Re}(bu\bar{v}) + c|v|^2 = \begin{pmatrix} \bar{u} & \bar{v} \end{pmatrix} \begin{pmatrix} a & b \\ \bar{b} & c \end{pmatrix} \begin{pmatrix} u \\ v \end{pmatrix}. \tag{14}$$

Let K be an imaginary quadratic number field, with discriminant D_K and ring of integers $\mathcal{O}_K$. For example, if $K = \mathbb{Q}(i)$, then $D_{\mathbb{Q}(i)} = -4$ and $\mathcal{O}_{\mathbb{Q}(i)} = \mathbb{Z}[i]$ is the ring of Gaussian integers. If the coefficients of the Hermitian form f satisfy $a, c \in \mathbb{R} \cap \mathcal{O}_K = \mathbb{Z}$ and $b \in \mathcal{O}_K$, then we say that f is *integral* (over K). It is easy to check that the values of the restriction of an integral binary Hermitian form to $\mathcal{O}_K \times \mathcal{O}_K$ are rational integers. If the *discriminant* $\Delta(f) = |b|^2 - ac$ of f is positive, then we say that f is *indefinite*, which is equivalent to saying that f takes both positive and negative values.

A binary Hermitian form naturally gives rise to a quaternary quadratic form. The representations of integers by positive definite quaternary quadratic forms have been studied for a long time (including Lagrange's four square theorem, see also the work of Ramanujan as in [Klo]).

In the case of indefinite forms, the counting problem is again complicated by the presence of an infinite group of automorphs: The group $\mathrm{SL}_2(\mathcal{O}_K)$ acts on the right by precomposition on the set of (indefinite) integral binary Hermitian forms, and the stabiliser of such a form f under this action is, analogously to the case of binary quadratic forms treated in Section 5.1, called the *group of automorphs* of the form and denoted by $\mathrm{SU}_f(\mathcal{O}_K)$. The Bianchi group $\mathrm{PSL}_2(\mathcal{O}_K)$ acts discretely on the upper halfspace model of $\mathbb{H}^3_\mathbb{R}$, with finite covolume. Now, the image in $\mathrm{PSL}_2(\mathcal{O}_K)$ of the group of automorphs of a fixed indefinite integral binary Hermitian form f is a Fuchsian subgroup that preserves a real hyperbolic plane $C(f)$ whose boundary at infinity is the circle

$$C_\infty(f) = \{[u : v] \in \mathbb{P}^1(\mathbb{C}) = \partial_\infty \mathbb{H}_\mathbb{R}^3 \ : \ f(u, v) = 0\}.$$

The group of automorphs $\mathrm{SU}_f(\mathcal{O}_K)$ is an arithmetic group that acts on $C(f)$ with finite covolume.

The cusps of a Bianchi group $\mathrm{PSL}_2(\mathcal{O}_K)$ are in a natural bijective correspondence with the ideal classes of K, see for example Theorem 2.4 in Chapter 7 of [EGM]. Let $x, y \in \mathcal{O}_K$ be not both zero, so that $[x : y] \in \mathbb{P}^1(\mathbb{C})$ is a cusp of $\mathrm{PSL}_2(\mathcal{O}_K)$. Then, if $y = 0$, the horoball $\mathcal{H}$ that consists of those points in the upper halfspace model $\mathbb{H}_\mathbb{R}^3$ whose vertical coordinate is at least 1 is precisely invariant, and if $y \neq 0$, then there is some $\tau > 0$ such that the horoball $\mathcal{H}$ centered at $\frac{x}{y}$ of Euclidean height τ is precisely invariant. Analogously with the case of indefinite binary quadratic forms, for any $g \in \mathrm{SL}_2(\mathcal{O}_K)$, the hyperbolic distance between $\mathcal{H}$ and $C_\infty(f \circ g) = g^{-1} C_\infty(f)$ is $\ln \frac{|f \circ g(x,y)|}{\tau |y|^2 \sqrt{\Delta(f)}}$, and, as in the case of binary quadratic forms, we found a connection between representing integers by f and the counting problem of Section 3.6.

We define a counting function of the representation of integers for each nonzero fractional ideal $\mathfrak{m}$ of K. For every $u, v \in K$, let $\langle u, v \rangle$ be the $\mathcal{O}_K$-module they generate. For every $s > 0$, we consider the integer $\psi_{f,\mathfrak{m}}(s)$ given by

$$\mathrm{Card} \ _{\mathrm{SU}_f(\mathcal{O}_K)} \backslash \big\{ (u, v) \in \mathfrak{m} \times \mathfrak{m} \ : \ (N\mathfrak{m})^{-1} |f(u, v)| \leq s, \quad \langle u, v \rangle = \mathfrak{m} \big\}.$$

Generalising the argument used for binary quadratic forms (see Section 5.1), we can again use the generalisation of Equation (11) to obtain an asymptotic expression for $\psi_{f,\mathfrak{m}}(s)$.

Theorem 5 (Parkkonen-Paulin [PaP3][PaP8, Theo. 5.2]) *There exists $\kappa > 0$ such that, as s tends to $+\infty$,*

$$\psi_{f,\mathfrak{m}}(s) = \frac{\pi \ \mathrm{Covol}(\mathrm{SU}_f(\mathcal{O}_K))}{2 \ |D_K| \ \zeta_K(2) \ \Delta(f)} \ s^2 (1 + \mathrm{O}(s^{-\kappa})). \quad \square$$

Here $\mathrm{Covol}(\mathrm{SU}_f(\mathcal{O}_K))$ is the area of the quotient of the hyperbolic plane $C(f)$ in $\mathbb{H}_\mathbb{R}^3$ by the group of automorphs of f, and ζ_K is Dedekind's zeta function of K. In the proof, after applying Equation (11), we use the fact that there is an explicit formula (essentially due to Humbert) for the volume

$$\mathrm{Vol}\left(\mathrm{PSL}_2(\mathcal{O}_K) \backslash \mathbb{H}_\mathbb{R}^3\right) = \frac{1}{4\pi^2} |D_K|^{3/2} \zeta_K(2).$$

See [Sar] for a proof of this formula using Eisenstein series, and Section 8.8 and Section 9.6 of [EGM] for further proofs. The following corollary follows immediately by taking $\mathfrak{m} = \mathcal{O}_K$: If $\mathcal{P}_K$ is the set of relatively prime pairs of integers of K, then

$$
\begin{aligned}
\text{Card } _{\text{SU}_f(\mathcal{O}_K)}\backslash\big\{(u, v) \in \mathcal{P}_K \ : \ |f(u, v)| \le s\big\} \\
= \frac{\pi \ \text{Covol}(\text{SU}_f(\mathcal{O}_K))}{2 \, |D_K| \, \zeta_K(2) \, \Delta(f)} \ s^2 \, (1 + \text{O}(s^{-\kappa})),
\end{aligned}
$$

as s tends to $+\infty$.

In general, one could compute the covolume of the group of automorphs $\text{SU}_f(\mathcal{O}_K)$ with the aid of Prasad's formula in [Pra]. Maclachlan and Reid [MaR] computed the covolumes of all stabilisers in $\text{PSL}(\mathbb{Q}(i))$ of Euclidean halfspheres in the upper halfspace model of $\mathbb{H}^3_{\mathbb{R}}$ centered at 0 with Euclidean radius $\sqrt{D}$, where D is a rational integer. This result can be used to obtain an even more explicit expression of the asymptotic formula of Theorem 5 when $K = \mathbb{Q}(i)$: A constant $\iota(f) \in \{1, 2, 3, 6\}$ is defined as follows. If $\Delta(f) \equiv 0 \mod 4$, let $\iota(f) = 2$. If the coefficients a and c of the form f as in Equation (14) are both even, let $\iota(f) = 3$ if $\Delta(f) \equiv 1 \mod 4$, and let $\iota(f)$ be the remainder modulo 8 of $\Delta(f)$ if $\Delta(f) \equiv 2 \mod 4$. In all other cases, let $\iota(f) = 1$. The class number of $\mathbb{Q}(i)$ is 1, and there is just one counting function to be considered. We prove in [PaP3, Coro. 3], and [PaP8, Coro. 5.3] for the error term, that if $K = \mathbb{Q}(i)$, there exists $\kappa > 0$ such that, as s tends to $+\infty$,

$$
\psi_{f, \mathbb{Z}[i]}(s) = \frac{\pi^2}{8 \, \iota(f) \, \zeta_{\mathbb{Q}(i)}(2)} \prod_{p | \Delta(f)} \left(1 + \left(\frac{-1}{p}\right) p^{-1}\right) \ s^2 \, (1 + \text{O}(s^{-\kappa})).
$$

Here p ranges over the odd positive rational primes and $\left(\frac{-1}{p}\right)$ is the Legendre symbol of -1 modulo p. We refer to [PaP3] for more details and more applications, including counting representations satisfying congruence conditions, and to [PaP8, Section 5] for their error terms.

5.3 Counting quadratic irrational in orbits of modular groups

The group $\text{PSL}_2(\mathbb{Z})$ acts transitively on the rational real numbers, but not transitively on the irrational algebraic real numbers of a given degree. Hence, counting results (for appropriate complexities) of algebraic irrationals within an orbit of $\text{PSL}_2(\mathbb{Z})$ is an interesting problem, and we give some solutions in [PaP4] in the quadratic case. Similar problems occur for quadratic irrational complex numbers under the action of (congruence subgroups of) Bianchi groups, and we illustrate them by the following result.

Let $\phi = \frac{1+\sqrt{5}}{2}$ be the Golden Ratio, and $\phi^{\sigma} = \frac{1-\sqrt{5}}{2}$ its Galois conjugate. Let K be an imaginary quadratic number field, with discriminant $D_K \ne -4$ (in order to simplify the statement in this survey), Dedekind's zeta function ζ_K

and ring of integers $\mathcal{O}_K$. We define as the complexity of a quadratic irrational α with Galois conjugate α^σ the quantity

$$h(\alpha) = \frac{2}{|\alpha - \alpha^\sigma|}.$$

(See [PaP4, Section 4.1] for algebraic versions and explanations). Let $\mathfrak{a}$ be a nonzero ideal in $\mathcal{O}_K$, and let $\Gamma_0(\mathfrak{a})$ be the Hecke congruence subgroup

$$\left\{ \pm \begin{pmatrix} a & b \\ c & d \end{pmatrix} \in \mathrm{PSL}_2(\mathcal{O}_K) \; : \; c \in \mathfrak{a} \right\}.$$

Corollary 6 (Parkkonen-Paulin [PaP4, Coro. 4.7], [PaP8, Coro. 4.3]) *There exists $\kappa > 0$ such that, as s tends to $+\infty$, the cardinality of*

$$\{\alpha \in \Gamma_0(\mathfrak{a}) \cdot \phi \mod \mathcal{O}_K \; : \; h(\alpha) \leq s\}$$

is equal to

$$\frac{2\pi^2 \, k_\mathfrak{a} \, \ln \phi}{|D_K| \, \zeta_K(2) \, N(\mathfrak{a}) \prod_{\mathfrak{p}|\mathfrak{a}} \left(1 + \frac{1}{N(\mathfrak{p})}\right)} \, s^2 \, (1 + O(s^{-\kappa})),$$

where $k_\mathfrak{a}$ is the smallest $k \in \mathbb{N} - \{0\}$ such that the 2k-th term of Fibonacci's sequence belongs to $\mathfrak{a}$, and $\mathfrak{p}$ ranges over the prime ideals in $\mathcal{O}_K$. □

5.4 Counting representations of integers by binary Hamiltonian forms

A *quaternion algebra* over a field F is a four-dimensional central simple algebra over F. A real quaternion algebra (that is, a quaternion algebra over $\mathbb{R}$) is isomorphic either to the algebra of real 2×2 matrices over $\mathbb{R}$ or to Hamilton's quaternion algebra $\mathbb{H}$ over $\mathbb{R}$, with basis elements $1, i, j, k$ as a $\mathbb{R}$-vector space, with unit element 1 and $i^2 = j^2 = -1, ij = -ji = k$. We define the *conjugate* of $x = x_0 + x_1 i + x_2 j + x_3 k$ in $\mathbb{H}$ by $\overline{x} = x_0 - x_1 i - x_2 j - x_3 k$, its *reduced trace* by $\mathrm{tr}(x) = x + \overline{x}$, and its *reduced norm* by $\mathrm{n}(x) = x\overline{x} = \overline{x}x$. We refer for instance to [Vig] for generalities on quaternion algebras.

A *binary Hamiltonian form* is a map $f : \mathbb{H} \times \mathbb{H} \to \mathbb{R}$ with

$$f(u, v) = a \, \mathrm{n}(u) + \mathrm{tr}(\overline{u} \, b \, v) + c \, \mathrm{n}(v),$$

whose *coefficients* a and c are real, and b lies in $\mathbb{H}$. The *matrix* $M(f)$ of f is the Hermitian matrix $\begin{pmatrix} a & b \\ \overline{b} & c \end{pmatrix}$, so that $f(u, v) = (\overline{u} \;\; \overline{v}) \begin{pmatrix} a & b \\ \overline{b} & c \end{pmatrix} \begin{pmatrix} u \\ v \end{pmatrix}$. The *discriminant* of f is

$$\Delta(f) = \mathrm{n}(b) - ac,$$

and f is *indefinite* (that is, f takes both positive and negative values) if and only if $\Delta > 0$.

In this section, we will describe the results in [PaP5] on the representation of integers by indefinite binary Hamiltonian forms. The proof follows the same ideas as in the previous two sections but the noncommutativity of the quaternions adds several new features.

In order to generalise the results of the previous two subsections to the context of Hamiltonian forms, we have to introduce the correct analogs of the ring of integers and of Bianchi groups for quaternion algebras. We say that a quaternion algebra A over $\mathbb{Q}$ is *definite* (or ramified over $\mathbb{R}$) if the real quaternion algebra $A \otimes_{\mathbb{Q}} \mathbb{R}$ is isomorphic to $\mathbb{H}$. We fix an identification between $A \otimes_{\mathbb{Q}} \mathbb{R}$ and $\mathbb{H}$, so that A is a $\mathbb{Q}$-subalgebra of $\mathbb{H}$. The *reduced discriminant* D_A of A is the product of the primes $p \in \mathbb{N}$ such that the quaternion algebra $A \otimes_{\mathbb{Q}} \mathbb{Q}_p$ over $\mathbb{Q}_p$ is a division algebra. For example, the $\mathbb{Q}$-vector space $\mathbb{H}_{\mathbb{Q}} = \mathbb{Q} + \mathbb{Q}i + \mathbb{Q}j + \mathbb{Q}k$ generated by $1, i, j, k$ in $\mathbb{H}$ is Hamilton's quaternion algebra over $\mathbb{Q}$. It is the unique (up to isomorphism) definite quaternion algebra over $\mathbb{Q}$ with discriminant $D_A = 2$.

A $\mathbb{Z}$-*lattice* I in A is a finitely generated $\mathbb{Z}$-module generating A as a $\mathbb{Q}$-vector space. An *order* in a quaternion algebra A over $\mathbb{Q}$ is a unitary subring $\mathcal{O}$ of A which is a $\mathbb{Z}$-lattice, and the order is *maximal* if it is maximal with respect to inclusion among all orders of A. The *Hurwitz order* $\mathcal{O} = \mathbb{Z} + \mathbb{Z}i + \mathbb{Z}j + \mathbb{Z}\frac{1+i+j+k}{2}$ in $\mathbb{H}_{\mathbb{Q}}$ is maximal, and it is the unique maximal order in $\mathbb{H}_{\mathbb{Q}}$ up to conjugacy.

The Dieudonné determinant (see [Die, Asl]) Det is the group morphism from the group $\mathrm{GL}_2(\mathbb{H})$ of invertible 2×2 matrices with coefficients in $\mathbb{H}$ to $\mathbb{R}_+^*$, defined by

$$\mathrm{Det}\left(\begin{pmatrix} a & b \\ c & d \end{pmatrix}\right)^2 = \mathrm{n}(a\,d) + \mathrm{n}(b\,c) - \mathrm{tr}(a\,\overline{c}\,d\,\overline{b})$$

$$= \begin{cases} \mathrm{n}(ad - aca^{-1}b) & \text{if } a \neq 0 \\ \mathrm{n}(cb - cac^{-1}d) & \text{if } c \neq 0 \\ \mathrm{n}(cb - db^{-1}ab) & \text{if } b \neq 0. \end{cases}$$

We will denote by $\mathrm{SL}_2(\mathbb{H})$ the group of 2×2 matrices with coefficients in $\mathbb{H}$ with Dieudonné determinant 1, which equals the group of elements of (reduced) norm 1 in the central simple algebra $\mathcal{M}_2(\mathbb{H})$ over $\mathbb{R}$, see [Rei, Section 9a]. We refer for instance to [Kel] for more information on $\mathrm{SL}_2(\mathbb{H})$.

The group $\mathrm{SL}_2(\mathbb{H})$ acts linearly on the left on the right $\mathbb{H}$-module $\mathbb{H} \times \mathbb{H}$. Let $\mathbb{P}_r^1(\mathbb{H}) = (\mathbb{H} \times \mathbb{H} - \{0\})/\mathbb{H}^{\times}$ be the right projective line of $\mathbb{H}$, identified as usual with the Alexandrov compactification $\mathbb{H} \cup \{\infty\}$ where $[1 : 0] = \infty$ and $[x : y] = xy^{-1}$ if $y \neq 0$. The projective action of $\mathrm{SL}_2(\mathbb{H})$ on $\mathbb{P}_r^1(\mathbb{H})$, induced by its linear action on $\mathbb{H} \times \mathbb{H}$, is then the action by homographies on $\mathbb{H} \cup \{\infty\}$ defined by

$$\begin{pmatrix} a & b \\ c & d \end{pmatrix} \cdot z = \begin{cases} (az+b)(cz+d)^{-1} & \text{if } z \neq \infty, -c^{-1}d \\ ac^{-1} & \text{if } z = \infty, c \neq 0 \\ \infty & \text{otherwise.} \end{cases}$$

The linear action on the left on $\mathbb{H} \times \mathbb{H}$ of the group $\mathrm{SL}_2(\mathbb{H})$ induces an action on the right on the set of binary Hamiltonian forms f by precomposition.

The above action of $\mathrm{SL}_2(\mathbb{H})$ on $\mathbb{H} \cup \{\infty\}$ induces a faithful left action of $\mathrm{PSL}_2(\mathbb{H}) = \mathrm{SL}_2(\mathbb{H})/\{\pm \mathrm{id}\}$ on $\mathbb{H} \cup \{\infty\} = \partial_\infty \mathbb{H}^5_{\mathbb{R}}$. By Poincaré's extension procedure (see for instance [PaP1, Lem. 6.6]), this action extends to a left action of $\mathrm{SL}_2(\mathbb{H})$ on the upper halfspace model of $\mathbb{H}^5_{\mathbb{R}}$ with coordinates $(z, r) \in \mathbb{H} \times \,]0, +\infty[$, by

$$\begin{pmatrix} a & b \\ c & d \end{pmatrix} \cdot (z, r) = \left(\frac{(az+b)\overline{(cz+d)} + a\bar{c}r^2}{\mathrm{n}(cz+d) + r^2 \,\mathrm{n}(c)}, \frac{r}{\mathrm{n}(cz+d) + r^2 \,\mathrm{n}(c)} \right).$$

In this way, the group $\mathrm{PSL}_2(\mathbb{H})$ is identified with the group of orientation preserving isometries of $\mathbb{H}^5_{\mathbb{R}}$.

Given an order $\mathcal{O}$ in a definite quaternion algebra over $\mathbb{Q}$, a binary Hamiltonian form f is *integral* over $\mathcal{O}$ if its coefficients belong to $\mathcal{O}$. Note that such a form f takes integral values on $\mathcal{O} \times \mathcal{O}$. The *Hamilton-Bianchi group* $\mathrm{SL}_2(\mathcal{O}) = \mathrm{SL}_2(\mathbb{H}) \cap \mathcal{M}_2(\mathcal{O})$ preserves the set of indefinite binary Hamiltonian forms f that are integral over $\mathcal{O}$. The stabiliser in $\mathrm{SL}_2(\mathcal{O})$ of such a form f is its *group of automorphs* $\mathrm{SU}_f(\mathcal{O})$.

The Hamilton-Bianchi group $\mathrm{SL}_2(\mathcal{O})$ is a (nonuniform) arithmetic lattice in the connected real Lie group $\mathrm{SL}_2(\mathbb{H})$ (see for instance [PaP1, p. 1104] for details). The volume of the quotient real hyperbolic orbifold $\mathrm{SL}_2(\mathcal{O}) \backslash \mathbb{H}^5_{\mathbb{R}}$ has a nice expression in terms of the discriminant D_A, generalising Humbert's formula.

Theorem 7 (Emery, Parkkonen-Paulin [PaP5])

$$\mathrm{Covol}(\mathrm{SL}_2(\mathcal{O})) = \frac{\zeta(3) \prod_{p|D_A}(p^3 - 1)(p - 1)}{11520}. \quad \square$$

This result is proved in [PaP5] using two different methods: In the Appendix of that paper, Emery (who was the first to prove the theorem in full generality) uses Prasad's formula and we give a different proof using the theory of Eisenstein series for quaternions developped in [KrO], following Sarnak's proof in [Sar] for Bianchi groups.

With a, b, c the coefficients of f, let

$$\mathcal{C}_\infty(f) = \{[u : v] \in \mathbb{P}^1_r(\mathbb{H}) \; : \; f(u, v) = 0\} \quad \text{and}$$
$$\mathcal{C}(f) = \{(z, r) \in \mathbb{H} \times \,]0, +\infty[\; : \; f(z, 1) + a\,r^2 = 0\}.$$

In $\mathbb{P}^1_r(\mathbb{H}) = \mathbb{H} \cup \{\infty\}$, the set $\mathcal{C}_\infty(f)$ is the 3-sphere of center $-\frac{b}{a}$ and radius $\frac{\sqrt{\Delta(f)}}{|a|}$ if $a \neq 0$, and it is the union of $\{\infty\}$ with the real hyperplane $\{z \in \mathbb{H} \; : \; \mathrm{tr}(\bar{z}b) + c = 0\}$ of $\mathbb{H}$ otherwise. The arithmetic group $\mathrm{SU}_f(\mathcal{O})$ acts with finite covolume on the hyperbolic hyperplane $\mathcal{C}(f)$.

The action by homographies of $\mathrm{SL}_2(\mathcal{O})$ preserves the right projective space $\mathbb{P}^1_r(\mathcal{O}) = A \cup \{\infty\}$, which is the set of fixed points of the parabolic elements of $\mathrm{SL}_2(\mathcal{O})$ acting on $\mathbb{H}^5_\mathbb{R} \cup \partial_\infty \mathbb{H}^5_\mathbb{R}$. In order to describe the orbits of parabolic fixed points, we recall some basic definitions and facts on ideals in a quaternion algebra, see [Vig]. The *left order* $\mathcal{O}_\ell(I)$ of a $\mathbb{Z}$-lattice I is $\{x \in A \; : \; xI \subset I\}$. A *left fractional ideal* of $\mathcal{O}$ is a $\mathbb{Z}$-lattice of A whose left order is $\mathcal{O}$. A *left ideal* of $\mathcal{O}$ is a left fractional ideal of $\mathcal{O}$ contained in $\mathcal{O}$. Two left fractional ideals $\mathfrak{m}$ and $\mathfrak{m}'$ of $\mathcal{O}$ are isomorphic as left $\mathcal{O}$-modules if and only if $\mathfrak{m}' = \mathfrak{m}\,c$ for some $c \in A^\times$. A (left) *ideal class* of $\mathcal{O}$ is an equivalence class of left fractional ideals of $\mathcal{O}$ for this equivalence relation. We will denote by $_\mathcal{O}\mathcal{I}$ the set of ideal classes of $\mathcal{O}$. The *class number* h_A of A is the number of ideal classes of a maximal order $\mathcal{O}$ of A. It is finite and independent of the maximal order $\mathcal{O}$ (see for instance [Vig, p. 87-88]).

For every (u, v) in $\mathcal{O} \times \mathcal{O} - \{(0, 0)\}$, consider the two left ideals of $\mathcal{O}$

$$I_{u,v} = \mathcal{O}u + \mathcal{O}v, \quad K_{u,v} = \begin{cases} \mathcal{O}u \cap \mathcal{O}v & \text{if } uv \neq 0, \\ \mathcal{O} & \text{otherwise.} \end{cases}$$

The map

$$\mathrm{SL}_2(\mathcal{O}) \backslash \mathbb{P}^1_r(\mathcal{O}) \to (_\mathcal{O}\mathcal{I} \times {}_\mathcal{O}\mathcal{I}),$$

which associates, to the orbit of $[u : v]$ in $\mathbb{P}^1_r(\mathcal{O})$ under $\mathrm{SL}_2(\mathcal{O})$, the couple of ideal classes $([I_{u,v}], [K_{u,v}])$ is a bijection by [KrO, Satz 2.1, 2.2]. In particular, the number of cusps of $\mathrm{SL}_2(\mathcal{O})$ (or the number of ends of $\mathrm{SL}_2(\mathcal{O}) \backslash \mathbb{H}^5_\mathbb{R}$) is the square of the class number h_A of A.

The *norm* $\mathrm{n}(\mathfrak{m})$ of a left ideal $\mathfrak{m}$ of $\mathcal{O}$ is the greatest common divisor of the norms of the nonzero elements of $\mathfrak{m}$. In particular, $\mathrm{n}(\mathcal{O}) = 1$. The *norm* of a left fractional ideal $\mathfrak{m}$ of $\mathcal{O}$ is $\frac{\mathrm{n}(c\mathfrak{m})}{\mathrm{n}(c)}$ for any $c \in \mathbb{N} - \{0\}$ such that $c\mathfrak{m} \subset \mathcal{O}$.

Let $\mathcal{O}$ be a maximal order in A, and let $\mathfrak{m}$ be a left fractional ideal of $\mathcal{O}$, with norm $\mathrm{n}(\mathfrak{m})$. For every $s > 0$, we consider the integer $\psi_{f,\mathfrak{m}}(s)$ equal to

Card $_{\mathrm{SU}_f(\mathcal{O})} \backslash \{(u, v) \in \mathfrak{m} \times \mathfrak{m} \; : \; \mathrm{n}(\mathfrak{m})^{-1} |f(u, v)| \leq s, \quad \mathcal{O}u + \mathcal{O}v = \mathfrak{m}\},$

which is the number of nonequivalent m-primitive representations by f of rational integers with absolute value at most s. Analogously to the cases of binary quadratic and Hermitian forms, we have an explicit asymptotic result for this counting function.

Theorem 8 (Parkkonen-Paulin [PaP5, Theo. 1] [PaP8, Coro. 5.6]) *There exists $\kappa > 0$ such that, as s tends to $+\infty$, with p ranging over positive rational primes,*

$$\psi_{f,\mathfrak{m}}(s) = \frac{45 \, D_A \, \mathrm{Covol}(\mathrm{SU}_f(\mathcal{O}))}{2 \, \pi^2 \, \zeta(3) \, \Delta(f)^2 \, \prod_{p|D_A}(p^3 - 1)} s^4 \, (1 + O(s^{-\kappa})). \qquad \square$$

The proof of the above result again uses Corollaire 4.9 of [PaP4], and [PaP7, Section 4] (see Section 9) for the error term. One considers the $h_A{}^2$ different orbits of the parabolic fixed points xy^{-1} of $\mathrm{SL}_2(\mathcal{O})$ for which $\mathcal{O}x + \mathcal{O}y = \mathfrak{m}$, and connects the counting functions $\psi_{f,x,y}(s)$ defined by

$$\mathrm{Card} \; _{\mathrm{SU}_f(\mathcal{O})}\backslash\big\{(u, v) \in \mathrm{SL}_2(\mathcal{O})(x, y) \; : \; \mathfrak{n}(\mathcal{O}x + \mathcal{O}y)^{-1}|f(u, v)| \leq s\big\}$$

with the geometric counting function that counts the common perpendiculars between a Margulis cusp neighbourhood of the cusp corresponding to xy^{-1} and the totally geodesic immersed hypersurface corresponding to $\mathcal{C}(f)$. The counting function $\psi_{f,x,y}$ depends (besides f) only on the $\mathrm{SL}_2(\mathcal{O})$-orbit of $[x : y]$ in $\mathbb{P}^1_r(\mathcal{O})$, and summing over all such orbits gives the result. We refer to [PaP5] for more details and more general results that cover finite index subgroups of $\mathrm{SL}_2(\mathcal{O})$, and [PaP8, Section 5.1] for the error term.

6 Patterson, Bowen-Margulis and skinning measures

Let M be a complete nonelementary connected Riemannian manifold of dimension at least 2, with pinched negative sectional curvature $-b^2 \leq K \leq -1$. Let $\widetilde{M} \to M$ be a universal Riemannian cover, and let Γ be its covering group. Let $x_0 \in \widetilde{M}$ and let $\delta = \delta_\Gamma \in \,]0, +\infty[$ be the critical exponent of Γ.

6.1 Patterson densities and Bowen-Margulis measures

Let $r > 0$. A family $(\mu_x)_{x \in \widetilde{M}}$ of nonzero finite measures on $\partial_\infty \widetilde{M}$ whose support is the limit set $\Lambda\Gamma$ is a *Patterson density of dimension r* for Γ if it is Γ-equivariant, that is, if it satisfies

$$\gamma_* \mu_x = \mu_{\gamma x} \tag{15}$$

for all $\gamma \in \Gamma$ and $x \in \widetilde{M}$, and if the pairwise Radon-Nikodym derivatives of the measures μ_x for $x \in \widetilde{M}$ exist and satisfy

$$\frac{d\mu_x}{d\mu_y}(\xi) = e^{-r\beta_\xi(x,y)} \tag{16}$$

for all $x, y \in \widetilde{M}$ and $\xi \in \partial_\infty \widetilde{M}$.

If *Poincaré's series*

$$\mathcal{P}_\Gamma(s) = \sum_{\gamma \in \Gamma} e^{-sd(x_0, \gamma x_0)}$$

diverges at $s = \delta$, then Γ is said to be of *divergence type*. In particular, this holds when $\widetilde{M}$ is a symmetric space and Γ is geometrically finite by [Sul2, CoI], see [DaOP] for many more general results. For groups of divergence type, there exists (see for instance [Rob2, Coro. 1.8]), up to multiplication by a constant, one and only one Patterson density $(\mu_x)_{x \in \widetilde{M}}$ of dimension δ for Γ: For every $x \in \widetilde{M}$, the measure μ_x is the weak-* limit of

$$\frac{1}{\mathcal{P}_\Gamma(s)} \sum_{\gamma \in \Gamma} e^{-s \, d(x, \gamma x_0)} \Delta_{\gamma x_0}$$

as $s \to \delta$, see [Pat, Kai], where Δ_y is the unit mass Dirac measure at any point $y \in \widetilde{M}$.

Let $(\mu_x)_{x \in \widetilde{M}}$ be a Patterson density of dimension δ for Γ. The *Bowen-Margulis measure* $\widetilde{m}_{\mathrm{BM}}$ for Γ on $T^1\widetilde{M}$ is defined, using Hopf's parametrisation, by

$$d\widetilde{m}_{\mathrm{BM}}(v) = \frac{d\mu_{x_0}(v_-)d\mu_{x_0}(v_+)dt}{d_{x_0}(v_-, v_+)^{2\delta}}$$
$$= e^{-\delta(\beta_{v_-}(\pi(v), x_0) + \beta_{v_+}(\pi(v), x_0))}d\mu_{x_0}(v_-)d\mu_{x_0}(v_+)dt,$$

see [Sul1, Sul2, Kai]. The Bowen-Margulis measure is independent of the base point x_0, and its support is (in Hopf's parametrisation) $(\Lambda\Gamma \times \Lambda\Gamma - \Delta) \times \mathbb{R}$, where Δ is the diagonal in $\Lambda\Gamma \times \Lambda\Gamma$. It is invariant under the geodesic flow, the antipodal map and the action of Γ, and thus it defines a measure m_{BM} on T^1M which is invariant under the geodesic flow of M and the antipodal map.

When the Bowen-Margulis measure m_{BM} is finite, the group Γ is of divergence type (see for instance [Rob2, p. 19]). Hence denoting the total mass of a measure m by $\|m\|$, the probability measure $\frac{m_{\mathrm{BM}}}{\|m_{\mathrm{BM}}\|}$ is then uniquely defined. When the sectional curvature of M has bounded derivatives, it is the unique probability measure of maximal entropy of the geodesic flow (see [OtP]). When finite, the Bowen-Margulis measure m_{BM} on T^1M is mixing for the geodesic flow, under the mild assumption conjecturally always satisfied, that the geodesic flow is topologically mixing (or that the set of the lengths of the closed geodesics in M is not contained in a discrete subgroup of $\mathbb{R}$), see [Bab1]. This condition holds for instance if M is locally symmetric, or if M is compact, or if Γ contains a parabolic element or if $\Lambda\Gamma$ is not totally disconnected, see for instance [Dal]. In this review, we assume that m_{BM} is finite.

6.2 Skinning measures

Let $\widetilde{C}$ be a nonempty closed convex subset of $\widetilde{M}$. We define in [PaP6] the *skinning measure* $\widetilde{\sigma}_{\widetilde{C}}$ of Γ on $\partial^1_+\widetilde{C}$, using the homeomorphism $w \mapsto w_+$ from $\partial^1_+\widetilde{C}$ to $\partial_\infty\widetilde{M} - \partial_\infty\widetilde{C}$, by

$$d\widetilde{\sigma}_{\widetilde{C}}(w) = e^{-\delta\,\beta_{w(+\infty)}(\pi(w),\,x_0)}\,d(P^+_{\widetilde{C}})_*(\mu_{x_0}|_{\partial_\infty\widetilde{M}-\partial_\infty\widetilde{C}})(w)$$
$$= e^{-\delta\,\beta_{w_+}(P_{\widetilde{C}}(w_+),\,x_0)}\,d\mu_{x_0}(w_+). \qquad (17)$$

We also consider $\widetilde{\sigma}_{\widetilde{C}}$ as a measure on $T^1\widetilde{M}$ with support contained in $\partial^1_+\widetilde{C}$. The skinning measure $\widetilde{\sigma}_{\widetilde{C}}$ is independent of the base point x_0, satisfies $\widetilde{\sigma}_{\gamma\widetilde{C}} = \gamma_*\widetilde{\sigma}_{\widetilde{C}}$ for every isometry γ of $\widetilde{M}$ and its support is $\{w \in \partial^1_+\widetilde{C} : w_+ \in \Lambda\Gamma\} = P^+_{\widetilde{C}}(\Lambda\Gamma - \Lambda\Gamma \cap \partial_\infty\widetilde{C})$. For any $x \in \widetilde{M}$, up to identifying the unit tangent sphere $T^1_x\widetilde{M}$ at x with the boundary at infinity $\partial_\infty\widetilde{M}$ by the map $v \mapsto v_+$, we have $\widetilde{\sigma}_{\{x\}} = \mu_x$.

The skinning measure has been defined by Oh and Shah [OS1, Section 1.2] for the outer unit normal bundles of spheres, horospheres and totally geodesic subspaces in real hyperbolic spaces, see also [HeP3, Lem. 4.3] for a closely related measure. The terminology comes from McMullen's proof of the contraction of the skinning map (capturing boundary information for surface subgroups of 3-manifold groups) introduced by Thurston to prove his hyperbolisation theorem.

When $\widetilde{C}$ is a horoball, the skinning measure of $\widetilde{C}$ is well known. In fact, the outer unit normal bundle $\partial^1_+\widetilde{C}$ of $\widetilde{C}$ is a leaf of the strong unstable foliation of the geodesic flow and the skinning measure $\widetilde{\sigma}_{\widetilde{C}}$ is the conditional measure of the Bowen-Margulis measure on this leaf, see for example [Mar2, Rob2]. The skinning measure of a horoball has also appeared as a measure on $\partial_\infty\widetilde{M}$ with the point at infinity ξ of the horoball removed in [Cos, Tuk, AM] in the constant curvature case and in [HeP2] under the name *Patterson measure on* $\partial_\infty\widetilde{M} - \{\xi\}$ in the general case. Furthermore, using the upper halfspace model of $\mathbb{H}^n_\mathbb{R}$, Oh and Shah consider in [OS1] a measure ω_Γ defined in $\mathbb{R}^{n-1} = \partial_\infty\mathbb{H}^n_\mathbb{R} - \{\infty\}$ by

$$d\omega_\Gamma(\xi) = e^{\delta\beta_\xi(x,\,(\xi,1))}d\mu_x(\xi).$$

Noticing that $(\xi, 1) = P_{\widetilde{C}}(\xi)$ if $\widetilde{C}$ is the horoball in $\mathbb{H}^n_{\mathbb{R}}$ that consists of the points whose vertical coordinate is at least 1, it follows that ω_Γ is the image of the skinning measure of $\widetilde{C}$ under the map $P_{\widetilde{C}}^{-1} : (\xi, 1) \mapsto \xi$.

For later use in Section 8, we introduce some convenient notation. Let $w \in T^1 \widetilde{M}$. When $\widetilde{C} = HB_-(w)$ is the unstable horoball of w, the conditional measure of the Bowen-Margulis measure on the strong unstable leaf $W^{\mathrm{su}}(w)$ of w is denoted by

$$\mu_{W^{\mathrm{su}}(w)} = \widetilde{\sigma}_{HB_-(w)},$$

and similarly, we denote by

$$\mu_{W^{\mathrm{ss}}(w)} = \iota_* \widetilde{\sigma}_{HB_+(w)}$$

the conditional measure of the Bowen-Margulis measure on the strong stable leaf $W^{\mathrm{ss}}(w)$ of w. These two measures are independent of the element w of a given strong unstable leaf and given strong stable leaf, respectively. We also define the conditional measure of the Bowen-Margulis measure on the stable leaf $W^{\mathrm{s}}(w)$ of w, using the homeomorphism $(v', t) \mapsto v = g^t v'$ from $W^{\mathrm{ss}}(w) \times \mathbb{R}$ to $W^{\mathrm{s}}(w)$, by

$$d\mu_w^{\mathrm{s}}(v) = e^{-\delta_\Gamma t} \, d\mu_{W^{\mathrm{ss}}(w)}(v')dt.$$

Let $\widetilde{C}$ be a proper nonempty closed convex subset of $\widetilde{M}$ such that the Γ-orbit of $\widetilde{C}$ is locally finite, and let C be its image in M. Since $\widetilde{\sigma}_{\widetilde{C}}$ is invariant under the stabiliser $\Gamma_{\widetilde{C}}$ of $\widetilde{C}$ in Γ, the measure $\widetilde{\sigma} = \sum_{\gamma \in \Gamma / \Gamma_{\widetilde{C}}} \gamma_* \widetilde{\sigma}_{\widetilde{C}}$ is a Γ-invariant locally finite Borel positive measure on $T^1 \widetilde{M}$ (independent of the choice of representatives of elements of $\Gamma / \Gamma_{\widetilde{C}}$), whose support is contained in the Γ-orbit of $\partial_+^1 \widetilde{C}$. Hence $\widetilde{\sigma}$ induces a locally finite Borel positive measure σ_C on $T^1 M = \Gamma \backslash T^1 \widetilde{M}$, called the *skinning measure* of the properly immersed closed convex subset C, whose support is contained in $\partial_+^1 C$.

Oh and Shah proved in particular that $\|\sigma_C\|$ is finite if $\widetilde{M}$ is geometrically finite with constant curvature -1 and either $\widetilde{C}$ is a horoball centered at a parabolic fixed point or $\delta_\Gamma > 1$ and $\widetilde{C}$ is a codimension 1 totally geodesic submanifold. See [OS3, Section 5] for a precise, more general statement in higher codimension. Extending this result in variable curvature (with a different proof), we give a sharp criterion in [PaP6, Theo. 9] for the finiteness of the skinning measure, by studying its decay in the cusps of M. This decay is analogous to the decay of the Bowen-Margulis measure in the cusps, which was first studied by Sullivan [Sul2] who called it the fluctuating density property (see also [SV] and [HeP2, Theo. 4.1]). The criterion, as in the case of the Bowen-Margulis measure in [DaOP], is a separation property of critical exponents.

6.3 Disintegration of the Bowen-Margulis measure

Let $\widetilde{C}$ be a proper nonempty closed convex subset of $\widetilde{M}$. Define

$$U_{\widetilde{C}} = \{v \in T^1\widetilde{M} : v_+ \notin \partial_\infty\widetilde{C}\}, \tag{18}$$

which is a nonempty open subset of $T^1\widetilde{M}$, invariant under the geodesic flow.

Let $f_{\widetilde{C}} : U_{\widetilde{C}} \to \partial^1_+\widetilde{C}$ be the composition of the map from $U_{\widetilde{C}}$ onto $\partial_\infty\widetilde{M} - \partial_\infty\widetilde{C}$ sending v to v_+ and the homeomorphism $P^+_{\widetilde{C}}$ from $\partial_\infty\widetilde{M} - \partial_\infty\widetilde{C}$ to $\partial^1_+\widetilde{C}$. The map $f_{\widetilde{C}}$ is a continuous fibration, invariant under the geodesic flow. The fiber of $f_{\widetilde{C}}$ above $w \in \partial^1_+\widetilde{C}$ is exactly the stable leaf

$$W^s(w) = \{v \in T^1\widetilde{M} : v_+ = w_+\}.$$

See [PaP6, PaP7] for further properties of $f_{\widetilde{C}}$, including the fact that $f_{\widetilde{C}}$ is a Hölder fibration when the sectional curvature of M has bounded derivatives.

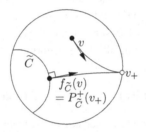

The following disintegration result of the Bowen-Margulis measure over the skinning measure of $\widetilde{C}$ is the crucial tool for the proof in [PaP7] of our general counting result, see Section 8.

Proposition 9 (Parkkonen-Paulin [PaP6]) *Let $\widetilde{C}$ be a proper nonempty closed convex subset of $\widetilde{M}$. The restriction to $U_{\widetilde{C}}$ of the Bowen-Margulis measure $\widetilde{m}_{\mathrm{BM}}$ disintegrates by the fibration $f_{\widetilde{C}} : U_{\widetilde{C}} \to \partial^1_+\widetilde{C}$, over the skinning measure $\widetilde{\sigma}_{\widetilde{C}}$ of $\widetilde{C}$, with conditional measure $e^{\delta\,\beta_{w_+}(\pi(w),\,\pi(v))}\,d\mu^s_w(v)$ on the fiber $f^{-1}_{\widetilde{C}}(w) = W^s(w)$ of $w \in \partial^1_+\widetilde{C}$:*

$$d\widetilde{m}_{\mathrm{BM}}(v) = \int_{w \in \partial^1_+\widetilde{C}} e^{\delta\,\beta_{w_+}(\pi(w),\,\pi(v))}\,d\mu^s_w(v)\,d\widetilde{\sigma}_{\widetilde{C}}(w). \qquad \square$$

7 Finite volume hyperbolic manifolds

In this section, we consider the special case when $\widetilde{M} = \mathbb{H}^n_{\mathbb{R}}$, Γ is a discrete group of isometries of $\widetilde{M}$ and $M = \Gamma \backslash \widetilde{M}$ has finite volume, and we relate the measures defined in Section 6 with more classical measures. For every $p \in \mathbb{N}$, we denote by λ_p the standard Lebesgue measure of $\mathbb{R}^p$.

Under the assumptions of this section, there exists a unique Patterson density $(\mu_x)_{x \in \mathbb{H}_\mathbb{R}^n}$ of dimension $n - 1$ for Γ normalised to have total mass $\text{Vol}(\mathbb{S}^{n-1})$ for every $x \in \mathbb{H}_\mathbb{R}^n$, which we call the spherical density and which we now describe.

In the unit ball model of $\mathbb{H}_\mathbb{R}^n$ with origin 0, the measure μ_0 of the *spherical density* $(\mu_x)_{x \in \mathbb{H}_\mathbb{R}^n}$ is the Lebesgue measure of $\mathbb{S}^{n-1} = \partial_\infty \mathbb{H}_\mathbb{R}^n$ and (see for instance [BriH, p. 273])

$$\frac{d\mu_x}{d\mu_0}(\xi) = e^{-(n-1)\beta_\xi(x,0)} = \left(\frac{1 - \|x\|^2}{\|x - \xi\|^2} \right)^{n-1}.$$

In the upper halfspace model with point at infinity ∞, using the standard inversion mapping the ball model to the upper halfspace model, the *spherical density* $(\mu_x)_{x \in \mathbb{H}_\mathbb{R}^n}$ has the expression, for every $\xi \in \mathbb{R}^{n-1} = \partial_\infty \mathbb{H}_\mathbb{R}^n - \{\infty\}$,

$$d\mu_x(\xi) = \left(\frac{2x_n}{\|x - \xi\|^2} \right)^{n-1} d\lambda_{n-1}(\xi), \tag{19}$$

where x_n is the vertical coordinate of any $x \in \mathbb{H}_\mathbb{R}^n$.

In the unit ball model of $\mathbb{H}_\mathbb{R}^n$, the visual distance d_0 seen from the origin 0 (see Section 2.3) coincides with half the chordal distance (see for example [Bou]). In the upper halfspace model, an easy computation shows that the Busemann cocycle of $\mathbb{H}_\mathbb{R}^n$ is

$$\beta_\xi(x, y) = \ln \left(\frac{y_n}{x_n} \frac{\|x - \xi\|^2}{\|y - \xi\|^2} \right) \tag{20}$$

for all $x, y \in \mathbb{H}_\mathbb{R}^n$ and all $\xi \in \mathbb{R}^{n-1}$. By Equation (3), for any base point $x \in \mathbb{H}_\mathbb{R}^n$ and all $\xi, \eta \in \mathbb{R}^{n-1}$, using the point $u = \left(\frac{\xi+\eta}{2}, \frac{\|\xi-\eta\|}{2} \right)$ as a chosen point on the geodesic line with endpoints ξ, η, we get an expression for the visual distance seen from x:

$$d_x(\xi, \eta) = \frac{x_n \|\xi - \eta\|}{\|x - \xi\| \|x - \eta\|}.$$

Thus, in the upper halfspace model, for any $v \in T^1 \mathbb{H}_\mathbb{R}^n$ such that $v_\pm \neq \infty$, we have

$$d\widetilde{m}_{\text{BM}}(v) = \frac{2^{2(n-1)} d\lambda_{n-1}(v_-) \, d\lambda_{n-1}(v_+) \, dt}{\|v_+ - v_-\|^{2(n-1)}}, \tag{21}$$

where t is the signed distance from the closest point to ∞ on the geodesic line $]v_-, v_+[$ to $\pi(v)$.

It is known that the Liouville measure, normalised to be a probability measure, is the probability measure of maximal entropy for the geodesic flow in constant curvature and finite volume. Thus, the Bowen-Margulis measure coincides (up to a positive multiplicative constant) with the Liouville measure. We now determine the proportionality constant.

Proposition 10 *Let M be a finite volume complete hyperbolic manifold of dimension $n \geq 2$, $d\,\mathrm{Vol}_{T^1 M}$ its Liouville measure, and dm_{BM} its Bowen-Margulis measure, constructed using the spherical Patterson density. Then*

$$m_{\mathrm{BM}} = 2^{n-1}\,\mathrm{Vol}_{T^1 M}\,.$$

In particular,

$$\|m_{\mathrm{BM}}\| = 2^{n-1}\,\mathrm{Vol}(\mathbb{S}^{n-1})\,\mathrm{Vol}(M). \tag{22}$$

Proof. We use the upper halfspace model

$$\mathbb{H}^n_{\mathbb{R}} = \{x = (\overline{x}, x_n) \in \mathbb{R}^{n-1} \times \mathbb{R} \,:\, x_n > 0\}.$$

We parametrise the unit tangent sphere at any point $x \in \mathbb{H}^n_{\mathbb{R}}$ by the positive endpoint $v_+ \in \mathbb{R}^{n-1} \cup \{\infty\}$ of a unit tangent vector $v \in T^1_x \mathbb{H}^n_{\mathbb{R}}$. This gives a parametrisation of the vectors $v \in T^1 \mathbb{H}^n_{\mathbb{R}}$ by the pairs $(x, v_+) \in \mathbb{H}^n_{\mathbb{R}} \times (\mathbb{R}^{n-1} \cup \{\infty\})$. Recall that the Liouville measure disintegrates as

$$d\,\mathrm{Vol}_{T^1 \mathbb{H}^n_{\mathbb{R}}}(v) = \int_{x \in \mathbb{H}^n_{\mathbb{R}}} d\,\mathrm{Vol}_{T^1_x \mathbb{H}^n_{\mathbb{R}}}(v)\, d\,\mathrm{Vol}_{\mathbb{H}^n_{\mathbb{R}}}(x).$$

Hence in the full-measure subset where $v_+ \neq \infty$, the Liouville measure may be written

$$d\,\mathrm{Vol}_{T^1 \mathbb{H}^n_{\mathbb{R}}}(v) = \frac{(2x_n)^{n-1}\, d\lambda_{n-1}(v_+)}{\|x - v_+\|^{2(n-1)}} \frac{d\lambda_n(x)}{x_n^n} \tag{23}$$

$$= \frac{2^{n-1} d\lambda_{n-1}(\overline{x})\, d\lambda_{n-1}(v_+) dx_n}{\|x - v_+\|^{2(n-1)}\, x_n}. \tag{24}$$

In order to relate the formulas (21) and (24), let us give the expression of the coordinates $(\overline{x}, x_n, v_+)$ in terms of the coordinates (v_-, v_+, t).

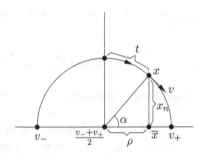

Let α be the angle between the segments $\left[\frac{v_-+v_+}{2}, v_+\right]$ and $\left[\frac{v_-+v_+}{2}, x\right]$. Let ρ be the algebraic distance from $\frac{v_-+v_+}{2}$ to $\overline{x}$ on the line through v_- and v_+ oriented from v_- to v_+. We have

$$\overline{x} = \frac{v_-+v_+}{2} + \rho \frac{v_+-v_-}{\|v_+-v_-\|}$$

and by a formula of [Bea, p. 147],

$$\sinh t = \frac{1}{\tan\alpha} = \frac{\rho}{x_n}.$$

Since $\rho^2 + x_n^2 = \|\frac{v_+-v_-}{2}\|^2$, we hence have

$$x_n = \frac{\|v_+-v_-\|}{2\cosh t} \quad \text{and} \quad \overline{x} = \frac{v_-+v_+}{2} + \frac{v_+-v_-}{2}\tanh t.$$

Writing $\overline{x} = (\overline{x}^1, \ldots, \overline{x}^{n-1})$ and $v_\pm = (v_\pm^1, \ldots, v_\pm^{n-1})$ and differentiating the above equations with v_+ constant, we have, for $i = 1\ldots, n-1$,

$$dx_n = -\frac{\sinh t}{2\cosh^2 t}\|v_+-v_-\| \, dt - \frac{1}{2\cosh t}\sum_{j=1}^{n-1}\frac{v_+^j-v_-^j}{\|v_+-v_-\|}\, dv_-^j$$

and

$$d\overline{x}^i = \frac{1-\tanh t}{2}\, dv_-^i + \frac{v_+^i-v_-^i}{2\cosh^2 t}\, dt.$$

Therefore an easy computation, using the facts that $\overline{x} - v_+ = \frac{1-\tanh t}{2}(v_--v_+)$ and $\|x-v_+\|^2 = \|\overline{x}-v_+\|^2 + x_n^2 = \frac{1-\tanh t}{2}\|v_+-v_-\|^2$, shows that

$$d\overline{x}^1 \wedge \cdots \wedge d\overline{x}^{n-1} \wedge dx_n$$

$$= \frac{\|v_+-v_-\|}{2\cosh t}\left(\frac{1-\tanh t}{2}\right)^{n-1} dv_-^1 \wedge \cdots \wedge dv_-^{n-1} \wedge dt$$

$$= x_n \left(\frac{\|x-v_+\|^2}{\|v_+-v_-\|^2}\right)^{n-1} dv_-^1 \wedge \cdots \wedge dv_-^{n-1} \wedge dt.$$

The result then follows from the formulas (21) and (24). □

Let now C be either a Margulis cusp neighbourhood in M or a totally geodesic immersed submanifold of M with finite volume, and let us relate the skinning measure of C to the usual Riemannian measure on the outer unit normal bundle of C. Note that the Riemannian measure $\text{Vol}_{\partial_+^1 C}$ disintegrates with respect to the base point fibration $\partial_+^1 C \to \partial C$ over the Riemannian measure of ∂C, with measure on the fiber of $x \in \partial C$ the spherical measure on the outer unit normal vectors to C at x:

$$d \operatorname{Vol}_{\partial^1_+ C}(v) = \int_{x \in \partial C} d \operatorname{Vol}_{\partial^1_+ C \cap T^1_x M}(v) \, d \operatorname{Vol}_{\partial C}(x). \tag{25}$$

Homogeneity considerations show that the skinning measure σ_C coincides up to a multiplicative constant with the Riemannian measure $\operatorname{Vol}_{\partial^1_+ C}$. We now compute the constant.

Proposition 11 *Let M be a finite volume complete hyperbolic manifold of dimension $n \geq 2$. We use the spherical Patterson density to define the skinning measures.*

(1) If C is a Margulis cusp neighbourhood, then

$$\sigma_C = 2^{n-1} \operatorname{Vol}_{\partial^1_+ C} .$$

(2) If C is a finite volume totally geodesic properly immersed submanifold of M, then

$$\sigma_C = \operatorname{Vol}_{\partial^1_+ C} .$$

In particular, if C is a Margulis cusp neighbourhood of M, then (see for instance [Hers, p. 473] for the last equality)

$$\|\sigma_C\| = 2^{n-1} \operatorname{Vol}(\partial^1_+ C) = 2^{n-1} \operatorname{Vol}(\partial C) = 2^{n-1}(n-1) \operatorname{Vol}(C),$$

and if C is a finite volume totally geodesic properly immersed submanifold of dimension $k \in \{1, \ldots, n-1\}$ of M, then

$$\|\sigma_C\| = \operatorname{Vol}(\mathbb{S}^{n-k-1}) \operatorname{Vol}(C).$$

Proof. (1) Consider the horoball $\widetilde{C}$ in the upper halfspace model of $\mathbb{H}^n_{\mathbb{R}}$ that consists of the points whose vertical coordinate is at least 1. Fix a base point $x_0 = (0, 1) \in \mathbb{R}^{n-1} \times \,]0, +\infty[$. Note that the closest point to $\xi \in \mathbb{R}^{n-1}$ in $\widetilde{C}$ is $P_{\widetilde{C}}(\xi) = (\xi, 1) \in \mathbb{R}^{n-1} \times \,]0, +\infty[$. Using the definition of the skinning measure for the first equality and the formulas (19) and (20) for the second one, we hence have

$$d\widetilde{\sigma}_{\widetilde{C}}(w) = e^{-(n-1)\beta_{w_+}(P_{\widetilde{C}}(w_+),\, x_0)} \, d\mu_{x_0}(w_+)$$

$$= (\|x_0 - w_+\|^2)^{n-1} \left(\frac{2}{\|x_0 - w_+\|^2} \right)^{n-1} d\lambda_{n-1}(w_+)$$

$$= 2^{n-1} d\lambda_{n-1}(w_+).$$

Since $\partial \widetilde{C} = \{(\overline{x}, 1) \,:\, \overline{x} \in \mathbb{R}^{n-1}\}$ is a codimension one submanifold of $\mathbb{H}^n_{\mathbb{R}}$, whose induced Riemannian metric is isometric to the Euclidean metric on $\mathbb{R}^{n-1}$ by the map $(\overline{x}, 1) \mapsto \overline{x}$, the result follows.

(2) Let $1 \le k \le n - 1$. In the upper halfspace model of $\mathbb{H}^n_{\mathbb{R}}$ with base point $x_0 = (0, \dots, 0, 1)$, consider the k-dimensional totally geodesic subspace,

$$\widetilde{C} = \{x = (x_1, \dots, x_n) \in \mathbb{H}^n_{\mathbb{R}} : x_1 = \cdots = x_{n-k} = 0\},$$

which is isometric to $\mathbb{H}^k_{\mathbb{R}}$ and has Riemannian volume form given by

$$d \operatorname{Vol}_{\widetilde{C}} = \frac{d\lambda_{k-1}(x_{n-k+1}, \dots, x_{n-1}) d\lambda_1(x_n)}{x_n^k}.$$

For any $\xi = (\xi^1, \xi^2) \in \mathbb{R}^{n-k} \times \mathbb{R}^{k-1} = \mathbb{R}^{n-1} = \partial_\infty \mathbb{H}^n_{\mathbb{R}} - \{\infty\}$, the closest point to ξ in $\widetilde{C}$ is $P_{\widetilde{C}}(\xi) = (0, \xi^2, \|\xi^1\|) \in \mathbb{R}^{n-k} \times \mathbb{R}^{k-1} \times]0, +\infty[= \mathbb{H}^n_{\mathbb{R}}$. Note that $\pi(w)_n = \|w^1_+\|$ and $\|\pi(w) - w_+\|^2 = 2\pi(w)_n \|w^1_+\|$ for every $w \in \partial^1_+ \widetilde{C}$. Recall that the map $w \mapsto w_+$ from $\partial^1_+ \widetilde{C}$ to $\partial_\infty \mathbb{H}^n_{\mathbb{R}} - \partial_\infty \widetilde{C} = \mathbb{R}^{n-1} - (\{0\} \times \mathbb{R}^{k-1})$ is a homeomorphism.

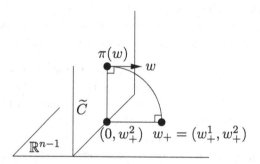

Using the definition of the skinning measure for the first equality and the formulas (19) and (20) for the second one, we hence get

$$d\widetilde{\sigma}_{\widetilde{C}}(w) = e^{-(n-1)\beta_{w_+}(P_{\widetilde{C}}(w_+), x_0)} d\mu_{x_0}(w_+)$$

$$= \left(\frac{\pi(w)_n}{1} \frac{\|x_0 - w_+\|^2}{\|\pi(w) - w_+\|^2}\right)^{n-1} \left(\frac{2}{\|x_0 - w_+\|^2}\right)^{n-1} d\lambda_{n-1}(w_+)$$

$$= \frac{d\lambda_{n-1}(w_+)}{\|w^1_+\|^{n-1}}.$$

On the other hand, by Equation (25), we have

$$d \operatorname{Vol}_{\partial^1_+ \widetilde{C}}(w) = d \operatorname{Vol}_{\mathbb{S}^{n-k-1}}\left(\frac{w^1_+}{\|w^1_+\|}\right) \frac{d\lambda_{k-1}(w^2_+) \, d\lambda_1(\|w^1_+\|)}{\|w^1_+\|^k}.$$

Using spherical coordinates on the first factor of $\mathbb{R}^{n-1} = \mathbb{R}^{n-k} \times \mathbb{R}^{k-1}$, we have

$$d\lambda_{n-1}(w_+) = \|w^1_+\|^{n-k-1} d \operatorname{Vol}_{\mathbb{S}^{n-k-1}}\left(\frac{w^1_+}{\|w^1_+\|}\right) d\lambda_1(\|w^1_+\|) \, d\lambda_{k-1}(w^2_+).$$

Hence $\widetilde{\sigma}_{\widetilde{C}} = \operatorname{Vol}_{\partial^1_+ \widetilde{C}}$, and the result follows by taking quotients. $\qquad \square$

8 The main counting result of common perpendiculars

Let M be a nonelementary complete connected Riemannian manifold with dimension at least 2 and pinched sectional curvature at most -1. Let $\widetilde{M} \to M$ be a universal Riemannian cover of M, with covering group Γ. Let δ be the critical exponent of Γ. We assume that the Bowen-Margulis measure m_{BM} of M is finite and mixing for the geodesic flow.

Theorem 12 (Parkkonen-Paulin [PaP7]) *Let C_- and C_+ be two properly immersed closed convex subsets of M. Assume that their skinning measures σ_{C_-} and σ_{C_+} are finite and nonzero. Then, as $s \to +\infty$,*

$$\mathcal{N}_{C_-,C_+}(s) \sim \frac{\|\sigma_{C_-}\|\,\|\sigma_{C_+}\|}{\delta\,\|m_{\mathrm{BM}}\|}\,e^{\delta s}.$$

As in Herrmann's result (see Equation (9) in Section 3.4) or Oh-Shah's result (see the end of Section 3.6), the endpoints of the common perpendiculars are evenly distributed simultaneously on C_- and on C_+, in the following sense.

Theorem 13 (Parkkonen-Paulin [PaP7]) *Let C_- and C_+ be two properly immersed closed convex subsets of M. Let Ω^- and Ω^+ be relatively compact subsets of $\partial_+^1 C_-$ and $\partial_+^1 C_+$, respectively. Assume that $\sigma_{C_-}(\Omega^-) \neq 0$, $\sigma_{C_+}(\Omega^+) \neq 0$ and $\sigma_{C_-}(\partial\Omega^-) = \sigma_{C_+}(\partial\Omega^+) = 0$. Then, as $s \to +\infty$, the number $\mathcal{N}_{\Omega^-,\,\Omega^+}(s)$ of common perpendiculars of C_- and C_+, with lengths at most s, and with initial vector in Ω^- and terminal vector in $\iota\,\Omega^+$, satisfies*

$$\mathcal{N}_{\Omega^-,\,\Omega^+}(s) \sim \frac{\sigma_{C_-}(\Omega^-)\,\sigma_{C_+}(\Omega^+)}{\delta\,\|m_{\mathrm{BM}}\|}\,e^{\delta s}.$$

When $C_- = \{x\}$, $C_+ = \{y\}$, are singletons, with $\widetilde{x}, \widetilde{y}$ lifts of x, y to $\widetilde{M}$, we recover Roblin's result in [Rob2] that

$$\mathcal{N}_{C_-,C_+}(s) \sim \frac{\|\mu_{\widetilde{x}}\|\,\|\mu_{\widetilde{x}}\|}{\delta\,\|m_{\mathrm{BM}}\|}\,e^{\delta s} = \frac{\|\sigma_{\{x\}}\|\,\|\sigma_{\{y\}}\|}{\delta\,\|m_{\mathrm{BM}}\|}\,e^{\delta s}.$$

Let us give a brief sketch of proof of these results, which uses directly the mixing property of the geodesic flow. This will, in particular, allow us in Section 9 to give estimates on the error terms in the presence of exponential decay of correlations. We refer to [PaP6] for complete proofs, and we only give here a reading guide, the actual proofs require a much more technical approach.

By definition, C_- and C_+ are the images in M of two proper nonempty closed convex subsets $\widetilde{C}_-$ and $\widetilde{C}_+$ in $\widetilde{M}$, whose Γ-orbits are locally finite.

We introduce dynamical neighbourhoods of $\partial_+^1 C_-$ and $\partial_+^1 C_+$, and we define bump functions supported in them, to which we will apply the mixing property. We fix $\eta > 0$ small enough and $R > 0$ big enough.

For every $w \in T^1 \widetilde{M}$, let $V_{w,R}$ be the ball of center w and radius R for Hamenstädt's distance $d_{W^{ss}(w)}$ on the strong stable leaf $W^{ss}(w)$ of w (see Section 2.4). For every proper nonempty closed convex subset $\widetilde{D}$ in $\widetilde{M}$ whose Γ-orbit is locally finite, let $\mathcal{V}_{\eta,R}(\widetilde{D})$ be the union for all $w \in \partial^1_+ \widetilde{D}$ and $s \in] - \eta, \eta[$ of the sets $g^s V_{w,R}$. These dynamical neighbourhoods $\mathcal{V}_{\eta,R}(\widetilde{D})$ of $\partial^1_+ \widetilde{D}$ are natural under isometries, hence, with D the image of $\widetilde{D}$ in M, they allow to define nice neighbourhoods $\mathcal{V}_{\eta,R}(D)$ of $\partial^1_+ D$, that scale nicely under the geodesic flow: $g^t \mathcal{V}_{\eta,R}(\widetilde{D}) = \mathcal{V}_{\eta,e^{-t}R}(\mathcal{N}_t \widetilde{D})$ for every $t \geq 0$.

Let $h_{\eta,R} : T^1 \widetilde{M} \to [0, +\infty]$ be the measurable Γ-invariant map defined by $w \mapsto \frac{1}{2\eta \, \mu_{W^{ss}(w)}(V_{w,R})}$. The constant $R > 0$ is chosen big enough so that the above denominator is nonzero if $w \in \partial^1_+ \widetilde{C}_\pm$ (see [PaP6, Lem. 7]). We denote by $\mathbb{1}_A$ the characteristic function of a subset A. Let $\widetilde{\phi}_{\eta,\widetilde{D}} : T^1 \widetilde{M} \to [0, +\infty]$ be the map defined by (using the convention $\infty \times 0 = 0$)

$$\widetilde{\phi}_{\eta,\widetilde{D}}(v) = \sum_{\gamma \in \Gamma / \Gamma_{\widetilde{D}}} h_{\eta,R} \circ f_{\gamma \widetilde{D}}(v) \, \mathbb{1}_{\mathcal{V}_{\eta,R}(\gamma \widetilde{D})}(v),$$

where $\Gamma_{\widetilde{D}}$ is the stabiliser of $\widetilde{D}$ in Γ and $h_{\eta,R} \circ f_{\gamma \widetilde{D}}(v) \, \mathbb{1}_{\mathcal{V}_{\eta,R}(\gamma \widetilde{D})}(v) = 0$ if $v \notin U_{\gamma \widetilde{D}}$, since $\mathcal{V}_{\eta,R}(\gamma \widetilde{D}) \subset U_{\gamma \widetilde{D}}$. The function $\widetilde{\phi}_{\eta,\widetilde{D}}$ is invariant under Γ, hence defines by taking the quotient by Γ a test function $\phi_{\eta,D} : T^1 M \to [0, +\infty]$. Now define $\phi_\eta^- = \phi_{\eta,C_-}$ and $\phi_\eta^+ = \phi_{\eta,C_+} \circ \iota$. The invariance of the Bowen-Margulis measure by the antipodal map and the disintegration result of Proposition 9 allow to prove (see [PaP6, Prop. 18]) that

$$\int_{T^1 M} \phi_\eta^\pm \, dm_{\mathrm{BM}} = \|\sigma_{C_\pm}\|, \tag{26}$$

and that $\phi_\eta^\pm \, dm_{\mathrm{BM}} \xrightarrow{\ *\ } \sigma_{C_\pm}$ as η goes to 0.

The main trick in the proof is to estimate in two ways the integral

$$\mathcal{I}_\eta(t) = \int_{T^1 M} \phi_\eta^- \circ g^{-t/2} \, \phi_\eta^+ \circ g^{t/2} \, dm_{\mathrm{BM}}.$$

On one hand, by Equation (26) and the mixing property of the geodesic flow, the integral $\mathcal{I}_\eta(t)$ converges, for every fixed $\eta > 0$, to $\frac{\|\sigma_{C_-}\| \, \|\sigma_{C_-}\|}{\|m_{\mathrm{BM}}\|}$ as $t \to +\infty$.

On the other hand, a vector $v \in T^1 M$, with a fixed lift $\widetilde{v}$ to $T^1 \widetilde{M}$, belongs to the support of $\phi_\eta^- \circ g^{-t/2} \, \phi_\eta^+ \circ g^{t/2}$ if and only if $g^{-t/2} v$ belongs to the support of ϕ_η^- and $g^{t/2} v$ belongs to the support of ϕ_η^+, that is, if and only if there exist $\gamma^\pm \in \Gamma$, $s^\pm \in] - \eta, \eta[$, $w^\pm \in \gamma^\pm \partial^1_+ \widetilde{C}_\pm$ and $v^\pm \in V_{w^\pm, R}$ such that $\widetilde{v} = g^{\frac{t}{2} + s^-} v^- = g^{-\frac{t}{2} - s^+} \iota v^+$. For every $\epsilon > 0$, by the properties of negative curvature, this implies, if η is small enough, and uniformly in t big enough, that $\pi(\widetilde{v})$ is not far from the midpoint of a common perpendicular arc between

$\gamma^-\widetilde{C}_-$ and $\gamma^+\widetilde{C}_+$, of length close to t, and that $g^{t/2}\gamma^-\partial_+^1\widetilde{C}_-$ is close to a piece of strong unstable leaf at $\widetilde{v}$, and $g^{-t/2}\gamma^+\iota\,\partial_+^1\widetilde{C}_+$ is close to a piece of strong stable leaf at $\widetilde{v}$ (see [PaP7, Lem. 7]). Furthermore, each such midpoint contributes to the integral $\mathcal{I}_\eta(t)$ by an amount which is, as η is small and uniformly in t big enough, almost $\frac{e^{-\delta t}}{2\eta}$. By a Cesaro type of argument, the results follows, by integrating $e^{\delta t}$.

To pass from Theorem 12 to Theorem 13, we replace $\partial_+^1 C_-$ and $\partial_+^1 C_+$ by Ω^- and Ω^+, the endpoints of the common perpendicular constructed above being close to $\gamma_-\Omega^-$ and $\gamma_+\Omega^+$, which have measure 0 boundary.

We end this section by completing the list of examples given in Section 3, adding the following two cases. They follow (see [PaP7]) by applying the main Theorem 12, the remarks following the statement of Proposition 11, and Equation (22).

Corollary 14 *Let M be a finite volume complete hyperbolic manifold of dimension $n \geq 2$.*

(1) If C_- and C_+ are properly immersed finite volume totally geodesic submanifolds of M of dimensions k_- and k_+ in $[1, n-1]$, respectively, then, as $s \to +\infty$,

$$\mathcal{N}_{C_-,C_+}(s) \sim \frac{\mathrm{Vol}(\mathbb{S}^{n-k_--1})\,\mathrm{Vol}(\mathbb{S}^{n-k_+-1})}{2^{n-1}(n-1)\,\mathrm{Vol}(\mathbb{S}^{n-1})}\,\frac{\mathrm{Vol}(C_-)\,\mathrm{Vol}(C_+)}{\mathrm{Vol}(M)}\,e^{(n-1)s}.$$

(2) If $\mathcal{H}_-$ and $\mathcal{H}_+$ are Margulis cusp neighbourhoods in M, then, as $s \to +\infty$,

$$\mathcal{N}_{\mathcal{H}_-,\mathcal{H}_+}(s) \sim \frac{2^{n-1}(n-1)\,\mathrm{Vol}(\mathcal{H}_-)\,\mathrm{Vol}(\mathcal{H}_+)}{\mathrm{Vol}(\mathbb{S}^{n-1})\,\mathrm{Vol}(M)}\,e^{(n-1)s}. \quad \square$$

In particular, if C_- and C_+ are closed geodesics of M of lengths ℓ_- and ℓ_+, respectively, then the number $\mathcal{N}(s)$ of common perpendiculars (counted with multiplicity) between C_- and C_+ of length at most s satisfies, as $s \to +\infty$,

$$\mathcal{N}(s) \sim \frac{\pi^{\frac{n}{2}-1}\left(\Gamma\left(\frac{n-1}{2}\right)\right)^2}{2^{n-2}(n-1)\,\Gamma\left(\frac{n}{2}\right)}\,\frac{\ell_-\,\ell_+}{\mathrm{Vol}(M)}\,e^{(n-1)s}.$$

When M is a closed hyperbolic surface (in particular $n=2$) and $C_- = C_+$, this formula has been obtained by Martin-McKee-Wambach [MMW] by trace formula methods. Obtaining the case $C_- \neq C_+$, as well as error terms, seems difficult by these methods.

9 Spectral gaps, exponential decay of correlations and error terms

Let M be a nonelementary complete connected Riemannian manifold with dimension at least 2 and pinched sectional curvature at most -1 having bounded derivatives. Let $\widetilde{M} \to M$ be a universal Riemannian cover of M, with covering group Γ. Let δ be the critical exponent of Γ. We assume that the Bowen-Margulis measure m_{BM} of M is finite and mixing for the geodesic flow. We denote by $\overline{m}_{\mathrm{BM}} = \frac{m_{\mathrm{BM}}}{\|m_{\mathrm{BM}}\|}$ its normalisation to a probability measure.

In this section, we give error terms in our main counting result, when the geodesic flow is exponentially mixing. Recall that there are two types of exponential mixing results.

Firstly, when M is locally symmetric with finite volume, then the boundary at infinity of $\widetilde{M}$, the strong unstable, unstable, stable, and strong stable foliations of M are smooth. Hence talking about C^ℓ-smooth leafwise defined functions on $T^1 M$ makes sense. We will denote by $C_c^\ell(T^1 M)$ the vector space of C^ℓ-smooth functions on $T^1 M$ with compact support and by $\|\psi\|_\ell$ the Sobolev $W^{\ell,2}$-norm of any $\psi \in C_c^\ell(T^1 M)$. Note that now the Bowen-Margulis measure of $T^1 M$ is the unique (up to a multiplicative constant) locally homogeneous smooth measure on $T^1 M$ (hence it coincides with the Liouville measure up to a multiplicative constant which we computed in Section 7 in constant curvature).

Given $\ell \in \mathbb{N}$, we will say that the geodesic flow on $T^1 M$ is *exponentially mixing for the Sobolev regularity* ℓ (or that it has *exponential decay of ℓ-Sobolev correlations*) if there exist $c, \kappa > 0$ such that for all $\phi, \psi \in C_c^\ell(T^1 M)$ and $t \in \mathbb{R}$, we have

$$\left| \int_{T^1 M} \phi \circ g^{-t}\, \psi\, d\overline{m}_{\mathrm{BM}} - \int_{T^1 M} \phi\, d\overline{m}_{\mathrm{BM}} \int_{T^1 M} \psi\, d\overline{m}_{\mathrm{BM}} \right| \leq c\, e^{-\kappa |t|}\, \|\psi\|_\ell\, \|\phi\|_\ell.$$

When Γ is a torsion free arithmetic lattice in the isometry group of $\widetilde{M}$, this property, for some $\ell \in \mathbb{N}$, follows from [KlM1, Theo. 2.4.5], with the help of [Clo, Theo. 3.1] to check its spectral gap property, and of [KlM2, Lem. 3.1] to deal with finite cover problems.

Secondly, when $\widetilde{M}$ is assumed to be as in the beginning of this section, then the boundary at infinity, the strong unstable, unstable, stable, and strong stable foliations are only Hölder smooth (as explained in Section 2.5), hence the appropriate regularity on functions on $\widetilde{M}$ is the Hölder one. For every $\alpha \in\,]0, 1[$, we denote by $C_c^\alpha(X)$ the space of α-Hölder-continuous real-valued functions with compact support on a metric space (X, d), endowed with the

Hölder norm

$$\|f\|_\alpha = \|f\|_\infty + \sup_{x,\,y \in X,\, x \neq y} \frac{|f(x) - f(y)|}{d(x,y)^\alpha}.$$

Given $\alpha \in \,]0, 1[$, we will say that the geodesic flow on $T^1 M$ is *exponentially mixing for the Hölder regularity* α (or that it has *exponential decay of α-Hölder correlations*) if there exist $\kappa, c > 0$ such that for all $\phi, \psi \in C_c^\alpha(T^1 M)$ and $t \in \mathbb{R}$, we have

$$\left| \int_{T^1 M} \phi \circ g^{-t}\, \psi \, d\overline{m}_{\text{BM}} - \int_{T^1 M} \phi \, d\overline{m}_{\text{BM}} \int_{T^1 M} \psi \, d\overline{m}_{\text{BM}} \right| \leq c\, e^{-\kappa |t|} \|\phi\|_\alpha \, \|\psi\|_\alpha.$$

This holds for compact manifolds M when M is two-dimensional by [Dol], when M is 1/9-pinched by [GLP, Coro. 2.7], when m_{BM} is the Liouville measure by [Liv], and when M is locally symmetric by [Moo].

Using smoothening processes of the functions $\phi_\eta^\pm$ introduced in the sketch of proof of Section 8, we obtain the following error terms in our main counting result Theorem 12.

Theorem 15 (Parkkonen-Paulin [PaP7]) *Let C_- and C_+ be two properly immersed closed convex subsets of M. Assume that their skinning measures σ_{C_-} and σ_{C_+} are finite and nonzero. Assume either that M is compact and the geodesic flow is exponentially mixing for the Hölder regularity, or that M is locally symmetric with finite volume, that $C^\pm$ have smooth boundary and the geodesic flow is exponentially mixing for the Sobolev regularity. Then there is some $\kappa > 0$ such that, as $s \to +\infty$,*

$$\mathcal{N}_{C_-, C_+}(s) = \frac{\|\sigma_{C_-}\| \, \|\sigma_{C_+}\|}{\delta \, \|m_{\text{BM}}\|} \, e^{\delta s} \left(1 + \mathrm{O}(e^{-\kappa s})\right).$$

This error term is also valid for the effective counting Theorem 13. This result gives in particular the exponential control in the error terms in the list of examples given in Section 3, as well as in Corollary 14.

As an application of Theorem 15, using Humbert's formula and the area of the fundamental domain of $\mathcal{O}_K$ in $\mathbb{R}^2$ (see for example [EGM, p. 318]), we get a version of Cosentino's asymptotic estimate (10) on the number of common perpendiculars from the Margulis cusp neighbourhood corresponding to the horoball of points with vertical coordinates at most 1 to itself in $\text{PSL}(\mathcal{O}_K) \backslash \mathbb{H}^3_\mathbb{R}$, valid for all discriminants:

$$\mathcal{N}(s) = \frac{\pi \, |\mathcal{O}_K^\times|^2}{4\sqrt{|D_K|}\, \zeta_K(2)} \, e^{2s} \left(1 + \mathrm{O}(e^{-\kappa s})\right),$$

when $s \to +\infty$. We refer to [PaP8] for further generalisations.

10 Gibbs measures and counting arcs with weights

Let M be a nonelementary complete connected Riemannian manifold with dimension at least 2 and pinched sectional curvature at most -1. Let $\widetilde{M} \to M$ be a universal Riemannian cover of M, with covering group Γ, and $x_0 \in \widetilde{M}$.

When counting geodesic arcs, it is sometimes useful to give them a higher weight if they are passing through a given region of M, and even more precisely, through a given region in position and direction. The trick is to introduce a *potential*, that is a Hölder-continuous map $F : T^1 M \to \mathbb{R}$. To shorten the exposition, we will assume in this survey that F is bounded and *reversible*, that is, that $F \circ \iota = F$. These assumptions are not necessary, up to the appropriate modifications, see [BrPP]. Given a piecewise smooth path $c : [a, b] \to M$, one defines its *weighted length* for the potential F as

$$\int_c F = \int_a^b F \circ \dot{c}(t)\, dt.$$

We are now going to adapt the material of Section 6 to the weighted case, see for instance [Led1, Led2, Ham2, Cou, Sch, Moh, PauPS, BrPP] with an emphasis on the last two ones for more information.

Let $\widetilde{F} = F \circ Tp : T^1\widetilde{M} \to \mathbb{R}$ be the lift of F by the universal cover $p : \widetilde{M} \to M$. For every x, y in $\widetilde{M}$, if $c : [0, d(x, y)] \to \widetilde{M}$ is the geodesic path from x to y, let $\int_x^y \widetilde{F} = \int_0^{d(x,y)} \widetilde{F} \circ \dot{c}(t)\, dt$. The *critical exponent* of the potential F is

$$\delta_F = \lim_{n \to +\infty} \frac{1}{n} \ln \sum_{\gamma \in \Gamma,\, d(x,\gamma y) \le n} e^{\int_x^{\gamma y} \widetilde{F}},$$

see [PauPS, Theo. 4.2] for the existence and finiteness of the above limit and its independence on $x, y \in \widetilde{M}$. Replacing the previous critical exponent δ (to which it is equal if $F = 0$), the critical exponent δ_F of the potential F will give the exponential growth rate in the counting of weighted common perpendiculars.

Similarly, the Busemann cocycle $\beta_\xi(x, y)$ needs to be replaced. The (normalized) *Gibbs cocycle* associated to the potential F is the function $C = C^F : \partial_\infty \widetilde{M} \times \widetilde{M} \times \widetilde{M} \to \mathbb{R}$ defined by

$$(\xi, x, y) \mapsto C_\xi(x, y) = \lim_{t \to +\infty} \int_y^{\xi_t} (\widetilde{F} - \delta_F) - \int_x^{\xi_t} (\widetilde{F} - \delta_F),$$

where $t \mapsto \xi_t$ is any geodesic ray with endpoint $\xi \in \partial_\infty \widetilde{M}$. The Gibbs cocycle is well defined by the Hölder-continuity of F. It satisfies obvious equivariance and cocycle properties: For all $x, y, z \in \widetilde{M}$, and for every isometry γ of $\widetilde{M}$, we have

$$C_{\gamma\xi}(\gamma x, \gamma y) = C_\xi(x, y) \quad \text{and} \quad C_\xi(x, z) + C_\xi(z, y) = C_\xi(x, y). \quad (27)$$

Similarly, the Bowen-Margulis measure needs to be replaced. A family $(\mu_x)_{x \in \widetilde{M}}$ of finite Borel measures on $\partial_\infty \widetilde{M}$, whose support is the limit set $\Lambda\Gamma$ of Γ, is a *Patterson density for the potential F* (of dimension δ_F) if

$$\gamma_* \mu_x = \mu_{\gamma x}$$

for all $\gamma \in \Gamma$ and $x \in \widetilde{M}$, and if the following Radon-Nikodym derivative exists for all $x, y \in \widetilde{M}$ and satisfies, for all $\xi \in \partial_\infty \widetilde{M}$,

$$\frac{d\mu_x}{d\mu_y}(\xi) = e^{-C_\xi(x, y)}.$$

Let $(\mu_x)_{x \in \widetilde{M}}$ be such a Patterson density. The *Gibbs measure* on $T^1\widetilde{M}$ for the potential F is the measure $\widetilde{m}_F$ on $T^1\widetilde{M}$ given by

$$d\widetilde{m}_F(v) = e^{C_{v_-}(x_0, \pi(v)) + C_{v_+}(x_0, \pi(v))} \, d\mu_{x_0}(v_-) \, d\mu_{x_0}(v_+) \, dt,$$

using Hopf's parametrisation. The Gibbs measure $\widetilde{m}_F$ is independent of the base point $x_0 \in \widetilde{M}$ used in its definition, and it is invariant under the actions of the group Γ and the geodesic flow. Thus, it defines a measure m_F on T^1M which is invariant under the geodesic flow, called the *Gibbs measure* on T^1M for the potential F. When the Gibbs measure m_F is finite, there exists a unique (up to a multiplicative constant) Patterson density for the potential F; the probability measure $\frac{m_F}{\|m_F\|}$ is uniquely defined; it is the unique probability measure of maximal pressure for the geodesic flow and the potential F when the sectional curvature of M has bounded derivatives; see [PauPS] for proofs of these claims. When finite, the Gibbs measure on T^1M is mixing if the geodesic flow is topologically mixing, see [Bab1].

Let D be a properly immersed closed convex subset of M, and let $\widetilde{D}$ be a proper nonempty closed convex subset of $\widetilde{M}$, whose Γ-orbit is locally finite and whose image in M is D. We also need to adapt the skinning measures to the presence of the potential F. The *skinning measure* of $\widetilde{D}$ for the potential F is the measure $\widetilde{\sigma}_{\widetilde{D}}^F$ on $\partial_+^1 \widetilde{D}$, defined, using the homeomorphism $v \mapsto v_+$ from $\partial_+^1 \widetilde{D}$ to $\partial_\infty \widetilde{M} - \partial_\infty \widetilde{D}$, by

$$d\widetilde{\sigma}_{\widetilde{D}}^F(v) = e^{C_{v_+}(x_0, P_{\widetilde{D}}(v_+))} \, d\mu_{x_0}(v_+).$$

It is independent of the base point x_0, and satisfies $\gamma_*(\widetilde{\sigma}_{\widetilde{D}}^F) = \widetilde{\sigma}_{\gamma\widetilde{D}}^F$ for every $\gamma \in \Gamma$. Let $\Gamma_{\widetilde{D}}$ be the stabiliser in Γ of $\widetilde{D}$. The Γ-invariant locally finite Borel positive measure $\sum_{\gamma \in \Gamma/\Gamma_{\widetilde{D}}} \gamma_* \widetilde{\sigma}_{\widetilde{D}}^F$ defines, through the covering $T^1\widetilde{M} \to T^1M$, a locally finite measure σ_D^F, called the *skinning measure* of D for the potential F. See [BrPP] for further information on the skinning measures with potential.

Let D_-, D_+ be two properly immersed closed convex subsets of M. For every $s \geq 0$, recall that $\text{Perp}_{D_-,D_+}(s)$ is the set of the common perpendiculars from D_- to D_+ having lengths at most s. The *weighted counting function* of common perpendiculars between D_- and D_+ (counted with multiplicities) for the potential F is

$$\mathcal{N}_{D_-,D_+,F}(s) = \sum_{c \in \text{Perp}_{D_-,D_+}(s)} m(c)\, e^{\int_c F}.$$

Using the same scheme of proof as explained in Section 8, we have the following asymptotic result.

Theorem 16 (Parkkonen-Paulin [BrPP]) *Let M be a nonelementary complete connected Riemannian manifold with pinched sectional curvature $-b^2 \leq K \leq -1$, and let $F : T^1 M \to \mathbb{R}$ be a (bounded, reversible) Hölder-continuous map. Let δ_F be the critical exponent of the potential F. Assume that the Gibbs measure m_F is finite and mixing for the geodesic flow. Let D_- and D_+ be two properly immersed closed convex subsets of M. Assume that $\sigma_{D_-}^F$ and $\sigma_{D_+}^F$ are finite and nonzero. Then, as $s \to +\infty$,*

$$\mathcal{N}_{D_-,D_+,F}(s) \sim \frac{\|\sigma_{D_-}^F\| \, \|\sigma_{D_+}^F\|}{\delta_F \, \|m_F\|} \, e^{\delta_F s}.$$

We have error terms in the presence of exponential decay of correlations, and the endpoints of the common perpendiculars are evenly distributed, that is, we may restrict to counting the common perpendiculars with endpoints in measurable subsets Ω_- and Ω_+, with finite nonzero skinning measures for the potential F and negligible boundary, of $\partial_+^1 D_-$ and $\partial_+^1 D_+$, respectively. We refer to [BrPP] for precise statements and proofs.

Jouni Parkkonen
Department of Mathematics and Statistics, P.O. Box 35
40014 University of Jyväskylä, FINLAND.
e-mail: jouni.t.parkkonen@jyu.fi

Frédéric Paulin
Département de mathématique, UMR 8628 CNRS, Bât. 425
Université Paris-Sud, 91405 ORSAY Cedex, FRANCE
e-mail: frederic.paulin@math.u-psud.fr

References

[AM] V. Ala-Mattila. *Geometric characterizations for Patterson-Sullivan measures of geometrically finite Kleinian groups.* Ann. Acad. Sci. Fenn. Math. Diss. **157**, 2011.

[Asl] H. Aslaksen. *Quaternionic determinants.* Math. Intelligencer **18** (1996) 57–65.

[Bab1] M. Babillot. *On the mixing property for hyperbolic systems.* Israel J. Math. **129** (2002) 61–76.

[Bab2] M. Babillot. *Points entiers et groupes discrets : de l'analyse aux systèmes dynamiques.* in "Rigidité, groupe fondamental et dynamique", Panor. Synthèses **13**, 1–119, Soc. Math. France, 2002.

[Bas] A. Basmajian. *The orthogonal spectrum of a hyperbolic manifold.* Amer. Math. J. **115** (1993) 1139–1159.

[Bea] A. F. Beardon. *The geometry of discrete groups.* Grad. Texts Math. **91**, Springer-Verlag, 1983.

[BHP] K. Belabas, S. Hersonsky, and F. Paulin. *Counting horoballs and rational geodesics.* Bull. Lond. Math. Soc. **33** (2001), 606–612.

[BFL] Y. Benoist, P. Foulon, and F. Labourie. *Flots d'Anosov à distributions stable et instable différentiables.* J. Amer. Math. Soc. **5** (1992) 33–74.

[Bou] M. Bourdon. *Structure conforme au bord et flot géodésique d'un CAT(−1) espace.* L'Ens. Math. **41** (1995) 63–102.

[Bowd] B. Bowditch. *Geometrical finiteness with variable negative curvature.* Duke Math. J. **77** (1995) 229–274.

[Bowe] R. Bowen. *The equidistribution of closed geodesics.* Amer. J. Math. **94** (1972), 413–423.

[Brid] M. Bridgeman. *Orthospectra of geodesic laminations and dilogarithm identities on moduli space.* Geom. Topol. **15** (2011) 707–733.

[BriK] M. Bridgeman and J. Kahn. *Hyperbolic volume of manifolds with geodesic boundary and orthospectra.* Geom. Funct. Anal. **20** (2010) 1210–1230.

[BriH] M. R. Bridson and A. Haefliger. *Metric spaces of non-positive curvature.* Grund. math. Wiss. **319**, Springer Verlag, 1999.

[Brin] M. Brin. *Ergodicity of the geodesic flow.* Appendix in W. Ballmann, *Lectures on spaces of nonpositive curvature*, DMV Seminar **25**, Birkhäuser, 1995, 81–95.

340

[BrPP] A. Broise-Alamichel, J. Parkkonen, and F. Paulin. *Equidistribution and counting under equilibrium states and in quantum graphs, with applications to non-Archimedean Diophantine approximation.* In preparation.

[Cal] D. Calegari. *Bridgeman's orthospectrum identity.* Topology Proc. **38** (2011) 173–179.

[Clo] L. Clozel. *Démonstration de la conjecture τ.* Invent. Math. **151** (2003) 297–328.

[Coh] H. Cohn. *A second course in number theory.* Wiley, 1962, reprinted as *Advanced number theory*, Dover, 1980.

[CoI] K. Corlette and A. Iozzi. *Limit sets of discrete groups of isometries of exotic hyperbolic spaces.* Trans. Amer. Math. Soc. **351** (1999) 1507–1530.

[Cos] S. Cosentino. *Equidistribution of parabolic fixed points in the limit set of Kleinian groups.* Erg. Theo. Dyn. Syst. **19** (1999) 1437–1484.

[Cou] Y. Coudene. *Gibbs measures on negatively curved manifolds.* J. Dynam. Control Syst. **9** (2003) 89–101.

[Dal] F. Dal'Bo. *Remarques sur le spectre des longueurs d'une surface et comptage.* Bol. Soc. Bras. Math. **30** (1999) 199–221.

[DaOP] F. Dal'Bo, J.-P. Otal, and M. Peigné. *Séries de Poincaré des groupes géométriquement finis.* Israel J. Math. **118** (2000) 109–124.

[Die] J. Dieudonné. *Les déterminants sur un corps non commutatif.* Bull. Soc. Math. France, **71** (1943) 27–45.

[Dol] D. Dolgopyat. *On decay of correlation in Anosov flows.* Ann. of Math. **147** (1998) 357–390.

[DRS] W. Duke, Z. Rudnick, and P. Sarnak. *Density of integer points on affine homogeneous varieties.* Duke Math. J. **71** (1993) 143–179.

[EGM] J. Elstrodt, F. Grunewald, and J. Mennicke. *Groups acting on hyperbolic space: Harmonic analysis and number theory.* Springer Mono. Math., Springer Verlag, 1998.

[EM] A. Eskin and C. McMullen. *Mixing, counting, and equidistribution in Lie groups.* Duke Math. J. **71** (1993) 181–209.

[Fed] H. Federer. *Curvature measures.* Trans. Amer. Math. Soc. **93** (1959) 418–491.

[Ghy] E. Ghys. *Flots d'Anosov dont les feuilletages stables sont différentiables.* Ann. Sci. Ec. Norm. Sup. **20** (1987) 251–270.

[GLP] P. Giulietti, C. Liverani, and M. Pollicott. *Anosov flows and dynamical zeta functions.* Ann. of Math. **178** (2013) 687–773.

[Gro] W. Grotz. *Mittelwert der Eulerschen φ-Funktion und des Quadrates der Dirichletschen Teilerfunktion in algebraischen Zahlkörpern.* Monatsh. Math. **88** (1979) 219–228.

[Ham1] U. Hamenstädt. *A new description of the Bowen-Margulis measure.* Erg. Theo. Dyn. Syst. **9** (1989) 455–464.

[Ham2] U. Hamenstädt. *Cocycles, Hausdorff measures and cross ratios.* Erg. Theo. Dyn. Syst. **17** (1997) 1061–1081.

[HaW] G. H. Hardy and E. M. Wright. *An introduction to the theory of numbers.* Oxford Univ. Press, sixth ed., 2008.

[Herr] O. Herrmann. *Über die Verteilung der Längen geodätischer Lote in hyperbolischen Raumformen.* Math. Z. **79** (1962) 323–343.

[Hers] S. Hersonsky. *Covolume estimates for discrete groups of hyperbolic isometries having parabolic elements.* Michigan Math. J. **40** (1993) 467–475.

[HeP1] S. Hersonsky and F. Paulin. *On the rigidity of discrete isometry groups of negatively curved spaces.* Comm. Math. Helv. **72** (1997) 349–388.

[HeP2] S. Hersonsky and F. Paulin. *Counting orbit points in coverings of negatively curved manifolds and Hausdorff dimension of cusp excursions.* Erg. Theo. Dyn. Syst. **24** (2004) 803–824.

[HeP3] S. Hersonsky and F. Paulin. *On the almost sure spiraling of geodesics in negatively curved manifolds.* J. Diff. Geom. **85** (2010) 271–314.

[HiP] M. Hirsch and C. Pugh. *Smoothness of horocycle foliations.* J. Diff. Geom. **10** (1975) 225–238.

[Hop] E. Hopf. *Ergodic theory and the geodesic flow on surfaces of constant negative curvature.* Bull. Amer. Math. Soc. **77** (1971) 863–877.

[Hub] H. Huber. *Zur analytischen Theorie hyperbolischen Raumformen und Bewegungsgruppen.* Math. Ann. **138** (1959) 1–26.

[HuK] S. Hurder and A. Katok. *Differentiability, rigidity and Godbillon-Vey classes for Anosov flows.* Publ. Math. IHES. **72** (1990) 5–61.

[Hux] M. N. Huxley. *Exponential sums and lattice points III.* Proc. London Math. Soc. **87** (2003) 591–609.

[Kai] V. Kaimanovich. *Invariant measures of the geodesic flow and measures at infinity on negatively curved manifolds.* Ann. Inst. Henri Poincaré, Phys. Théo. **53** (1990) 361–393.

[KaH] A. Katok and B. Hasselblatt. *Introduction to the modern theory of dynamical systems.* Ency. Math. App. **54**, Camb. Univ. Press, 1995.

[Kel] R. Kellerhals. *Quaternions and some global properties of hyperbolic 5-manifolds.* Canad. J. Math. **55** (2003) 1080–1099.

[KlM1] D. Kleinbock and G. Margulis. *Bounded orbits of nonquasiunipotent flows on homogeneous spaces.* Sinai's Moscow Seminar on Dynamical Systems, 141–172, Amer. Math. Soc. Transl. Ser. **171**, Amer. Math. Soc. 1996.

[KlM2] D. Kleinbock and G. Margulis. *Logarithm laws for flows on homogeneous spaces.* Invent. Math. **138** (1999) 451–494.

[Klo] H. D. Kloosterman. *On the representation of numbers in the form* $ax^2 + by^2 + cz^2 + dt^2$. Acta Math. **49** (1927) 407–464.

[Kon] A. Kontorovich. *The hyperbolic lattice point count in infinite volume with applications to sieves.* Duke Math. J. **149** (2009) 1–36.

[KoO] A. Kontorovich and H. Oh. *Apollonian circle packings and closed horospheres on hyperbolic 3-manifolds.* J. Amer. Math. Soc. **24** (2011) 603–648.

[Kor] J. Korevaar. *Tauberian theory.* Grund. math. Wiss. **329**, Springer Verlag, 2010.

[KrO] V. Krafft and D. Osenberg. *Eisensteinreihen für einige arithmetisch definierte Untergruppen von* $SL_2(\mathbb{H})$. Math. Z. **204** (1990) 425–449.

[Lan] E. Landau. *Elementary number theory.* Chelsea Pub. 1958.

[LaP] P. D. Lax and R. S. Phillips. *The asymptotic distribution of lattice points in Euclidean and non-Euclidean spaces.* J. Funct. Anal. **46** (1982) 280–350.

[Led1] F. Ledrappier. *Structure au bord des variétés à courbure négative.* Sém. Théorie Spec. Géom. Grenoble **13**, Année 1994–1995, 97–122.

[Led2] F. Ledrappier. *A renewal theorem for the distance in negative curvature.* In "Stochastic Analysis" (Ithaca, 1993), p. 351–360, Proc. Symp. Pure Math. **57** (1995), Amer. Math. Soc.

[Liv] C. Liverani. *On contact Anosov flows.* Ann. of Math. **159** (2004) 1275–1312.

[MaR] C. Maclachlan and A. Reid. *Parametrizing Fuchsian subgroups of the Bianchi groups.* Canad. J. Math. **43** (1991) 158-181.

[Mar1] G. Margulis. *Applications of ergodic theory for the investigation of manifolds of negative curvature.* Funct. Anal. Applic. **3** (1969) 335–336.

[Mar2] G. Margulis. *On some aspects of the theory of Anosov systems.* Mono. Math., Springer Verlag, 2004.

[MMW] K. Martin, M. McKee, and E. Wambach. *A relative trace formula for a compact Riemann surface.* Int. J. Number Theory **7** (2011) 389–429; see webpage of first author for errata.

[Mey] R. Meyerhoff. *The ortho-length spectrum for hyperbolic 3-manifolds.* Quart. J. Math. Oxford **47** (1996) 349–359.

[Moh] O. Mohsen. *Le bas du spectre d'une variété hyperbolique est un point selle.* Ann. Sci. École Norm. Sup. **40** (2007) 191–207.

[Moo] C. Moore. *Exponential decay of correlation coefficients for geodesic flows.* In "Group representations, ergodic theory, operator algebras, and mathematical physics" (Berkeley, 1984), 163–181, Math. Sci. Res. Inst. Publ. **6**, Springer, 1987.

[Mos] G. D. Mostow. *Strong rigidity of locally symetric spaces.* Ann. Math. Studies **78**, Princeton Univ. Press, 1973.

[OS1] H. Oh and N. Shah. *The asymptotic distribution of circles in the orbits of Kleinian groups.* Invent. Math. **187** (2012) 1–35.

[OS2] H. Oh and N. Shah. *Equidistribution and counting for orbits of geometrically finite hyperbolic groups.* J. Amer. Math. Soc. **26** (2013) 511–562.

[OS3] H. Oh and N. Shah. *Counting visible circles on the sphere and Kleinian groups.* In "Geometry, Topology, and Dynamics in Negative Curvature", 272–287, London Math. Soc. Lecture Notes in Math. **425**, Cambridge University Press, 2016.

[OtP] J.-P. Otal and M. Peigné. *Principe variationnel et groupes Kleiniens.* Duke Math. J. **125** (2004) 15–44.

[PaP1] J. Parkkonen and F. Paulin. *Prescribing the behaviour of geodesics in negative curvature.* Geom. & Topo. **14** (2010) 277–392.

[PaP2] J. Parkkonen and F. Paulin. *Equidistribution, counting and arithmetic applications.* Oberwolfach Report **29** (2010) 35–37.

[PaP3] J. Parkkonen and F. Paulin. *On the representations of integers by indefinite binary Hermitian forms.* Bull. London Math. Soc. **43** (2011) 1048–1058.

[PaP4] J. Parkkonen and F. Paulin. *Équidistribution, comptage et approximation par irrationnels quadratiques.* J. Mod. Dyn. **6** (2012) 1–40.

[PaP5] J. Parkkonen and F. Paulin. *On the arithmetic and geometry of binary Hamiltonian forms.* Appendix by Vincent Emery. Algebra & Number Theory **7** (2013) 75–115.

[PaP6] J. Parkkonen and F. Paulin. *Skinning measures in negative curvature and equidistribution of equidistant submanifolds.* Erg. Theo. Dyn. Syst. **34** (2014) 1310–1342.

[PaP7] J. Parkkonen and F. Paulin. *Counting common perpendicular arcs in negative curvature.* Preprint [arXiv:1305.1332], to appear in Erg. Theo. Dyn. Syst.

[PaP8] J. Parkkonen and F. Paulin. *On the arithmetic of crossratios and generalised Mertens' formulas.* Numéro Spécial "Aux croisements de la géométrie hyperbolique et de l'arithmétique", F. Dal'Bo, C. Lecuire eds, Ann. Fac. Scien. Toulouse **23** (2014) 967–1022.

[PaPo] W. Parry and M. Pollicott. *An analog of the prime number theorem for closed orbits of Axiom A flows.* Ann. of Math. **118** (1983) 573–591.

[Par] W. Parry. *Bowen's equidistribution theory and the Dirichlet density theorem.* Erg. Theo. Dyn. Syst. **4** (1984) 117–134.

[Pat] S. J. Patterson. *The limit set of a Fuchsian group.* Acta. Math. **136** (1976) 241–273.

[PauPS] F. Paulin, M. Pollicott, and B. Schapira. *Equilibrium states in negative curvature.* Astérisque **373**, Soc. Math. France 2015.

[Pol] M. Pollicott. *A symbolic proof of a theorem of Margulis on geodesic arcs on negatively curved manifolds.* Amer. J. Math. **117** (1995) 289–305.

[Pra] G. Prasad. *Volumes of S-arithmetic quotients of semi-simple groups.* Publ. Math. IHES **69** (1989) 91–117.

[Rei] I. Reiner. *Maximal orders.* Academic Press, 1972.

[Rob1] T. Roblin. *Sur la fonction orbitale des groupes discrets en courbure négative.* Ann. Inst. Fourier **52** (2002) 145–151.

[Rob2] T. Roblin. *Ergodicité et équidistribution en courbure négative.* Mémoires Soc. Math. France, **95** (2003).

[Sar] P. Sarnak. *The arithmetic and geometry of some hyperbolic three-manifolds.* Acta Math. **151** (1983) 253–295.

[Sch] B. Schapira. *On quasi-invariant transverse measures for the horospherical foliation of a negatively curved manifold.* Erg. Theo. Dyn. Syst. **24** (2004) 227–257.

[Sha] R. Sharp. *Periodic orbits of hyperbolic flows.* In G. A. Margulis, "On some aspects of the theory of Anosov systems", Springer Verlag, 2004.

[Sto] L. Stoyanov. *Spectra of Ruelle transfer operators for axiom A flows.* Nonlinearity **24** (2011) 1089–1120.

[SV] B. Stratmann and S. L. Velani. *The Patterson measure for geometrically finite groups with parabolic elements, new and old.* Proc. London Math. Soc. **71** (1995) 197–220.

[Sul1] D. Sullivan. *The density at infinity of a discrete group of hyperbolic motions.* Publ. Math. IHES **50** (1979) 172–202.

[Sul2] D. Sullivan. *Entropy, Hausdorff measures old and new, and the limit set of geometrically finite Kleinian groups.* Acta Math. **153** (1984) 259–277.

[Tuk] P. Tukia. *The Poincaré series and the conformal measure of conical and Myrberg limit points.* J. Analyse Math. **62** (1994) 241–259.

[Vig] M. F. Vignéras. *Arithmétique des algèbres de quaternions.* Lect. Notes in Math. **800**, Springer Verlag, 1980.

[Wal] R. Walter. *Some analytical properties of geodesically convex sets.* Abh. Math. Sem. Univ. Hamburg **45** (1976) 263–282.

10

Lattices in hyperbolic buildings

ANNE THOMAS[1]

Introduction

This survey is intended as a brief introduction to the theory of hyperbolic buildings and their lattices. Hyperbolic buildings are negatively curved geometric objects which also have a rich algebraic and combinatorial structure, and the study of these buildings and the lattices in their automorphism groups involves a fascinating mixture of techniques from many different areas of mathematics.

Roughly speaking, a hyperbolic building is obtained by gluing together many hyperbolic spaces which are tiled by polyhedra. For the precise definition, together with background on general buildings and known constructions of hyperbolic buildings, see Section 1 below.

Given a hyperbolic building Δ, we write $G = \mathrm{Aut}(\Delta)$ for the group of automorphisms, or cellular isometries, of Δ. When the building Δ is locally finite, the group G equipped with the compact-open topology is naturally a locally compact topological group, and so has a Haar measure μ. In this topology on G, a subgroup $\Gamma < G$ is discrete if and only if it acts on Δ with finite cell stabilisers. A *lattice* in G is a discrete subgroup $\Gamma < G$ such that $\mu(\Gamma \backslash G) < \infty$, and a lattice Γ is *cocompact* (or *uniform*) if $\Gamma \backslash G$ is compact. The Haar measure μ on G may be normalised so that the covolume $\mu(\Gamma \backslash G)$ of a lattice $\Gamma < G$ is given by the formula

$$\mu(\Gamma \backslash G) = \sum \frac{1}{|\operatorname{Stab}_\Gamma(v)|} \tag{1}$$

where the sum is taken over a set of representatives for the orbits of the vertices of Δ under the action of Γ. A discrete subgroup $\Gamma < G$ is then a lattice if and only if this sum converges, and is a cocompact lattice if and only if this sum has finitely many terms.

[1] The author is supported in part by ARC Grant No. DP110100440.

For brevity, we will refer to lattices in the automorphism groups of hyperbolic buildings as lattices in hyperbolic buildings. In Section 2 we describe known constructions of lattices in hyperbolic buildings, then discuss many different questions concerning such lattices. Much of the study of lattices in hyperbolic buildings is motivated by the well-developed theory of lattices in semisimple Lie groups.

We recommend that the reader consult also the survey Farb–Hruska–Thomas [24], which discusses more general polyhedral complexes and their automorphism groups and lattices. In order to avoid repetition, we have concentrated here on questions concerning lattices which are particularly pertinent to hyperbolic buildings, and/or where there has been progress since [24] was written. We also provide greater detail than [24] on hyperbolic Coxeter groups and constructions of hyperbolic buildings.

Acknowledgements

The author thanks the organizers of "Geometry, Topology and Dynamics in Negative Curvature" for the opportunity to attend such a well-run and interesting conference, the London Mathematical Society for travel support and an anonymous referee for helpful comments. The author is indebted to her coauthors on [24] and to many of the researchers cited below for numerous rewarding discussions.

1 Buildings and hyperbolic buildings

In this section we recall definitions and results concerning both general and hyperbolic buildings. We begin with a summary of the relevant theory of Coxeter groups and Coxeter polytopes in Section 1.1, and some background on polyhedral complexes in Section 1.2. Buildings and examples of spherical and Euclidean buildings are discussed in Section 1.3 before we focus on hyperbolic buildings in Section 1.4. Some references for the theory of buildings are Abramenko–Brown [1], Brown [11] and Ronan [53].

1.1 Coxeter groups and Coxeter polytopes

We mostly follow the reference Davis [19], particularly Chapter 6, and concentrate on the hyperbolic case.

Recall that a *Coxeter group* is a group W with finite generating set S and presentation of the form

$$W = \langle s \in S \mid (st)^{m_{st}} = 1 \rangle$$

where $s, t \in S$, $m_{ss} = 1$ for all $s \in S$ and if $s \neq t$ then $m_{st} = m_{ts}$ is an integer ≥ 2 or $m_{st} = \infty$, meaning that the product st has infinite order. The pair (W, S) is called a *Coxeter system*. A Coxeter system (W, S) is *right-angled* if for each $s, t \in S$ with $s \neq t$, $m_{st} \in \{2, \infty\}$. Note that $m_{st} = 2$ if and only if $st = ts$.

Let $\mathbb{X}^n$ be the n–dimensional sphere, n–dimensional Euclidean space or n–dimensional (real) hyperbolic space. Many important examples of Coxeter groups arise as discrete reflection groups acting on $\mathbb{X}^n$, as follows. Let P be a convex polyhedron in $\mathbb{X}^n$ with all dihedral angles integer submultiples of π. Such a P is called a *Coxeter polytope*. Let $W = W(P)$ be the group generated by the set $S = S(P)$ of reflections in the codimension one faces of P. Then (W, S) is a Coxeter system and W is a discrete subgroup of the isometry group of $\mathbb{X}^n$ (see [19, Theorem 6.4.3]). Moreover, the action of W tessellates $\mathbb{X}^n$ by copies of P. For example, let P be a right-angled hyperbolic p–gon, $p \geq 5$. Then the corresponding Coxeter system is right-angled with p generators, one for each side of P, so that if s and t are reflections in distinct sides then $m_{st} = 2$ when these sides are adjacent, and otherwise $m_{st} = \infty$.

Suppose that $\mathbb{X}^n$ is the sphere or Euclidean space. Then Coxeter polytopes $P \subset \mathbb{X}^n$ exist and have been classified in every dimension, and the corresponding Coxeter systems (W, S) are the spherical or affine Coxeter systems, respectively (see [19, Table 6.1] for the classification).

If $\mathbb{X}^n$ is n–dimensional hyperbolic space $\mathbb{H}^n$, then there is no complete classification of Coxeter polytopes. Vinberg's Theorem [65] establishes that compact hyperbolic Coxeter polytopes can exist only in dimension $n \leq 29$, although at the time of writing the highest dimension in which an example is known is $n = 8$ (due to Bugaenko [12]). Finite volume hyperbolic Coxeter polytopes have also been investigated, with for example Prokhorov [50] proving these can exist only in dimension $n \leq 995$. For $n \leq 6$ (respectively, $n \leq 19$), there are infinitely many essentially distinct compact (respectively, finite volume) hyperbolic Coxeter polytopes (Allcock [2]). In dimension 3, Andreev's Theorem [3] classifies compact hyperbolic Coxeter polytopes, but in dimensions $n \geq 4$ only special cases have been considered, and there seems little hope of a complete list. Hyperbolic Coxeter polytopes which are simplices exist in dimensions $n \leq 4$ only, and their classification is given in [19, Table 6.2]. Right-angled compact hyperbolic polytopes also exist in dimensions $n \leq 4$ only, and there are infinitely many examples in each dimension $n \leq 4$ (see [66]). A right-angled example in dimension 3 is the dodecahedron, which tessellates $\mathbb{H}^3$ as depicted on the cover of Thurston's book [59], and a right-angled example in dimension 4 is the 120–cell, which has 120 dodecahedral faces. For other special cases of compact hyperbolic Coxeter polytopes, see for example the work of Esselmann [23], Felikson–Tumarkin [25, 26, 61], Kaplinskaja [40] and

Tumarkin [60, 62, 63], and for an overview of results in the finite volume case, see the introduction to [2].

Let $W = W(P)$ be the Coxeter group generated by reflections in the faces of a hyperbolic Coxeter polytope P. If P is compact then W is a word-hyperbolic group, that is, a group which is hyperbolic in the sense of Gromov. (For background on word-hyperbolic groups, see [10]. Necessary and sufficient conditions for word-hyperbolicity of Coxeter groups were established by Moussong [19, Corollary 12.6.3].) On the other hand, some authors such as Humphreys [38] reserve "hyperbolic Coxeter group" for the case that P is a compact simplex. In this survey, when necessary we will refer to the discrete reflection group $W(P)$, where P is a (compact) hyperbolic Coxeter polytope, as a *(cocompact) geometric hyperbolic Coxeter group*.

1.2 Polyhedral complexes and links

Polyhedral complexes are generalisations of (geometric realisations of) simplicial complexes. Roughly speaking, they are obtained by gluing together polyhedra from the constant curvature space $\mathbb{X}^n$ (the sphere, Euclidean space or hyperbolic space), using isometries along faces. For the formal definition of a polyhedral complex, see for example [24, Section 2.1]. We sometimes refer to 2-dimensional polyhedral complexes as *polygonal complexes*.

The tessellation of $\mathbb{X}^n$ by copies of a Coxeter polytope P is a simple example of a polyhedral complex. A metric tree is a 1-dimensional Euclidean polyhedral complex, and a product of two such trees is a 2-dimensional Euclidean polygonal complex.

Let x be a vertex of an n-dimensional polyhedral complex X. The *link* of x, denoted $\mathrm{Lk}(x, X)$, is the spherical $(n-1)$-dimensional polyhedral complex obtained by intersecting X with an n-sphere of sufficiently small radius centred at x. For example, if X has dimension 2, then $\mathrm{Lk}(x, X)$ may be identified with the graph whose vertices correspond to endpoints of edges of X that are incident to x, and whose edges correspond to corners of faces of X incident to x. By rescaling so that for each x the n-sphere around x has radius 1, we induce a canonical metric on each link.

The importance of links is that they provide a local condition for nonpositive or negative curvature of polyhedral complexes, using the following result which combines several theorems of Gromov. For these theorems as well as background on the nonpositive curvature condition CAT(0), and the negative curvature condition CAT(-1), see [10].

Theorem 1 Gromov *Let X be a contractible polyhedral complex of piecewise constant curvature κ. If X has finitely many isometry types of cells, then*

X is CAT(κ) *if and only if for all vertices x of X, the link* Lk(x, X) *is a* CAT(1) *space. In particular, if X is a contractible Euclidean (respectively, hyperbolic) polygonal complex with finitely many isometry types of cells, then X is* CAT(0) *(respectively,* CAT(-1)) *if and only if every embedded loop in the graph* Lk(x, X) *has length at least* 2π.

An important special case of a polyhedral complex is a (k, L)–*complex*, which is a polygonal complex in which each face is a regular k–gon, for $k \geq 3$ an integer, and the link at each vertex is a fixed finite graph L. So long as k and L satisfy a simple condition, Ballmann–Brin [5] showed that a contractible CAT(0) (k, L)–complex may be constructed by a "free" inductive process of adding k–gons to the previous stage. Another construction of (k, L)–complexes for k even is the special case of the Davis–Moussong complex described in [24, Section 3.6].

1.3 Buildings

We will adopt the following "geometric" definition of a building, which is the most appropriate for the hyperbolic case. Other definitions, using simplicial complexes or chamber systems, may be found in, for example, [1] or [53].

Definition 2 Let $\mathbb{X}^n$ be respectively the n–dimensional sphere, n–dimensional Euclidean space or n–dimensional hyperbolic space. Let P be a Coxeter polytope in $\mathbb{X}^n$ and let (W, S) be the corresponding Coxeter system. A respectively *spherical, Euclidean or hyperbolic building of type* (W, S) is a polyhedral complex Δ equipped with a maximal family of subcomplexes, called *apartments*, so that each apartment is isometric to the tessellation of $\mathbb{X}^n$ by copies of P, called *chambers*, and so that:

(1) any two chambers of Δ are contained in a common apartment; and
(2) for any two apartments $\mathcal{A}$ and $\mathcal{A}'$, there exists an isometry $\phi : \mathcal{A} \to \mathcal{A}'$ which fixes $\mathcal{A} \cap \mathcal{A}'$.

The tessellation of a single copy of $\mathbb{X}^n$ by images of P satisfies Definition 2 and is sometimes called a *thin* building. We will mainly be interested in *thick* buildings, those where there is "branching", that is, where each codimension one face of each chamber is contained in at least three distinct chambers. Thick buildings may be thought of as obtained by gluing together many copies of the same tessellation of $\mathbb{X}^n$. A building is *right-angled* if it is of type (W, S) a right-angled Coxeter system.

We now discuss some important examples of spherical and Euclidean buildings. Hyperbolic buildings will be considered in Section 1.4 below.

Examples of spherical buildings

A first example of a spherical building is the complete bipartite graph $K_{q,q}$, which is thick so long as $q \geq 3$. The chambers are the edges of this graph, metrised as quarter-circles, and the apartments are the embedded loops of length 2π. This is a (right-angled) building of type (W, S) where

$$W = \langle s, t \mid s^2 = t^2 = (st)^2 = 1 \rangle$$

is the dihedral group of order 4, acting on the circle.

An example of a spherical building Δ which is not right-angled is the flag complex of the projective plane over the finite field $\mathbb{F}_q$ of order q, which may be constructed as follows. Let V be the vector space $\mathbb{F}_q \times \mathbb{F}_q \times \mathbb{F}_q$ over $\mathbb{F}_q$, let $\mathcal{P}$ be the collection of one-dimensional subspaces of V (the *points* of the projective plane) and let $\mathcal{L}$ be the collection of two-dimensional subspaces of V (the *lines*). A point $p \in \mathcal{P}$ is defined to be *incident* to a line $l \in \mathcal{L}$ if $p \subset l$. The building Δ is then the bipartite graph with vertex set $\mathcal{P} \sqcup \mathcal{L}$ and edges corresponding to incidence. See for instance [53, Chapter 1, Example 3] for the verification that Δ is indeed a building, of type

$$W = \langle s, t \mid s^2 = t^2 = (st)^3 = 1 \rangle$$

the symmetric group on three letters, which is isomorphic to the dihedral group of order 6, acting on the circle. In particular, the apartments of Δ are embedded cycles of 6 edges, and correspond to bases of V. The *standard apartment* is that corresponding to the standard basis for V.

The structure of this spherical building is strongly connected with the structure of the group $G = \mathrm{SL}_3(\mathbb{F}_q)$, in a way that generalises to many other pairings of buildings with groups. The action of G on V induces a natural action on the building Δ. The stabilisers of edges of Δ are the cosets of the upper-triangular subgroup $B < G$, the (standard) *Borel subgroup*, and the stabilisers of vertices are the *parabolic subgroups* of G. Thus the chambers of Δ can be identified with the cosets G/B, and the vertices of Δ can be identified with the disjoint union $G/P_1 \sqcup G/P_2$, where P_1 is the parabolic subgroup fixing the span of $(1, 0, 0)$ and P_2 that fixing the span of $(1, 0, 0)$ and $(0, 1, 0)$. The pointwise stabiliser of the standard apartment is the diagonal subgroup $T < G$, called the *torus*, and the setwise stabiliser of the standard apartment is the group of monomial matrices $N < G$, that is, matrices with exactly one non-zero entry in each row and each column. The normaliser of the torus T is the group N, with quotient $N/T \cong W$, and the group W is called the *Weyl group*.

The discussion in the previous paragraph could be summarised by saying that the group $G = \mathrm{SL}_3(\mathbb{F}_q)$ has a (B, N)–*pair*, also known as a *Tits system*; there is then a building associated to G, of type its Weyl group, which is constructed using cosets of important subgroups as indicated. (For the rather technical definition of a (B, N)–pair, see for example [53, Chapter 5].) Other spherical buildings are associated to other finite groups of Lie type, using their respective (B, N)–pairs.

All one-dimensional spherical buildings are *generalised m–gons*, meaning that they are graphs with diameter m edges and shortest embedded circuit containing $2m$ edges. A finite, thick generalised m–gon exists for $m \in \{2, 3, 4, 6, 8\}$ only (Feit–Higman, see [53, Theorem 3.4]). Generalised 2–gons are complete bipartite graphs. Generalised 3–gons are flag complexes of projective planes, and so there is no classification known. There are many examples of finite generalised 4–gons, but only one or two known examples of finite generalised 6– or 8–gons. For more on generalised m–gons, see for example [30, Chapter 5].

Examples of Euclidean buildings

A first example of a Euclidean building is the tree T_q of valence q, metrised so that each edge has length say 1. The chambers are the edges of the tree, and the apartments are the bi-infinite geodesics. This is a (right-angled) building of type (W, S) where

$$W = \langle s, t \mid s^2 = t^2 = 1 \rangle$$

is the infinite dihedral group acting on the real line, with the generating reflections s and t fixing points distance 1 apart. The product of trees $T_q \times T_q$ is a 2–dimensional right-angled Euclidean building with apartments the tessellation of the Euclidean plane by unit squares, and associated Coxeter system the direct product of two infinite dihedral groups.

Many Euclidean buildings (and all irreducible Euclidean buildings of dimension ≥ 3) are of "algebraic" origin, as in the following example. Let K be a nonarchimedean local field, such the p–adics $\mathbb{Q}_p$ or the field of formal Laurent series $\mathbb{F}_q((t))$. The group $G = \mathrm{SL}_n(K)$ has a (B, N)–pair with associated Weyl group W an affine Coxeter group, so that the action of W tessellates $(n - 1)$–dimensional Euclidean space. Thus the group G has an associated Euclidean building of dimension $(n - 1)$. For instance the building for $\mathrm{SL}_3(\mathbb{Q}_p)$ has apartments the tessellation of the Euclidean plane by equilateral triangles. For further details and references concerning Euclidean buildings, which are also known as *affine buildings*, see [24, Section 3.1].

Links of buildings

Let x be a vertex of an n–dimensional building. Then it is easy to verify that the link of x is a spherical building of dimension $(n - 1)$, with the induced apartment and chamber structure. For example, the link of each vertex in

$T_q \times T_q$ is the complete bipartite graph $K_{q,q}$, and the link of each vertex in the building for $SL_3(\mathbb{F}_q((t)))$ is the spherical building Δ for $SL_3(\mathbb{F}_q)$ from Section 1.3 above.

With the natural piecewise spherical structure, a spherical building is a CAT(1) space. This was shown by Davis [18], generalising a result of Gromov [32] for right-angled spherical Coxeter systems and of Moussong for all spherical Coxeter systems (see [19, Theorem 12.3.3]). By Theorem 1 above, it follows that Euclidean (respectively, hyperbolic) buildings are CAT(0) spaces (respectively, CAT(−1) spaces). The result that irreducible Euclidean buildings are CAT(0) was already well-known, and a proof can be found in [1, Section 11.2].

1.4 Hyperbolic buildings

We first discuss examples and constructions of hyperbolic buildings. Many of these constructions of hyperbolic buildings also yield lattices, as discussed in Section 2 below. We then briefly discuss the classification of hyperbolic buildings.

A hyperbolic building Δ of dimension 2 is sometimes called a *Fuchsian* building, since if Δ has type (W, S) then W may be regarded as a Fuchsian group. By the restrictions on the dimension of hyperbolic Coxeter polytopes discussed in Section 1.1 above, hyperbolic buildings with compact (respectively, finite volume) chambers can exist only in dimension $n \leq 29$ (respectively, $n \leq 995$), and right-angled hyperbolic buildings with compact chambers exist only in dimensions 2, 3 and 4.

A first example of a hyperbolic building is *Bourdon's building* $I_{p,q}$, defined and studied in [7]. Let P be a regular right-angled hyperbolic p–gon, with $p \geq 5$. Then P is a Coxeter polytope, and the building $I_{p,q}$ has type the associated right-angled Coxeter system. Thus $I_{p,q}$ is a right-angled Fuchsian building with apartments hyperbolic planes tessellated by copies of P. The link of each vertex of $I_{p,q}$ is the complete bipartite graph $K_{q,q}$, with $q \geq 2$, and so by Theorem 1 above, Bourdon's building is CAT(−1). Each edge of $I_{p,q}$ is contained in q chambers, thus $I_{p,q}$ is thick for $q \geq 3$. Bourdon's building can be thought of as a hyperbolic version of the product of trees $T_q \times T_q$. However it is not globally a product space. It is a (k, L)–complex with $k = p$ and $L = K_{q,q}$, and as descibed in [7] may be constructed using the Ballmann–Brin inductive process (see Section 1.2 above).

A slightly more general example is the right-angled Fuchsian building $I_{p,\mathbf{q}}$ where $\mathbf{q} = (q_s)_{s \in S}$ is a p–tuple of integers $q_s \geq 2$, indexed by the generators of the Coxeter system (W, S) associated to the regular right-angled p–gon P.

The edges of $I_{p,\mathbf{q}}$ are assigned types $s \in S$, and the vertices then inherit types $\{s, t\} \subset S$ with $m_{st} = 2$, so that the natural action of W on each apartment is type-preserving. The parameters q_s record that each edge of type s is contained in q_s chambers, which makes the building *regular*. Each vertex of type $\{s, t\}$ has link the complete bipartite graph K_{q_s, q_t}.

Any regular right-angled building may be constructed using complexes of groups. (Note that not all right-angled buildings are hyperbolic, since for example a product of trees is not a hyperbolic building.) For the general theory of complexes of groups see [10], and for a summary of this theory in the context of polyhedral complexes see [24, Section 2.3]. The construction for $I_{p,\mathbf{q}}$ that we now sketch appears in Bourdon [8, Example 1.5.1(a)], was known earlier to Davis and Meier and is equivalent to a special case of constructions given in [8, Section 5] and [19, Example 18.1.10]. Let the p–gon P and parameters $\mathbf{q}$ be as in the previous paragraphs. For each $s \in S$ let G_s be a group of order q_s. A complex of groups over P with universal cover the building $I_{p,\mathbf{q}}$ is obtained by assigning groups to the face, edges and vertices of P as follows. The face group is trivial, the group on the edge of type s is G_s and the group on the vertex of type $\{s, t\}$ with $m_{st} = 2$ is the direct product $G_s \times G_t$. Very roughly speaking, the universal cover is obtained by "unfolding" this orbifold-like data.

An example of a Fuchsian building which is not right-angled and is constructed using complexes of groups is as follows. Let P be a regular hyperbolic k–gon, with $k \geq 6$ even and all dihedral angles $\frac{\pi}{3}$. The universal cover of the following complex of groups over P, from Gaboriau–Paulin [28, Section 3.1], is a Fuchsian building with apartments hyperbolic planes tessellated by copies of P, and the link of every vertex the spherical building Δ described in Section 1.3 above. Let $G = \mathrm{SL}_3(\mathbb{F}_q)$ and let B, P_1 and P_2 be the Borel and parabolic subgroups of G as in Section 1.3 above. The face group is B, the edge groups alternate between P_1 and P_2 and all vertex groups are G. In [28, Section 3.4] Gaboriau–Paulin also use complexes of groups to construct some Fuchsian buildings with non-compact chambers. In [7, Example 1.5.3] Bourdon uses complexes of groups to construct Fuchsian buildings with right-angled triangles as chambers.

Some additional constructions of Fuchsian buildings are as follows. Suppose L is a one–dimensional spherical building and $k \geq 3$ is even. Then a Davis–Moussong (k, L)–complex (see Section 1.2 above) may be metrised as a hyperbolic building with all links L and all chambers k–gons. Vdovina [64] constructed various Fuchsian buildings with even-sided chambers as universal covers of finite polygonal complexes whose links are one-dimensional spherical buildings, with not necessarily the same link at each vertex, while Kangaslampi–Vdovina [39] using similar techniques constructed Fuchsian buildings with chambers n–gons, $n \geq 3$, and links generalised 4–gons. Bourdon [7, Example 1.5.2] obtained certain Fuchsian buildings by "hyperbolising" affine buildings.

Recall from Section 1.3 above that some spherical and Euclidean buildings are obtained as buildings for groups which have (B, N)–pairs. In similar fashion, some hyperbolic buildings arise as buildings for Kac–Moody groups. A Kac–Moody group Λ over a finite field $\mathbb{F}_q$ may be thought of as an infinite-dimensional analogue of an algebraic group over a nonarchimedean local field. The group Λ has twin (B, N)–pairs, which yield isomorphic twin buildings Δ_+ and Δ_-, with the group Λ acting diagonally on the product $\Delta_+ \times \Delta_-$. When the Weyl group W of Λ is a (cocompact) geometric hyperbolic Coxeter group, then the associated buildings $\Delta_\pm$ are hyperbolic buildings (with compact chambers). For example, the building $I_{p,q+1}$ may be realised as a Kac–Moody building when q is a power of a prime. For further details, see Carbone–Garland [15] and Rémy [51].

Apart from right-angled and Kac–Moody buildings, there are very few known constructions of hyperbolic buildings of dimension greater than 2. Haglund–Paulin [36] have constructed some three–dimensional hyperbolic buildings using "tree-like" decompositions of the corresponding Coxeter systems, while Davis [20] gives covering-theoretic constructions of some three-dimensional hyperbolic buildings, including some where not all links are the same.

As discussed in Section 1.1 above, there is no complete classification of geometric hyperbolic Coxeter systems, and so there seems little hope of classifying general hyperbolic buildings. Even for Fuchsian buildings, where the associated Coxeter systems are classified, the possible links L are generalised m–gons, which have not been classified. Indeed even after fixing a Coxeter system (W, S) and the link L at each vertex, there may be uncountably many hyperbolic buildings of type (W, S) with links L (see for example [28, Theorem 3.6]). There are however cases in which "local data" does determine the building. For example, Bourdon's building $I_{p,q}$ is the unique simply-connected polygonal complex such that all faces are right-angled hyperbolic p–gons, and all links are $K_{q,q}$ [7, 54]. For more on the question of uniqueness, see [24, Section 2.3].

2 Lattices

We now discuss lattices in hyperbolic buildings. The state of the theory is such that for many hyperbolic buildings, the basic question of the existence of lattices in their automorphism groups is open. We thus begin by describing known constructions of lattices in Section 2.1, then discuss a range of other questions concerning lattices in Section 2.2.

2.1 Constructions of lattices

Let Δ be a hyperbolic building, and recall the characterisation of lattices in $\mathrm{Aut}(\Delta)$ from the introduction. It will be seen that much more is known about constructions of cocompact lattices in $\mathrm{Aut}(\Delta)$ than about constructions of non-cocompact lattices.

If Δ is the universal cover of a finite polyhedral complex, as for example in the constructions of Fuchsian buildings due to Vdovina [64], then the fundamental group of that finite polyhedral complex is a cocompact lattice in $\mathrm{Aut}(\Delta)$, since it acts freely and cocompactly on Δ.

Now suppose Δ is the universal cover of a complex of finite groups, over a finite underlying polyhedral complex Y. For instance, one of the constructions of Bourdon's building $I_{p,q}$ is as the universal cover of a complex of finite groups over a right-angled hyperbolic p–gon P. Let Γ be the fundamental group of this complex of groups. Roughly speaking, Γ is an amalgam of the finite groups associated to the cells of Y. Then Γ is a cocompact lattice in $\mathrm{Aut}(\Delta)$, since Γ acts on Δ with finite stabilisers and compact quotient Y.

In [27], Futer–Thomas constructed cocompact lattices in $\mathrm{Aut}(I_{p,q})$ which are fundamental groups of complexes of groups over Y a (tessellated) surface. They also showed that for some p and g, whether there exists a cocompact lattice $\Gamma < \mathrm{Aut}(I_{p,q})$ so that the quotient by the action of Γ is a genus g surface depends upon the value of q. This is the only known case where the values of the parameters p and q affect the existence of lattices in $\mathrm{Aut}(I_{p,q})$.

Complexes of finite groups may also be used to construct non-cocompact lattices. In this case, the underlying complex Y is infinite, and the assigned finite groups must have orders growing fast enough that the series in Equation (1) above converges. For example, Thomas [55] obtained many cocompact and non-cocompact lattices for right-angled buildings by constructing a functor from graphs of groups, with tree lattices as their fundamental groups, to complexes of groups, with right-angled building lattices as their fundamental groups. More elaborate complexes of groups were used by Thomas to construct both cocompact and non-cocompact lattices for certain Fuchsian buildings in [56] and Davis–Moussong complexes in [57].

When Δ is a Davis–Moussong (k, L)–complex, then there is an associated Coxeter group $W(k, L)$, which is *not* the type of the building Δ, and the group $W(k, L)$ may be regarded as a cocompact lattice in $\mathrm{Aut}(\Delta)$. In [8, Example 1.5.2] Bourdon gives a lattice construction which "lifts" lattices for

Euclidean buildings to cocompact and non-cocompact lattices for certain Fuchsian buildings.

Now suppose that the hyperbolic building Δ is the building for a Kac–Moody group Λ. Recall that Λ acts diagonally on the product $\Delta_+ \times \Delta_-$, where $\Delta_\pm \cong \Delta$. Carbone–Garland [15] and independently Rémy [51] showed that the stabiliser in Λ of any point in the negative building Δ_- is a non-cocompact lattice in the automorphism group of the positive building Δ_+. These stabilisers may also be considered as lattices in the *complete Kac–Moody group* $\hat\Lambda = \hat\Lambda_+$, which is a totally disconnected locally compact group acting on Δ_+, and is obtained by completing Λ using one of several methods. Some of the lattices in $\hat\Lambda$ constructed in Gramlich–Horn–Mühlherr [31] are for hyperbolic buildings. We do not know of any other constructions of lattices in complete Kac–Moody groups whose associated buildings are hyperbolic.

2.2 Questions about lattices

As mentioned in the introduction, many questions concerning lattices in hyperbolic buildings are motivated by comparison with known results concerning lattices in semisimple Lie groups. This background and motivation for the questions below is treated much more thoroughly in the corresponding sections of [24], to which we refer the reader. In general, a lot more is known about cocompact than about non-cocompact lattices in hyperbolic buildings. There are also cases where the behaviour of compact and non-cocompact lattices is dramatically different. It does appear that there is greater "rigidity" when the hyperbolic building has (compact) simplicial chambers, perhaps because the associated Coxeter system then has the property, like all irreducible affine and finite Coxeter systems, that all m_{st} are finite.

Classification

Once it is known that the automorphism group of a hyperbolic building Δ admits lattices, an immediate next question is to classify the lattices in $\mathrm{Aut}(\Delta)$ up to conjugacy. This has only been done in special cases. For instance in [39], Kangaslampi–Vdovina classify the torsion-free groups which act simply transitively on the vertices of Fuchsian buildings with triangular chambers and links the smallest generalised 4–gon, and in [16], Carbone–Kangaslampi–Vdovina classify all such groups with torsion.

Commensurability and commensurators

Lattices may also be classified up to commensurability. Recall that two subgroups $\Gamma_1, \Gamma_2 < G$ are *commensurable* if there exists $g \in G$ so that

$g\Gamma_1 g^{-1} \cap \Gamma_2$ has finite index in both $g\Gamma_1 g^{-1}$ and Γ_2. Haglund [33, Theorem 1.1] proved that for $p \geq 6$, all cocompact lattices in $\mathrm{Aut}(I_{p,q})$ are commensurable. In contrast, there are uncountably many commensurability classes of non-cocompact lattices in $\mathrm{Aut}(\Delta)$ for Δ a regular right-angled building (Thomas [46, Main Theorem 2(b)]).

The *commensurator* of a lattice $\Gamma < G$ is the subgroup consisting of elements $g \in G$ such that $g\Gamma g^{-1}$ and Γ are commensurable. Haglund [34] and independently Kubena–Thomas [43] proved that for Δ a regular right-angled building, the commensurator of a canonical cocompact lattice is dense in $G = \mathrm{Aut}(\Delta)$. The question of commensurators of non-cocompact lattices is wide open, even for $I_{p,q}$.

Covolumes

A basic question is to determine the set

$$\{\mu(\Gamma \backslash G) \mid \Gamma < G \text{ is a lattice}\}$$

of covolumes of lattices in G. Aspects of this question have been considered by Thomas for certain right-angled buildings [46], Fuchsian buildings [56] and Davis–Moussong complexes [57], but many cases remain open. For instance, it would be interesting to determine whether the set of covolumes of cocompact lattices in the Fuchsian buildings with triangular chambers considered in [39] has a positive lower bound.

Property (T) and finiteness properties

Ballmann–Świątkowski [6], Dymara–Januszkiewicz [21] and Żuk [70] have shown that the automorphism groups of many hyperbolic buildings Δ with simplicial chambers have Kazhdan's Property (T). On the other hand, Corollary 3 of [6] implies that the automorphism groups of Fuchsian buildings with chambers p–gons, $p \geq 4$, do not have Property (T). In higher dimensions, it follows from Niblo–Reeves [48, Theorem B] that the automorphism groups of right-angled buildings do not have Property (T), and Haglund–Paulin [35, Theorem 1.5] showed that the automorphism groups of hyperbolic buildings with chambers "even" polytopes do not have Property (T).

A group G has Property (T) if and only if all of its lattices have Property (T), and it is a well-known result of Kazhdan [42] that lattices with Property (T) are finitely generated. All cocompact lattices in a hyperbolic building are finitely generated since they are fundamental groups of finite complexes of finite groups. Infinite generation of some non-cocompact lattices for certain hyperbolic buildings was established by Thomas [57] and Thomas–Wortman [58]. It is not known whether for these buildings, there are any non-cocompact lattices which are finitely generated.

Very little is known about higher finiteness properties for lattices in hyperbolic buildings, apart from a recent result of Gandini [29] which bounds the homological finiteness length of non-cocompact lattices in Aut(X) for X a locally finite contractible polyhedral complex. As a corollary, such lattices are not finitely presentable. The examples of [57] and [58] show that Gandini's bound is not sharp.

Residual finiteness, linearity and simplicity

These questions are of particular interest for cocompact lattices in hyperbolic buildings with compact chambers, since such lattices are (finitely generated) word-hyperbolic groups. Recall that a group Γ is *residually finite* if for all $1 \neq \gamma \in \Gamma$, there exists a finite quotient $\Gamma \to \overline{\Gamma}$ such that $\overline{\gamma}$ is nontrivial. It is a theorem of Mal'cev [47] that every finitely generated linear group is residually finite. A major open conjecture of Gromov states that all word-hyperbolic groups are residually finite, while it is unknown whether every word-hyperbolic group is linear (see the introduction to Kapovich–Wise [41]). Word-hyperbolic groups are never simple [32, 49].

An important result, due to Wise [67], is the residual finiteness of cocompact lattices which are fundamental groups of complexes of finite groups over hyperbolic p–gons, for p large enough (depending upon the angles of the polygon). Combining this with [33, Theorem 1.1], Haglund proved that for $p \geq 6$, all cocompact lattices in Aut($I_{p,q}$) are linear and thus residually finite. The case $I_{5,q}$ is open. As noted in Kangaslampi–Vdovina [39], the residual finiteness and linearity of the cocompact lattices in Fuchsian buildings with triangular chambers is also open.

For non-cocompact lattices, nonlinear examples in Aut($I_{p,q}$) were obtained by Rémy [52] using Kac–Moody theory. A striking recent result of Caprace–Rémy [14] is the simplicity of many Kac–Moody groups Λ, which are non-cocompact lattices in the product of their twin buildings $\Delta_+ \times \Delta_-$.

Rigidity

Various rigidity questions for lattices in hyperbolic buildings overlap with questions about the structure of the building itself and its boundary, and so we discuss these issues together here. This section is only intended to be a list of some recent work in this area.

A careful proof of the folklore result that the visual boundary of a right-angled hyperbolic building is a Menger sponge was provided by Dymara–Osajda in [22]. Bourdon [7, Theorems 1.1 and 1.2] determined the conformal dimension of the visual boundary of $I_{p,q}$, and related this to its Hausdorff dimension. A version of Mostow rigidity was also established by Bourdon [7, Theorem 1.3] for cocompact lattices in $I_{p,q}$, using combinatorial Patterson–Sullivan measures. Bourdon–Pajot [9] established quasi-isometric rigidity for $I_{p,q}$, and Xie [68] generalised this result to all Fuchsian buildings.

Since hyperbolic buildings are CAT(-1) spaces, several rigidity results for divergence groups apply to lattices in hyperbolic buildings. As noted in the introduction to Burger–Mozes [13], the notion of a divergence group comes from Patterson–Sullivan theory for Kleinian groups. If Aut(Δ) acts cocompactly on the hyperbolic building Δ, as is the case for most known constructions, then any nonelementary lattice $\Gamma <$ Aut(Δ) is a divergence group [13, Corollary 6.5(2)]. Hersonsky–Paulin [37] generalised Mostow rigidity to divergence groups acting by isometries on many CAT(-1) spaces, including some hyperbolic buildings, and Burger–Mozes [13] established CAT(-1) super-rigidity results for divergence groups.

More recently, Daskalopoulos–Mese–Vdovina [17] studied harmonic maps from symmetric spaces into target spaces including hyperbolic buildings, and as an application proved super-rigidity results for the isometry groups of a class of complexes including hyperbolic buildings. Super-rigidity results for Kac–Moody groups Λ were obtained by Caprace–Rémy in [14].

An important ingredient in some classical rigidity results is the Howe–Moore property for unitary representations (see for example [69]), which concerns decay of matrix coefficients. This property was shown not to hold for Aut($I_{p,q}$) by Bader–Shalom [4, pp. 447–449], using Mackey theory.

Finally, the volume entropy of hyperbolic buildings considers the asymptotic growth of volumes of balls in the building (by analogy with volume entropy for Riemannian manifolds). This topic has been investigated by Hersonsky–Paulin [37], Leuzinger [45] and most thoroughly by Ledrappier–Lim [44], using the geodesic flow on apartments and measures on suitable boundaries.

Anne Thomas
School of Mathematics and Statistics F07
University of Sydney NSW 2006, AUSTRALIA
E-mail: anne.thomas@sydney.edu.au

References

[1] P. ABRAMENKO AND K. S. BROWN, *Buildings*, vol. 248 of Graduate Texts in Mathematics, Springer, New York, 2008. Theory and applications.

[2] D. ALLCOCK, *Infinitely many hyperbolic Coxeter groups through dimension 19*, Geom. Topol., 10 (2006), pp. 737–758 (electronic).

[3] E. M. ANDREEV, *Convex polyhedra in Lobačevskiĭ spaces*, Mat. Sb. (N.S.), 81 (123) (1970), pp. 445–478.

[4] U. BADER AND Y. SHALOM, *Factor and normal subgroup theorems for lattices in products of groups*, Invent. Math., 163 (2006), pp. 415–454.

[5] W. BALLMANN AND M. BRIN, *Polygonal complexes and combinatorial group theory*, Geom. Dedicata, 50 (1994), pp. 165–191.

[6] W. BALLMANN AND J. ŚWIATKOWSKI, *On L^2-cohomology and property (T) for automorphism groups of polyhedral cell complexes*, Geom. Funct. Anal., 7 (1997), pp. 615–645.

[7] M. BOURDON, *Immeubles hyperboliques, dimension conforme et rigidité de Mostow*, Geom. Funct. Anal., 7 (1997), pp. 245–268.

[8] M. BOURDON, *Sur les immeubles fuchsiens et leur type de quasi-isométrie*, Ergodic Theory Dynam. Systems, 20 (2000), pp. 343–364.

[9] M. BOURDON AND H. PAJOT, *Rigidity of quasi-isometries for some hyperbolic buildings*, Comment. Math. Helv., 75 (2000), pp. 701–736.

[10] M. R. BRIDSON AND A. HAEFLIGER, *Metric spaces of non-positive curvature*, vol. 319 of Grundlehren der Mathematischen Wissenschaften [Fundamental Principles of Mathematical Sciences], Springer-Verlag, Berlin, 1999.

[11] K. S. BROWN, *Buildings*, Springer Monographs in Mathematics, Springer-Verlag, New York, 1998. Reprint of the 1989 original.

[12] V. O. BUGAENKO, *Arithmetic crystallographic groups generated by reflections, and reflective hyperbolic lattices*, in Lie groups, their discrete subgroups, and invariant theory, vol. 8 of Adv. Soviet Math., Amer. Math. Soc., Providence, RI, 1992, pp. 33–55.

[13] M. BURGER AND S. MOZES, *CAT(-1)-spaces, divergence groups and their commensurators*, J. Amer. Math. Soc., 9 (1996), pp. 57–93.

[14] P.-E. CAPRACE AND B. RÉMY, *Simplicity and superrigidity of twin building lattices*, Invent. Math., 176 (2009), pp. 169–221.

[15] L. CARBONE AND H. GARLAND, *Existence of lattices in Kac-Moody groups over finite fields*, Commun. Contemp. Math., 5 (2003), pp. 813–867.

[16] L. CARBONE, R. KANGASLAMPI, AND A. VDOVINA, *Groups acting simply transitively on hyperbolic buildings*, LMS J. Comput. Math. 15 (2012), pp. 101–112.

[17] G. DASKALOPOULOS, C. MESE, AND A. VDOVINA, *Superrigidity of hyperbolic buildings*, Geom. Funct. Anal., 21 (2011), pp. 905–919.

[18] M. W. DAVIS, *Buildings are* CAT(0), in Geometry and cohomology in group theory (Durham, 1994), vol. 252 of London Math. Soc. Lecture Note Ser., Cambridge Univ. Press, Cambridge, 1998, pp. 108–123.

[19] ———, *The geometry and topology of Coxeter groups*, vol. 32 of London Mathematical Society Monographs Series, Princeton University Press, Princeton, NJ, 2008.

[20] ———, *Examples of buildings constructed via covering spaces*, Groups Geom. Dyn., 3 (2009), pp. 279–298.

[21] J. DYMARA AND T. JANUSZKIEWICZ, *Cohomology of buildings and their automorphism groups*, Invent. Math., 150 (2002), pp. 579–627.

[22] J. DYMARA AND D. OSAJDA, *Boundaries of right-angled hyperbolic buildings*, Fund. Math., 197 (2007), pp. 123–165.

[23] F. ESSELMANN, *The classification of compact hyperbolic Coxeter d-polytopes with d + 2 facets*, Comment. Math. Helv., 71 (1996), pp. 229–242.

[24] B. FARB, C. HRUSKA, AND A. THOMAS, *Problems on automorphism groups of nonpositively curved polyhedral complexes and their lattices*, in Geometry, rigidity, and group actions, Chicago Lectures in Math., Univ. Chicago Press, Chicago, IL, 2011, pp. 515–560.

[25] A. FELIKSON AND P. TUMARKIN, *On hyperbolic Coxeter polytopes with mutually intersecting facets*, J. Combin. Theory Ser. A, 115 (2008), pp. 121–146.

[26] ———, *Coxeter polytopes with a unique pair of non-intersecting facets*, J. Combin. Theory Ser. A, 116 (2009), pp. 875–902.

[27] D. FUTER AND A. THOMAS, *Surface quotients of hyperbolic buildings*, Int. Math. Res. Not. IMRN 2012, no. 2, 437–477.

[28] D. GABORIAU AND F. PAULIN, *Sur les immeubles hyperboliques*, Geom. Dedicata, 88 (2001), pp. 153–197.

[29] G. GANDINI, *Bounding the homological finiteness length*, Bull. Lond. Math. Soc. 44 (2012), pp. 1209–1214.

[30] C. GODSIL AND G. ROYLE, *Algebraic graph theory*, vol. 207 of Graduate Texts in Mathematics, Springer-Verlag, New York, 2001.

[31] R. GRAMLICH, M. HORN, AND B. MÜHLHERR, *Abstract involutions of algebraic groups and of Kac-Moody groups*, J. Group Theory, 14 (2011), pp. 213–249.

[32] M. GROMOV, *Hyperbolic groups*, in Essays in group theory, vol. 8 of Math. Sci. Res. Inst. Publ., Springer, New York, 1987, pp. 75–263.

[33] F. HAGLUND, *Commensurability and separability of quasiconvex subgroups*, Algebr. Geom. Topol., 6 (2006), pp. 949–1024.

[34] ———, *Finite index subgroups of graph products*, Geom. Dedicata, 135 (2008), pp. 167–209.

[35] F. HAGLUND AND F. PAULIN, *Simplicité de groupes d'automorphismes d'espaces à courbure négative*, in The Epstein birthday schrift, vol. 1 of Geom. Topol. Monogr., Geom. Topol. Publ., Coventry, 1998, pp. 181–248 (electronic).

[36] ——, *Constructions arborescentes d'immeubles*, Math. Ann., 325 (2003), pp. 137–164.

[37] S. HERSONSKY AND F. PAULIN, *On the rigidity of discrete isometry groups of negatively curved spaces*, Comment. Math. Helv., 72 (1997), pp. 349–388.

[38] J. E. HUMPHREYS, *Reflection groups and Coxeter groups*, vol. 29 of Cambridge Studies in Advanced Mathematics, Cambridge University Press, Cambridge, 1990.

[39] R. KANGASLAMPI AND A. VDOVINA, *Cocompact actions on hyperbolic buildings*, Internat. J. Algebra Comput., 20 (2010), pp. 591–603.

[40] I. M. KAPLINSKAJA, *The discrete groups that are generated by reflections in the faces of simplicial prisms in Lobačevskiĭ spaces*, Mat. Zametki, 15 (1974), pp. 159–164.

[41] I. KAPOVICH AND D. T. WISE, *The equivalence of some residual properties of word-hyperbolic groups*, J. Algebra, 223 (2000), pp. 562–583.

[42] D. A. KAŽDAN, *On the connection of the dual space of a group with the structure of its closed subgroups*, Funkcional. Anal. i Priložen., 1 (1967), pp. 71–74.

[43] A. KUBENA AND A. THOMAS, *Density of commensurators for uniform lattices of right-angled buildings*, J. Group Theory 15 (2012), pp. 565–611.

[44] F. LEDRAPPIER AND S. LIM, *Volume entropy of hyperbolic buildings*, J. Mod. Dyn., 4 (2010), pp. 139–165.

[45] E. LEUZINGER, *Entropy of the geodesic flow for metric spaces and Bruhat-Tits buildings*, Adv. Geom., 6 (2006), pp. 475–491.

[46] S. LIM AND A. THOMAS, *Counting overlattices for polyhedral complexes*, Topology Proc., 36 (2010), pp. 229–247.

[47] A. MALCEV, *On isomorphic matrix representations of infinite groups*, Rec. Math. [Mat. Sbornik] N.S., 8 (50) (1940), pp. 405–422.

[48] G. NIBLO AND L. REEVES, *Groups acting on* CAT(0) *cube complexes*, Geom. Topol., 1 (1997), p. approx. 7 pp. (electronic).

[49] A. Y. OLSHANSKIĬ, SQ-*universality of hyperbolic groups*, Mat. Sb., 186 (1995), pp. 119–132.

[50] M. N. PROKHOROV, *Absence of discrete groups of reflections with a noncompact fundamental polyhedron of finite volume in a Lobachevskiĭ space of high dimension*, Izv. Akad. Nauk SSSR Ser. Mat., 50 (1986), pp. 413–424.

[51] B. RÉMY, *Construction de réseaux en théorie de Kac-Moody*, C. R. Acad. Sci. Paris Sér. I Math., 329 (1999), pp. 475–478.

[52] B. RÉMY, *Topological simplicity, commensurator super-rigidity and non-linearities of Kac-Moody groups*, Geom. Funct. Anal., 14 (2004), pp. 810–852. With an appendix by P. Bonvin.

[53] M. RONAN, *Lectures on buildings*, University of Chicago Press, Chicago, IL, 2009. Updated and revised.

[54] J. ŚWIĄTKOWSKI, *Trivalent polygonal complexes of nonpositive curvature and Platonic symmetry*, Geom. Dedicata, 70 (1998), pp. 87–110.

[55] A. THOMAS, *Lattices acting on right-angled buildings*, Algebr. Geom. Topol., 6 (2006), pp. 1215–1238.

[56] ——, *On the set of covolumes of lattices for Fuchsian buildings*, C. R. Math. Acad. Sci. Paris, 344 (2007), pp. 215–218.

[57] A. THOMAS, *Existence, covolumes and infinite generation of lattices for Davis complexes*, Groups Geom. Dyn. 6 (2012), pp. 765–801.

[58] A. Thomas and K. Wortman, *Infinite generation of non-cocompact lattices on right-angled buildings*, Algebr. Geom. Topol., 11 (2011), pp. 929–938.

[59] W. P. Thurston, *Three-dimensional geometry and topology. Vol. 1*, vol. 35 of Princeton Mathematical Series, Princeton University Press, Princeton, NJ, 1997. Edited by Silvio Levy.

[60] P. Tumarkin, *Compact hyperbolic Coxeter n-polytopes with n + 3 facets*, Electron. J. Combin., 14 (2007), pp. Research Paper 69, 36 pp. (electronic).

[61] P. Tumarkin and A. Felikson, *On bounded hyperbolic d-dimensional Coxeter polytopes with d + 4 hyperfaces*, Tr. Mosk. Mat. Obs., 69 (2008), pp. 126–181.

[62] P. V. Tumarkin, *Hyperbolic Coxeter polytopes in $\mathbb{H}^m$ with n + 2 hyperfacets*, Mat. Zametki, 75 (2004), pp. 909–916.

[63] ———, *Hyperbolic n-dimensional Coxeter polytopes with n + 3 facets*, Tr. Mosk. Mat. Obs., 65 (2004), pp. 253–269.

[64] A. Vdovina, *Combinatorial structure of some hyperbolic buildings*, Math. Z., 241 (2002), pp. 471–478.

[65] È. B. Vinberg, *Absence of crystallographic groups of reflections in Lobachevskiĭ spaces of large dimension*, Trudy Moskov. Mat. Obshch., 47 (1984), pp. 68–102, 246.

[66] È. B. Vinberg and O. V. Shvartsman, *Discrete groups of motions of spaces of constant curvature*, in Geometry, II, vol. 29 of Encyclopaedia Math. Sci., Springer, Berlin, 1993, pp. 139–248.

[67] D. T. Wise, *The residual finiteness of negatively curved polygons of finite groups*, Invent. Math., 149 (2002), pp. 579–617.

[68] X. Xie, *Quasi-isometric rigidity of Fuchsian buildings*, Topology, 45 (2006), pp. 101–169.

[69] R. J. Zimmer, *Ergodic theory and semisimple groups*, vol. 81 of Monographs in Mathematics, Birkhäuser Verlag, Basel, 1984.

[70] A. Żuk, *La propriété (T) de Kazhdan pour les groupes agissant sur les polyèdres*, C. R. Acad. Sci. Paris Sér. I Math., 323 (1996), pp. 453–458.

Printed in the United States
By Bookmasters